Gordon McComb's
Gadgeteer's Goldmine!
55 Space-Age Projects

Gordon McComb's Gadgeteer's Goldmine!
55 Space-Age Projects

Gordon McComb

TAB Books
Division of McGraw-Hill

New York San Francisco Washington, D.C. Auckland Bogotá
Caracas Lisbon London Madrid Mexico City Milan
Montreal New Delhi San Juan Singapore
Sydney Tokyo Toronto

pbk   12 13 14 15 12 FGR/FGR 9 0 9 8 7 6 5 4 3 2 1 0
hc   2 3   4   5   6   7   8   9 10 12 FGR/FGR 9 9 8 7 6 5 4 3 2 1

Library of Congress Cataloging-in-Publication Data

McComb, Gordon.
    [Gadgeteer's goldmine]
    Gordon McComb's gadgeteer's goldmine! : 55 space-age projects / by
Gordon McComb.
        p.   cm.
    Includes bibliographical references.
    ISBN 0-8306-8360-7   ISBN 0-8306-3360-X (pbk.)
    1. Electronics—Amateurs' manuals.   I. Title.
TK9965.M35   1990
621.381—dc20   90-35338
                                                CIP

Acquisitions Editor: Roland S. Phelps
Technical Editor: Lisa A. Doyle
Director of Production: Katherine G. Brown
Book Design: Jaclyn J. Boone

About the author

Gordon McComb has written two dozen books and over 1,000 magazine articles, most of them on science and technology. His writing has appeared in *Popular Science*, *Omni*, *High Technology*, *Radio-Electronics*, *Popular Electronics*, *Family Handyman*, *Video*, *Macworld*, *MacUser*, and dozens of others. He is an avid electronics experimenter and inventor. His other books for TAB include *The Laser Cookbook: 88 Practical Projects* (No. 3090) and *Troubleshooting and Repairing VCRs* (No. 2960).

To Marshall Lee McComb. Welcome aboard, kid.

Contents

Projects

Acknowledgments

Who says work can't be fun? Writing this book provided many rewards including the chance to exchange ideas with many other gadgeteers. I'm grateful for the help provided to me by Anthony Charlton of Allegro Systems, Dennis Meredith of Meredith Instruments, Roger Sonntag of General Science and Engineering, Jeff Korman of Fobtron Components, Tesla coil expert Russell Clift, and fellow author Forrest Mims III.

There were many more "behind the scenes" people that made a great impact in the preparation of this book. Thanks go to the TAB BOOKS team: Roland Phelps, who shared my vision of a book on kooky gadgets, Lisa Doyle for her excellent editing, and Tim Higgins for being willing to listen to my complaints.

Introduction

Thomas Alva Edison began tinkering with electricity, batteries, and other scientific paraphernalia of the middle 1800s when he was about 10 years old. He carried out his first experiments in the cellar of the Edison household in Port Huron, Michigan. One of Edison's first experiments was creating static electricity by frantically rubbing together the fur of two tomcats. The trouble: the tomcats were still alive, and despite the cat's tails being wired together, the experiment resulted in little more than deep scratches on Edison's hands and arms.

Edison's inquisitiveness with the nature around him was more than just the passing fancy of an overly curious child. The man's inventive spirit stayed with him during his growing years. By the age of 12, Edison sold candy and newspapers as a free-lance vendor on the Port Huron railroad, but on the three-hour trips to Detroit, he tinkered in a mobile laboratory he set up in a baggage car at the rear of the train. Young Thomas (actually, he was referred to as Al in those days) was finally booted from the train when one of his experiments went haywire and the baggage car caught fire and burned.

During the years Edison spent tinkering in quiet solitude in the rumbling baggage car, he got his first taste of the pursuit and adventure of *gadgeteering*: the art of experimenting with mechanics, electronics, chemistry, physics, and other branches of scientific inquiry to come up with new and better ways of doing things.

Though Thomas Edison is best known for his various breakthrough inventions—the electric light, the phonograph, and the motion picture camera—perhaps his greatest achievement is that he popularized the notion of the serious, professional inventor.

Before that time, inventors applied their art in a haphazard fashion, often working out of some dusty and dank basement and projecting to the public the image of the scientist gone mad. Edison was first to organize a centralized research and development lab, starting in earnest with his historic Menlo Park facilities.

When Edison announced a new breakthrough or innovation, the entire world listened. On more than one occasion, Edison's announcements even had a profound effect on Wall Street, causing stock market prices to swell or plummet, depending on the gravity of his news.

Edison never stopped his quest for gadgeteering, even past his 80th birthday. And no wonder. Few diversions offer so much exhilaration, excitement, and yearning, simply from the sight of a new creation throbbing to life or the breaking through of some scientific barrier. Edison knew other emotions as well, like agony, torment, and suffering when his creations sat lifeless and pallid. Yet such emotions were well worth the price if just one gadget—every now and then— proved to be a success.

Gadgeteer's Goldmine is written for all those who feel the giddiness that Edison did when he flipped the switch and found his wacky invention actually worked. This book presents over 55 practical and affordable gadgets you can build in your own home or workshop (or in a baggage car of a train, if that's your preference).

Some of the gadgets are decades old; they are included here to give you a solid foothold in the techniques and practice of gadgeteering. And besides, they're fun. Others are new and unique and are presented to help you forge your own ideas and methods. You are free to use the basic designs provided in the following pages and elaborate on them as you see fit.

In this one volume, you'll find outrageous plans for building high-voltage Tesla coils, shocking Van de Graaff generators, intelligent voice-controlled robots, and sophisticated see-in-the-dark nighttime viewers. You'll also learn how to detect dangerous levels of radiation, how to photograph your own ''physic'' aura, how to dazzle an audience with moving patterns of laser light, and much more.

WHO THIS BOOK IS FOR

Gadgeteer's Goldmine is written for a wide variety of readers. If you're into electronics, you'll enjoy the many circuits you can build, including one that lets you carry your voice over a beam of light, or the project that detects even the faintest earthquakes. A number of the projects are excellent springboards for science fairs. These include experimenting with solid-state lasers, testing materials for radioactivity, and analyzing the properties of various piezoelectric materials.

If you're into mechanics, you should find a great deal of pleasure and satisfaction in building a high-tension induction spark maker, a 250,000-volt Tesla coil, and a ''bionic'' ear for listening to faraway sounds.

In all cases, the designs used in *Gadgeteer's Goldmine* have been tested in prototype form, either by myself or by the people who have kindly allowed me to reproduce their ''pet projects'' for this book. I encourage you to improve on the basic designs, but you can rest assured that the projects have actually been constructed and field tested.

The projects in *Gadgeteer's Goldmine* include all the necessary information on how to construct a variety of high-tech, space-age gizmos and contraptions. Suggested alternative approaches, parts lists, and sources of electronic and mechanical components are also provided where appropriate. A complete list of sources, including mail-order companies that cater to the individual hobbyist, is in Appendix A.

HOW TO USE THIS BOOK

Gadgeteer's Goldmine is divided into 26 chapters. All chapters present one or more actual hands-on projects that you can duplicate in your own gadget creations. Each chapter is a stand-alone unit, and with few exceptions, you don't have to read them in a particular order. However, a few chapters present beginning or advanced information, and you'll want to read the preliminary chapter first. For example, Chapter 9, ''The science of lasers,'' offers basic information on lasers and laser safety, so you should read it first before going on to Chapters 10 through 17, which detail numerous projects and experiments in laser technology. Generally, each project is free-standing (it is not dependent on any other).

CONVENTIONS USED IN THIS BOOK

You need little advance information before you can jump headfirst into this book. But you should take note of a few conventions I've used in the description of electronic parts and in the schematic diagrams for the electronic circuits (yes, you do need to know how to read schematics).

TTL integrated circuits are referenced by their standard 74XX number. The ''LS'' identifier, such as 74LS04, is assumed. I built most of the circuits using LS TTL chips, but the projects should work with the other TTL-family ICs—the standard (non-LS) chips, as well as those with the S, ALS, and C identifiers. If you use a type of TTL chip other than LS, you should consider current consumption, fan-out, and other design criteria because these factors could affect the operation and performance of the circuit.

In some cases, however, a certain TTL-compatible IC is specified in the design. Unless the accompanying text recommends otherwise, you should use only the chip specified. Certain CMOS TTL-compatible chips offer the same functions as a sister IC, but the pinouts and operation might differ.

The chart in FIG. I-1 details the conventions used in the schematic diagrams. Note that unconnected wires are shown by crossing lines, a broken line, or a ''looped'' line. Connected wires are shown by a connecting dot.

Details on the specific parts used in the circuits are provided in the parts list tables that accompany each schematic. Refer to the parts list for information on resistor and capacitor type, tolerance, and wattage or voltage ratings. Some

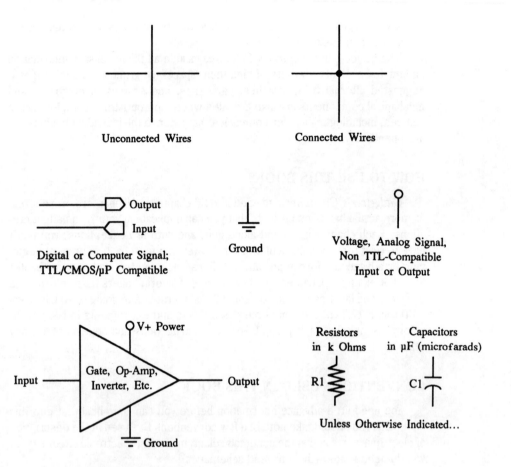

Fig. I-1. Conventions used in the schematic diagrams of this book.

parts lists also provide a source for complete kits or special components for help in building the project.

In all schematics, the parts are referenced by component type and number.

- IC# means an integrated circuit (IC).
- R# means a resistor or potentiometer (variable resistor).
- C# means a capacitor.
- D# means a diode, a zener diode, or sometimes a light-sensitive photo-diode.
- Q# means a transistor or sometimes a light-sensitive phototransistor.
- LED# means a light-emitting diode (unless otherwise noted, any visible or infrared LED will do).
- XTAL# means a crystal or ceramic resonator.
- S# means a switch, RL# means a relay, SPKR# means a speaker, and MIKE# means a microphone.

SAFETY FIRST

Would you believe that if misused or handled carelessly, some of the projects in *Gadgeteer's Goldmine* can be LETHAL? Several of the projects use or generate many thousands of volts. And while the voltage might be low in current, the shock could be enough to kill you under certain circumstances. Don't think of any project in *Gadgeteer's Goldmine* as a toy. Handle each one carefully and follow *all* standard safety precautions at *all times*. For safety's sake, keep the projects—whether they are finished or not—away from children.

1

Introduction to high-voltage devices

Everyone knows about Benjamin Franklin's shocking discovery of electricity in 1752. In fact, anyone flying a kite during an electrical storm can learn something about electricity in much the same way as old Ben. Realistically, though, it's not the best way to get an education in electricity.

The study of electrics goes way back before Ben Franklin's days. In fact, Franklin already knew about electricity when he flew his kite. He was inventing the lightning conductor, which obviously proved successful. Franklin based his assumptions about electricity, in part, on speculations made by the early Egyptians. The Egyptians of 2000 B.C. knew that a strange "unseen" force surrounded them that caused lightning, sparks, and other forms of electrical discharge.

The study of electricity has fascinated us for thousands of years. In the late 1800s, after the development of batteries, dynamos, and induction coils, the study of electrics soared. Experimenters the world over marveled at the shocks they received as they toyed with the new wonders of electricity.

Of particular interest to many innovators of the early electrical craze was high voltage—electrical charges in excess of 500 volts. High voltage is a little like the sound barrier: once you cross the threshold from low- to high-voltage potentials, electricity starts behaving in unusual ways.

This chapter details the nature of high voltage and how to work with it. There are special precautionary measures you need to follow if you want to work safely with high voltages. The text that follows can help you when experimenting with the various high-voltage projects such as the Jacob's ladder, Tesla coil, and Van de Graaff generator.

THE NATURE OF HIGH VOLTAGE

The term *voltage* is defined as an electromotive force. It is similar in nature to water flowing down a river. The water flows because of gravity; as the water flows, it creates a force that, depending on how rapid the water moves, can float leaves downstream or break off mighty limbs from a tree.

Voltage is measured by the difference in potential between two points or wires. Assuming one wire is at earth (ground) potential, which is theoretically zero, and the other wire has a charge of 10 volts, the voltage difference between the two wires is 10 volts. The measurement also takes into account the *velocity* or *pressure* of the voltage (expressed in *amperes*), and the amount of energy used up when the voltage gets to its destination (expressed in *watts*).

Low-voltage applications include flashlights, which operate on 3 to 6 volts, transistor radios, which require 9 volts, and most car electrical systems, which use 12 volts. These low-voltage applications are relatively easy to work with. Not only is the voltage direct (it doesn't change polarity over time), it is low enough that you don't feel it as a shock.

Medium voltage is found in the electrical system in your house. The 120-volt ac standard used in the United States can deliver a healthy shock, partly because it is a higher voltage than found on battery-operated equipment like flashlights and radios, and because it is alternating—it changes polarity once every 1/120 of a second (60 complete cycles per second).

High voltage is anything over about 500 volts. A number of common electrical devices require voltages this high and even higher. For example, Geiger counters, used in the detection of radiation, operate on 500 to 1,000 volts. Helium-neon gas lasers, like the kind used in supermarket check-out stands, run on 1,200 to 3,000 volts. And flyback transformers used with television CRTs generate thousands of volts of electricity. We'll be experimenting with these voltages and devices throughout this book. High voltages come in three basic forms: *dc, ac,* and *static.*

Dc high voltage

Dc high voltage is at a continuous voltage level and does not change polarities, as depicted in FIG. 1-1. Dc high voltage is the same as battery power, like in a flashlight battery, except that the output is 500 volts or more instead of 1.5 volts.

Ac high voltage

Ac high voltage changes polarity at least a few times each second (see FIG. 1-1). Typically, the polarity is 60 cycles per second, because the high voltage is derived from 120 Vac house current, which is 60 cycles per second.

Static high voltage

Dc and ac potential remain in conductors only as long as the energy source is applied. Remove the energy source, and the voltage disappears. Static voltage is trapped in an insulating body—like the inside of a glass jar—and will remain there until discharged (in actual practice, tiny leakage currents cause the trapped static to diminish over time).

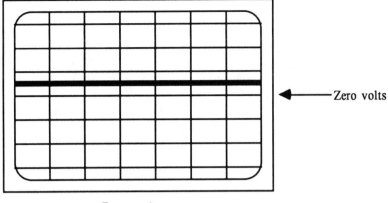

Dc waveform

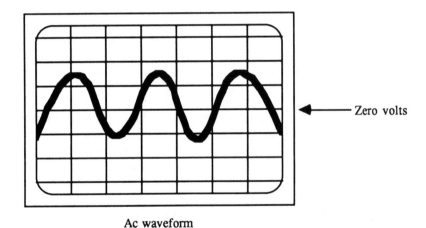

Ac waveform

Fig. 1-1. Dc voltage stays constant about the 0V (ground) point, whereas alternating current varies between positive and negative voltages with respect to the 0 V point.

A lightning bolt is an example of a high static voltage. A charge built up in a capacitor, which is an insulating body when it's connected to an active circuit, is also static electricity. A number of the projects in the chapter, and the chapters that follow, utilize the static charge left in a capacitor to power high-voltage devices.

Expressing high voltage

Very often, high voltages are in excess of 1,000, 10,000, even 100,000 volts. All those zeros can get in the way, so a more convenient method of expressing very high voltages is to use the *kilovolt*. One kilovolt, or kV, is equal to 1,000 volts, so 10 kV is 10,000 volts. Throughout this book we'll use the kV shorthand to express very high voltage.

THE CORONA EFFECT

High voltages exhibit a peculiar effect. When a high-voltage terminal is exposed to air, ions are jettisoned from the terminal into the surrounding air molecules. This causes a blue arcing or *corona effect*. Corona occurs at energy levels in excess of about 600 volts and is a common by-product of many of the projects in this book.

The charge of the ions is determined by the polarity of the high-voltage terminal. In many high-voltage power supplies, the negative terminal is at ground potential (at *or near* 0 volts), and the positive terminal carries the high voltage. In this instance, the ions jettisoned from the positive terminal are positively charged. When the high-voltage terminal is at a negative potential with respect to ground, the ions are negatively charged.

Although studies are still being made on the effects of positively and negatively charged ions on people, it is believed that:

- High concentrations of negative ions seem to promote alertness and mental agility. Over the long term, it is supposed to help respiratory conditions.
- High concentrations of positive ions are said to promote hostility and aggressiveness.

Heavy ion concentrations are also said to provide healing, relieve the pain of burns, and promote plant growth. While you are free to experiment with these theories (these projects aren't covered in *Gadgeteer's Goldmine*), you should use caution to avoid excessive exposure to the by-products of ion generation such as production of ozone gas and ultraviolet rays. Keep reading for more information on these topics.

Though the corona discharge is not particularly harmful, avoid exposure to it as much as possible. Why? The discharge produces ozone gas and ultraviolet rays, both of which—over a period of time—have an adverse effect on your health. Also, if the voltage is high enough, the discharge can turn into a mean spark, which can jump up to several inches through the air. The shock poses a serious hazard even if you don't actually touch the high voltage output terminal. Let's take a look at each of the dangers of corona discharge.

Ozone

Corona discharge produces toxic ozone, or O_3, which in even small quantities can cause nausea, vomiting, and irritation of the lungs. Ozone has a characteristically "sweet," almost saccharine-like smell (the same odor you smell after an electrical storm). Despite the benefits this gas offers in protecting us from radiation from the sun, direct exposure to it can be harmful.

You can minimize ozone build-up by keeping your lab well ventilated. If you're operating a Tesla coil or other apparatus that produces considerable ozone, aim a small fan at the arc discharge and keep some windows open.

Ultraviolet rays

The discharge arc produces shortwave ultraviolet light, which can cause considerable eye damage. Corona arcs can be especially harmful because they emit a small, extremely bright flash of ultraviolet light.

Avoid direct exposure to the spark gap by surrounding it in an opaque (metal or dark plastic) enclosure. At the very least, you can place a sheet of dark green glass (the kind used in arc welder's helmets) in front of the discharge. The glass will stop most of the harmful ultraviolet rays, but you'll be able to see a little bit of the spark.

Shock

Though air is an insulator, its dielectric barrier breaks down at higher voltages. That's what causes corona discharge, and eventually a spark, in the first place.

TABLE 1-1 shows the length of static or dc spark, depending on voltage. Note the spark length increases proportionally with the voltage. The higher the voltage, the longer the spark.

Table 1-1. Spark Length Versus Voltage

Voltage	Needle Points	1-inch Sphere	2-inch Sphere	10-inch Sphere
5 kV	0.17″	0.05″	0.06″	0.063″
10 kV	0.33″	0.11″	0.11″	0.126″
15 kV	0.50″	0.17″	0.17″	0.190″
20 kV	0.69″	0.23″	0.23″	0.250″
25 kV	0.87″	0.30″	0.30″	0.320″
30 kV	1.00″	0.37″	0.37″	0.390″
40 kV	1.50″	0.56″	0.51″	0.510″
50 kV	2.00″	0.79″	0.67″	0.650″
100 kV	6.10″	—	1.90″	1.370″
150 kV	10.30″	—	—	2.170″
200 kV	14.00″	—	—	4.060″
250 kV	17.30″	—	—	4.060″

For reference, high voltage pierces air that is dry and at room temperature and standard pressure at 1,143 volts per millimeter. You can convert millimeters to inches by dividing by 25.4; convert centimeters to inches by dividing by 2.54. For example, if the spark is 10 millimeters (1 centimeter or about 0.4 of an inch), the voltage is approximately 11,430 volts (assuming no humidity—voltages are higher with greater humidity).

You can see by the table that you don't need to actually touch the high-voltage terminal to receive a shock. At a voltage of, say, 50 kV, you need only come no closer than about two inches from the terminal to draw a spark—and a healthy shock—from it. *See the following section on the dangers of high-voltage electrocution and how to avoid it.* Be sure to read it; your life depends on it.

Minimizing corona discharge

Some circuits produce a corona discharge because of poor component layout. For example, the leads of a capacitor that stores a 2 kV charge should not be too close to the ground rail of your circuit. This type of discharge is unnecessary and in fact is usually deleterious to the operation of the circuit.

You can minimize unnecessary corona discharge by using a *corona dope* paint or high-voltage putty. Both paint and putty, shown in FIG. 1-2, are commonly available at electronics stores that cater to the TV repair tech. Cost is minimal and the paint or putty lasts for many applications. You might need to apply several layers of the paint or putty to completely eliminate the unwanted discharge.

Other circuits, especially those like a Tesla coil that use a high-voltage spark gap, by their nature produce some type of discharge. You can't use corona dope or high-voltage putty on circuits that require an active spark gap. Instead, provide a barrier to contain the spark gap in a plastic enclosure. The enclosure should be dark or opaque plastic to decrease the levels of ultraviolet radiation.

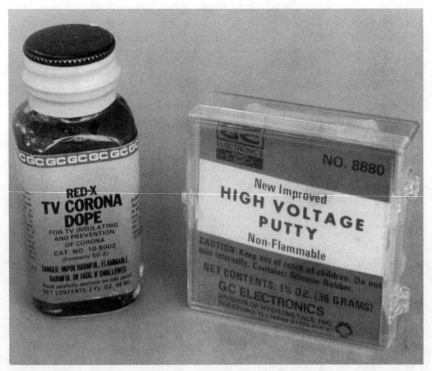

Fig. 1-2. High-voltage putty and corona dope help protect you from dangerous arcing and discharge.

THE DANGERS OF HIGH-VOLTAGE ELECTROCUTION

Most of the circuits detailed in this book develop high voltages at low current. Because of the low current, many electronics experimenters disregard the dangers of the higher potentials, relying on the maxim "It's the volts that jolts, but it's the mils (as in milliamps) that kills."

True, a circuit that produces a high voltage of even 10 kV at 50 microamps is not nearly as dangerous as a circuit that develops 100 volts at 50 amps, but you can still receive a nasty—and even fatal—shock from either one.

High voltage is *never* to be treated lightly, even if the current is so low it doesn't register on your volt-ohmmeter. You'll not soon forget a shock of 100,000 volts even if it is only a few microamps. The jolt will be enough to cause the muscles in your arms to contract and perhaps force you to grasp and hold the high-voltage terminal. Unless you're lucky enough to have someone nearby pull the plug or physically tackle you away from the high-voltage source, your body will not be able to cope with the extreme voltage. You'll be unconscious (or worse) in a matter of minutes. Follow these precautions when working with high voltages.

- Always turn off power when working with high-voltage circuits unless you are actively testing the circuit. However, for your own protection, avoid testing high-voltage circuits while they are in operation.
- If you must test a "live" circuit, use clip-on test leads rated at least twice the voltage you are measuring. That is, if the circuit develops 10 kV, the leads must be rated at least 20 kV.
- Always keep one hand in your pocket when testing a live circuit. Especially, don't hold a test probe in one hand, and a grounded line in the other. This prevents the high voltage from crossing your heart as it discharges from one hand to the other.
- Most multimeters cannot measure voltages in excess of 2 kV, requiring you to use a high-voltage probe. Most commercially manufactured probes are designed for TV use, so they are rated at no more than 40 kV. This should be sufficient for most of the applications you'll find in this book, but not for ultrahigh-voltage generators like Tesla coils and Van de Graaff generators. These cannot be directly measured by ordinary test gear.
- After turning the power off, always discharge high-voltage capacitors to ground. Place a 1- to 10-megohm resistor in the ground path to prevent a direct short. This drains the residual charge—which can be fatal in itself—left in the capacitors. The resistor prevents the capacitor from discharging too rapidly, which can cause the component to burst.
- Shield all high-voltage output terminals. Not only does this reduce exposure to the spark gap (and therefore dangerous ultraviolet rays), it lessens the chance of an accidental shock.
- Wear rubber footwear when working with high-voltage circuits. Or, place a heavy rubber mat under your chair and workbench.

- Remove dangling clothes and jewelry before working with high voltage. Don't place metal jewelry in your pocket.
- Keep low-voltage and high-voltage sections separate in all of your circuits. This is good design practice anyway because it reduces the likelihood of unwanted corona discharge, but it also helps to prevent you from being electrocuted when testing.
- Keep all flammables (circuit board cleaner, cigarette lighters, etc.) at least five feet from the workbench or high-voltage circuit. Do not allow open flame near circuits that produce large amounts of ozone.

Always remember one simple slogan when working with devices that produce more than 500 volts:

High voltage can KILL you!!

If need be, print this message on a large banner and hang it in front of your workbench. Never forget it. Your life depends on it.

Throughout this book, special precautions and warnings are provided for each project that uses or develops high voltages. Use these warnings as a guide in considering which projects are suitable for students and young experimenters.

2

Build a Jacob's ladder

The Jacob's ladder is perhaps the easiest high-voltage device to build, yet it's one of the most spectacular. If you've ever seen an old Frankenstein flick (and who hasn't), you've seen the Jacob's ladder in operation: Two slender wires snake upwards at slightly opposing angles to one another, making a funnel shape as they extend into the sky. Starting from the bottom or *neck* of the wires, a spark appears, and then slowly crawls upward, straddling the wires as it gets longer and longer. Just as the spark dies out at the top, another one takes its place at the bottom and the cycle repeats itself.

The Jacob's ladder device is named after the biblical character of Jacob, who in a vision (or a dream, depending on the version of the Bible you read) sees an angel climb up and down a ladder. The name of the spark device is said to have originated with the special-effects creators who developed the device for motion pictures.

HOW JACOB'S LADDER WORKS

The Jacob's ladder is little more than a high-voltage source and two wires that function as the output terminals. The output terminals are spaced close together at the base and further apart as the wires extend outward. When the ladder is turned on, a spark forms at the base. This spark heats and ionizes the air just above it, which substantially reduces the breakdown voltage (abbreviated BV in high-voltage texts) of the air.

Meanwhile, the breakdown voltage of the air at the point of the spark increases because the spark heats the air and creates a near vacuum. This differential causes the spark to move from the region of high-breakdown voltage to low, so the spark travels upward.

If the spark reaches the top of the electrodes, it bows out, extending its length until it can bow no more. The spark finally extinguishes, and a new, fresh spark starts at the base of the terminals. The process repeats itself over and over again.

Spark length and thickness is determined primarily by the output of the high-voltage power supply. The higher the voltage, the longer and thicker the spark. Output voltage is also partly determined by the thickness of the output terminal wires. For best results, use the thickest wires possible.

PARTS FOR JACOB'S LADDER

At the heart of the Jacob's ladder is an older-fashioned high-voltage neon sign transformer (see TABLE 2-1 for a complete parts list). These plug directly into the wall socket in your home and boost the voltage from 120 volts to 7 kV or more. The higher the voltage, the better the spark climb. The ideal neon sign transformer has an unloaded output of 10 kV or more (unloaded means nothing is connected to the output terminals; output voltage decreases as the resistance of the load increases).

Table 2-1. Parts List for Jacob's Ladder

1	Neon sign transformer, 10 kV or higher
2	Insulated standoffs (ceramic preferred), with mounting hardware
1	Enclosure
2	8$^{1}/_{2}$-inch lengths of $^{1}/_{8}$-inch-diameter stiff wire (such as welding rod or coat hanger rod)
1	Ac plug and line cord
1	SPST switch, rated at 10 amps or better at 120 Vac
1	Fuse holder
1	Fuse, rated at 10 amps or better at 120 Vac

Some newer neon sign power supply designs use a transformer with a lower voltage output, and then step up the voltage with a ladder network of capacitors and diodes (such a power supply is detailed in future chapters including Chapter 12, High-tech laser projects). These designs are not really suitable for use in a Jacob's ladder because the spark will fizzle out before it has a chance to travel very far up the wires.

Where do you find suitable neon sign transformers? Start at a local neon sign building and repair shop. Most will sell you one of the older type of transformers for less than $10. Be sure the transformer is in working condition; have the store test it out before you bring it home.

If you can't locate a suitable transformer at a neon sign shop, another local source is an electronic surplus house. You might also have some luck with mail-order sources, particularly those that deal regularly in unusual or special-purpose transformers. These include Fair Radio, H&R Sales, and American Design Components. Addresses for these and other sources are provided in Appendix A.

BUILDING JACOB'S LADDER

Most neon sign transformers come equipped with insulated standoffs for the anode (positive) and cathode (negative) terminals. If yours doesn't, you'll have to add them. The insulators should be the feed-through variety and be at least 1/2 inch in diameter by 1 or 2 inches long.

If your neon sign transformer already comes with suitable insulators, skip to the next paragraph. If it does not, attach your own insulators to the transformer with heavy wire or a #6 or #8 hardware bolt (length determined by the length of the insulator). Solder the wire or bolt it to the output terminals of the transformer if necessary. Make sure the connections are tight. When you are done, secure the insulators to the body of the transformer using RTV cement or epoxy. If the transformer won't easily accept the hardware and insulators, build a box to house it and add the output terminals on the top, as depicted in FIG. 2-1.

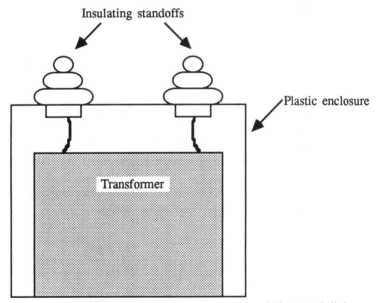

Fig. 2-1. Place the neon sign transformer and insulating standoffs in a plastic enclosure.

Once the insulators are in place, cut two pieces of #8 or #12 copper wire to 8 to 12 inches. With a pair of heavy-duty pliers or a wire-bending jig, bend the wires as shown in FIG. 2-2. The lower loop should be large enough to fit over the output terminals on the transformer. Carefully form a V, starting at the base of the wires and expanding at the top. You'll need to adjust the distance between the wires after you initially test the Jacob's ladder, so don't be too critical about spacing just yet. The figure provides some spacing dimensions you can try for starters.

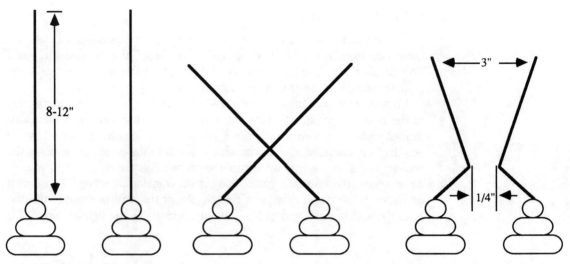

Fig. 2-2. Bend the wire terminals in a cross, then shape them into a ∨. Adjust the spacing to produce the longest, brightest spark.

Bear in mind that ideal spacing depends on the actual output of the transformer, the diameter and makeup of the output terminal wires, humidity, altitude, and perhaps even by the angle at which you hold your head. In any case, trial-and-error is the best approach in determining the best spacing of the wires on your Jacob's ladder.

Add a switch box and fuse as depicted in the schematic in FIG. 2-3. Be sure the power cord is long enough to conveniently reach an outlet. A length of 6 to 8 feet should be sufficient. Be certain you don't omit the fuse. It serves as the only safety device on the Jacob's ladder and is there to help prevent transformer overload and avoid a possible fire.

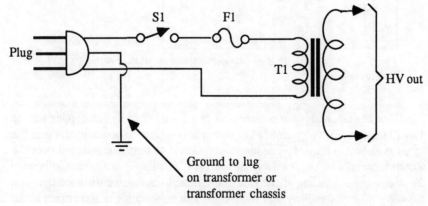

Fig. 2-3. Wiring diagram for the Jacob's ladder transformer.

TESTING JACOB'S LADDER

Before plugging the Jacob's ladder into an ac outlet, double-check your wiring. Be certain there are no shorts, either on the incoming 120 Vac line, or at the high-voltage output terminals.

During operation, *stay away from the ladder wires*. Keep your fingers, nose, and other bodily parts at least 5 inches from the transformer (except for the switch), output terminals, and ladder wires. Otherwise, you'll receive a nasty and potentially lethal shock. If the sparking gets out of control and it might not be safe to turn the ladder off at the switch, simply unplug it from the wall receptacle.

If the ladder passes inspection, flip the switch to the ON position, and plug in the power cord. If the transformer is wired correctly and the ladder wires are spaced the proper distance apart (which they probably are *not*), you should see a blue-white arc dance up the wires accompanied by an unmusical crackle of corona discharge.

If there is no spark at all, unplug the ladder and check your wiring. Try squeezing the wires closer together (but only a smidgen) and try again. If the spark appears at the bottom of the ladder, but proceeds only part way up the wires, unplug the unit and space the upper part of the wires closer together. At all times the pair of wires should create a V, or the spark won't climb.

Ideally, the spark should travel up the wires at about 2 to 3 inches per second. A spark that travels too fast up the ladder could be an indication that the wires are spaced too close together. Unplug the Jacob's ladder and spread the wires apart a bit.

After some tinkering, the wires should be spaced at the proper distance to create a glowing, climbing spark that travels at least $3/4$ to $4/5$ the way up the wire before extinguishing. You have obtained ideal wire spacing when the spark travels 2 to 3 inches per second, and reaches the top of the wires and bows out before extinguishing. Each spark climb will be slightly different. Some sparks might not reach the top while others do.

USING JACOB'S LADDER

The Jacob's ladder produces a fair amount of ozone, so prolonged use is not recommended, especially indoors. Power the ladder for only 5 minutes or less at a time. Not only will this help keep the transformer from prematurely burning out, it will let the ozone disperse.

For safety, build a Plexiglas funnel to house the entire Jacob's ladder. The funnel can be constructed out of acrylic tube stock (diameter determined by the outside dimension of your neon sign transformer). Place a lid on the top, and provide holes to let the ozone escape. The plastic shield helps prevent accidental shock and reduce some of the ultraviolet rays emitted by the crawling spark.

Plastics fabricators carry a wide variety of plastic parts. You might find it to be most cost effective to have the funnel built for you, following plans you provide, than to build it from scratch yourself. Or, you can construct the funnel from scrap pieces, which you can purchase at a plastics fabricator or at most plastic supply stores (note that the supply stores might require you purchase a minimum length of tube stock, so you could end up with more than you need).

Once the ladder is complete and working properly, what do you do with it? The Jacob's ladder device makes a great conversation piece, and is a wild addition to any party. Be sure the wires are sufficiently insulated to prevent guests from getting a nasty shock. Come Halloween, place the Jacob's ladder in a window by the front door, and watch the frightened faces as the kids walk up to your porch and wonder if trick-or-treating at your house is really worth it.

Note that the sparking output of Jacob's ladder can cause significant radio frequency interference (RFI). Listening to the radio or watching TV might be impossible while the ladder is in operation. Depending on the output of the ladder and the length of the terminal wires (high-voltage outputs and longer wires create more interference), the RFI could even travel to your neighbor's TVs and radios (don't be surprised if you get a complaint or two).

USING A VARIABLE TRANSFORMER

The neon sign transformer steps up the input voltage to the desired output voltage. This can be used to vary the voltage output of the ladder and therefore control the speed and size of the climbing arc.

Suppose your neon sign transformer is rated at 10 kV when supplied with 120 volts ac. You'll obtain the rated output only if the input voltage is 120 volts. If the input is more, the output will be more; if the input is less, the output will be less.

You can easily control the voltage applied to the neon sign transformer with a variable line transformer, often referred to as by the tradename *Variac*. The Variac lets you dial in the desired voltage—typically from 0 volts to 120 or 125 volts. Since the Jacob's ladder draws little current, you can use even a small Variac without worry of overload. In any case, you'll want to be sure the variable transformer you use incorporates a fuse. Remember: observe safety first.

Connect the Variac in line with the ac receptacle and the Jacob's ladder. Turn the dial all the way to zero, and turn the Jacob's ladder on. Slowly increase the voltage at the Variac until a spark forms at the base of the ladder wires— about 40 to 60 volts. Increase the voltage until the spark begins to climb. You might need to readjust the spacing of the wires when applying less than full voltage to the input terminals of the neon sign transformer.

3

Plasma sphere
experiments

You've probably seen a plasma sphere in high-tech catalogs or electronic gadget boutiques. In operation, gases inside a clear sphere are ionized by high frequency (ac) high voltage. Through something called the *Townsend Avalanche* effect, the electrically excited gases turn into a visible plasma—the so-called fourth state of matter—radiating from the central high-voltage terminal to the edge of the glass sphere.

The effect looks like a ball of lightning or tiny lightning bolts coursing through the sphere. While the device might look space-age and complex, it's rather simple to build. This chapter presents plans for building your own plasma sphere using inexpensive parts, most of which are available at salvage stores.

CONSTRUCTING THE POWER SUPPLY

The parts list for the plasma generator power supply is detailed in TABLE 3-1. The plasma generator uses a flyback transformer salvaged from an old television set. Flyback transformers are used to generate very high voltages for electrostatically charging the television picture tube. You can actually dismember an old TV to recover the flyback transformer, but an easier method is to buy the transformer as a surplus item. Many of the mail-order companies listed in Appendix A, such as Fair Radio, John Meshna & Associates, and Electronic Goldmine, routinely carry suitable flyback transformers.

Figure 3-1 shows a typical flyback transformer that's ideal for this project. You'll want an open-frame transformer similar to the one shown because you'll need to remove the primary winding and replace it with a few turns of magnet wire. The completely sealed, modular flyback transformers are unsuitable for this project because there is no easy way to rewind the primary.

Refer to FIG. 3-2 on rewinding the transformer flyback. First, remove the existing primary, if any, and clean off any residual glue or tape. Wind five turns

Table 3-1. Parts List for Plasma Globe Power Supply

R1	27 Ω, 5 watts
R2	120 to 140 Ω, 5 watts
R3	470 Ω
C1	2200 µF, polarized electrolytic
C2	1 µF, polarized electrolytic
IC1	LM7812 12-volt voltage regulator
Q1, Q2	2N3055 npn power transistors
LED1	Light-emitting diode
BR1	Bridge rectifier, 4 amps
T1	120-to-18-volt step-down transformer, 3 amps
T2	Flyback transformer (see text)
S1	SPST switch, rated at 10 amps or better at 120 Vac
Misc.	Heatsink for Q1 and Q2, heatsink for IC1, project box, fuse holder, fuse (10 amps or better at 120 Vac)

All resistors are 5 to 10 percent tolerance, 1/4 watt, unless otherwise noted. All capacitors are 10 to 20 percent tolerance, rated at 35 volts or more.

of 18 AWG insulated hookup wire, twist a 1/2-inch loop in the wire, and finish by winding five more turns. Start winding at the top of the transformer coil and be sure the windings on the top and bottom of the loop go in the same direction. That is, if you start by winding counterclockwise, also complete the winding in a counterclockwise fashion. While winding, you might need to tape or clamp the

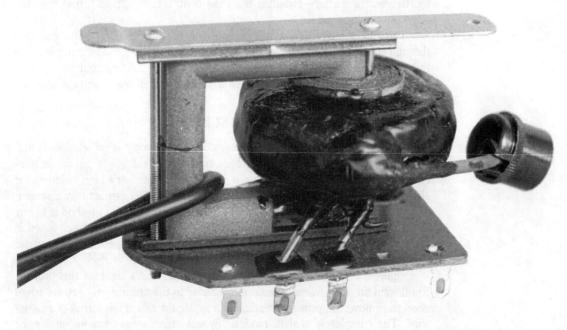

Fig. 3-1. A "typical" flyback transformer. This one has just a single loop of wire for the primary.

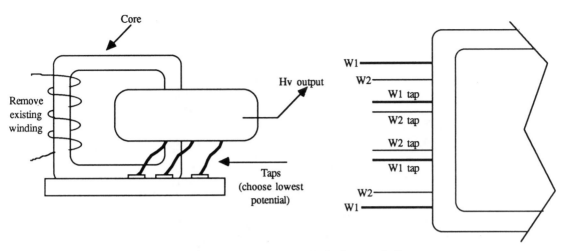

Fig. 3-2. How to rewire the flyback transformer with a new set of primary windings.

wires in place so they don't slip. With a second length of 18 AWG hookup wire, place another winding over the existing one, this time making a total of four turns. Wind two turns, make a 1/2-inch loop, and wind two more. The second winding should turn the same direction as the first (such as counterclockwise). Place the center loop of this new winding opposite the loop you previously made; however, the loops should not be more than about 1/4-inch apart.

Carefully remove 1/8 inch of insulation from the two center loops (these loops become center taps of the windings). Make sure the two loops are physically separated by about 1/4 inch. Wrap tape around the windings so that the ends and center loops are exposed but the rest is securely attached to the primary core. Your rewired flyback should look something like the one in FIG. 3-3.

Follow the schematic in FIG. 3-4 for completing the flyback transformer hookup. The two 2N3055 transistors *must* be mounted on a heatsink, using suitable insulators and heat-transfer paste. The heatsink should provide at least 30 square inches (front and back) of cooling area for *each* transistor. If you find the transistors get too hot, you'll need to provide larger heatsinks or add a small fan to circulate air around the transistors.

Resistors R1 and R2 are rated at 2 watts or more each. They, too, should be mounted in or on heatsinks. You can purchase high-wattage resistors through most electronics supply houses and through surplus, such as All Electronics (see Appendix A).

Most flyback transformers have a number of low-voltage taps. You should use the tap one that provides the lowest potential; that is, the terminal that is closest to 0 volts. You might not be able to determine this until the transformer is tested under power. Still, you might have some luck testing the taps while the transformer is unpowered using a volt-ohmmeter set to low ohms. Connect one

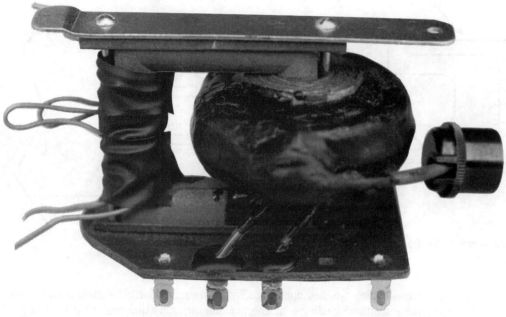

Fig. 3-3. Photo of a rewired flyback transformer, ready for action.

lead of the meter to the high-voltage output and the other lead to one of the taps. The ground tap should have the highest resistance.

Mount the modified flyback transformer in a large plastic box (avoid metal), and attach the heatsinks to the outside of the box. Lead lengths between transistor and transformer should be as short as possible. See FIG. 3-5 for construction details. Avoid mounting the heatsinks inside the enclosure, as that will prevent adequate ventilation. If you can't avoid internal installation for the transistors and heatsinks, you'll need to provide ventilation slots or holes or forced-air cooling with a small rotary fan.

Complete the power supply by adding the primary power transformer (it steps down 120 Vac to 12 Vac), the diode bridge, filter capacitor, and fuse. Double-check your wiring and be on the lookout for shorts. Check with a multimeter to be sure.

TESTING THE TRANSFORMER

You should test the transformer before building the plasma globe and completing the project. Connect a wire from the high-voltage output terminal and position it close (1/8 to 1/4 inch) from the ground terminal.

Plug the transformer into a wall socket and turn the power switch to ON. Immediately you should hear a crackling sound and you might see a spark jump

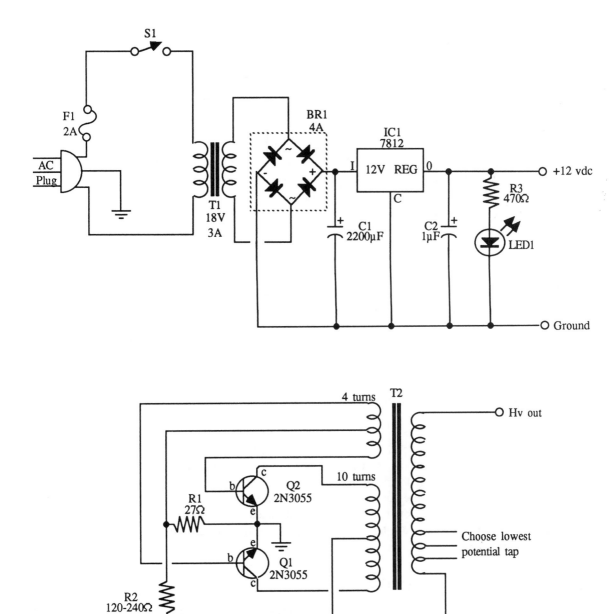

Fig. 3-4. Schematic diagram for the flyback transformer power supply. IC1 should be attached to a suitable heatsink, as with transistors Q1 and Q2.

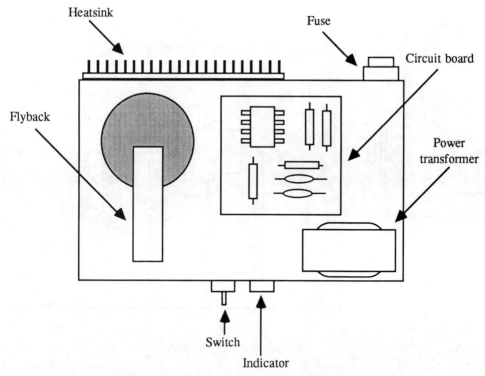

Heatsink

Fuse

Circuit board

Flyback

Power
transformer

Switch

Indicator

Fig. 3-5. Suggested parts layout for the base of the plasma generator.

from the output terminal. If you don't, turn the power switch to OFF and position the output terminal a little closer to or farther away from the ground post. *Do not* operate the power supply for more than 2 or 3 seconds.

Another method of testing the power supply is to connect a wire from the high-voltage output to the center or rim of an ordinary incandescent light bulb. Leave the ground unconnected. Turn the power supply on and the bulb should glow with an eerie purple haze.

CONSTRUCTING THE PLASMA GLOBE

The glass plasma globe is the toughest to make because it requires many steps. You might find it easier to purchase a ready-made globe (one source is Information Unlimited; address provided in Appendix A), or to use a large 4- or 5-inch clear decorator bulb. Use the following plans to construct your own globe. A parts list for the globe is provided in TABLE 3-2.

The plasma globe consists of two basic parts, a clear glass sphere and a metal high-voltage output terminal. To operate properly, the glass sphere must be sealed with the metal terminal centered inside it. Air must then be with-

Table 3-2. Parts List for Plasma Globe

1	3^1/$_2$-inch diameter, ^{1}/$_8$-inch thick acrylic plastic disc
1	5-inch #10 machine bolt, with assorted hardware
1	4- to 5-inch glass decorator globe (untinted preferred)
1	Round metal draw pull
1	^{1}/$_2$-inch-long schedule 125 PVC pipe (about 2^3/$_4$ inches)
1	^{1}/$_4$-inch NPT brass gas cock

drawn from the globe so there is at least a partial vacuum. The better the vacuum, the brighter and sharper the rays of plasma. Procedures for sucking the air out of the globe are presented later.

The best glass spheres to use are 6-inch-diameter decorative lamp globes. You can buy them at most any home improvement store for as little as $4 or $5. The clear or gold-tinted globes are better than the smoked variety; you'll be able to see the glowing plasma better.

Cut a 3^1/$_2$-inch-diameter circle out of a piece of ^{1}/$_8$-inch plastic (you might be able to buy the circle already cut for you). As shown in FIG. 3-6, drill a ^{1}/$_4$-inch hole in the center and a ^{1}/$_4$-inch hole one inch off to one side. The plastic disc serves as a base and lid for the glass globe, so if your globe has a larger or smaller opening, adjust the size of the disc accordingly.

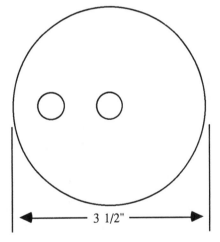

Fig. 3-6. Template for the plastic disc that serves as the base of the glass sphere.

3 1/2"

Thread two nuts and accompanying washers onto a 5-inch-long #10 bolt, as shown in FIG. 3-7. Insert the end of the bolt through the hole in the disc, and secure it in place by adding a washer and nut on the other side of the disc.

The output terminal is a 1-inch-diameter metal hardware drawer pull, available at most any hardware store. They come in both chrome- and brass-plated

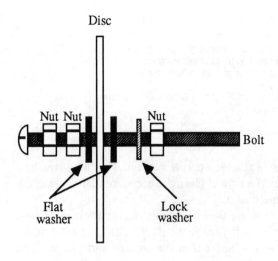

Fig. 3-7. Attach the bolt to the
disc with hardware as shown.

versions; I used the chrome-plated kind. The drawer pull is ideal for the output terminal because it is spherical shaped, already polished to a smooth finish, and it includes a threaded hole for #10 hardware.

Add a #10 nut and washer to the end of the bolt. Then secure the drawer pull to the top of the bolt. Temporarily fit the drawer pull and bolt into the glass globe. Visually check that the pull is in the center of the globe. If not, adjust the nuts on the disc end of the bolt to add or subtract some length.

Now measure the distance between the underside lip of the pull and the surface of the disc. Cut a piece of $1/2$-inch diameter PVC plastic tube to *exactly* this size (about $2^3/4$ inches). Remove the drawer pull.

Slip the plastic over the bolt and place the drawer pull back on. As you tighten the drawer pull, center the plastic so the bolt goes right through the middle. The finished assembly should look like FIG. 3-8. Be sure the drawer pull is firmly attached to the bolt and that the nut and washer act to anchor the metal drawer pull in place. Apply RTV silicone sealant around the rib of the plastic tube at both the drawer pull and disc to seal off the air inside the tube.

In the second hole in the plastic, insert a $1/4$-inch NPT brass gas cock. You might also be able to use a plastic or brass valve designed for aquarium use. Apply RTV silicone sealant around the threads of the gas cock or valve to ensure an airtight fit. It is vitally important that the seal around the threads is complete.

Clean the inside of the glass globe with mild detergent and water and let it dry completely. Insert the drawer pull terminal assembly into the end of the glass globe and press the plastic disc firmly against the opening in the globe. Coat the rim of the globe with epoxy to secure the plastic to the glass. Let set, and then liberally apply clear RTV silicone sealant all around the rim of the glass. You should try for one, continuous bead of silicone around the rim. Let the silicone sealant set completely, preferably overnight.

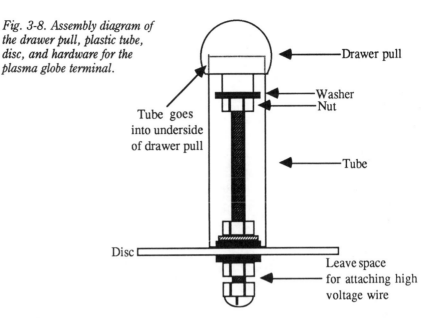

Fig. 3-8. Assembly diagram of the drawer pull, plastic tube, disc, and hardware for the plasma globe terminal.

Drawer pull

Washer
Nut

Tube goes into underside of drawer pull

Tube

Disc

Leave space for attaching high voltage wire

EVACUATING THE GLASS GLOBE

One of the hardest parts of building the globe is clearing the air from inside it. There are several methods you can use. In all cases, the recommended vacuum is 0.5 to 2.0 torr. One torr equals 1 mm of mercury, expressed as 1 mm/Hg, or $1/760$ of an atmosphere). Use a vacuum gauge for accuracy. If the vacuum gauge is graduated in mm/Hg, the desired vacuum range is 0.5 mm/Hg to 2.0 mm/Hg.

To evacuate the globe, use one of the following techniques.

- Use a motorized vacuum pump to suck out the air. Close the gas cock when you reach the desired vacuum.
- Use a hand-operated vacuum pump like that in FIG. 3-9, available through most school lab supplies as well as Edmund Scientific Company. The volume of air inside the globe is small, so you don't need to pump for long.
- Use a water-operated aspirator (also available at most school lab supply companies). The water aspirator connects to a water faucet and produces a vacuum by running water pressure. Most water aspirators don't come with a vacuum gauge, so you'll need to add one yourself.

To evacuate the globe, attach the vacuum apparatus (motorized vacuum pump, hand-operating vacuum pump, or water aspirator) to the gas cock mounted on the underside of the plastic disc. Open the gas cock and remove the air. Close the gas cock when the desired pressure is reached, but keep the vacuum apparatus connected. At 5- or 10-minute intervals, open the gas cock and

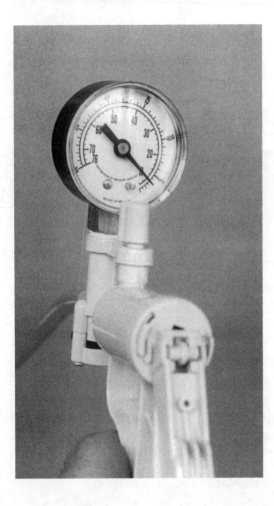

Fig. 3-9. A hand-operated vacuum pump, available through most lab supply outfits.

watch the vacuum gauge to see if air has reentered the globe. If it has, there's a leak.

Inspect the seals around the rim of the globe and the threads of the gas cock (or aquarium valve) and reapply silicone sealant, if necessary. Evacuate the air once again.

CONNECTING THE GLOBE TO THE POWER SUPPLY

The globe and power supply should be enclosed in a plastic cabinet for safety and looks. A parts list for the enclosure is provided in TABLE 3-3.

Cut a piece of 4-inch black ABS plastic pipe to three inches in length. Place the pipe around the base of the glove over the plastic disc. Apply epoxy to the inside of the pipe to secure the globe in place.

Connect a wire from the high-voltage output of the flyback transformer to the bolt on the underside of the plastic disc, as shown in FIG. 3-10. Use the nuts you previously added to firmly secure the wire to the bolt. Be sure the wire is

Table 3-3. Parts List for Enclosure

1	Project box for enclosure, approximately 8 by 8 by 4 inches
1	3-inch length of 4-inch-diameter black ABS plastic pipe for neck
2	$1/2$-inch by 1-inch L brackets
Misc.	Hardware for mounting, grommet for power cord

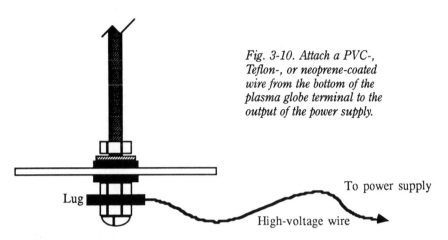

Fig. 3-10. Attach a PVC-, Teflon-, or neoprene-coated wire from the bottom of the plasma globe terminal to the output of the power supply.

held tightly in the nuts and is flush to the metal. Cut off any extra exposed pieces of wire. Otherwise, the wire could arc and reduce the effectiveness of the globe.

Attach the bottom of the plastic pipe to the top of the plasma generator case. The prototype used two small $1/2$-inch L-shaped brackets, as shown in FIG. 3-11.

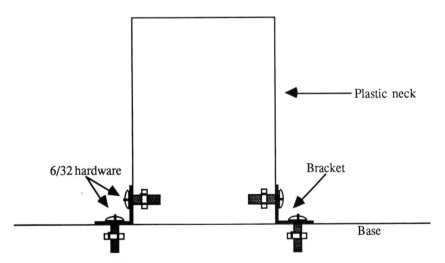

Fig. 3-11. Attach the plastic neck to the base of the generator with $6/32$ hardware and L brackets.

OPERATING THE PLASMA GENERATOR

Assuming everything is wired correctly and the air is sucked out of the globe, turning the plasma generator on should reveal a vivid display of deep blue streams of light within the glass sphere. You can direct the streams by placing your fingers on the glass globe. You can make the plasma streams dance around the glove by moving your hand around the periphery of the glass.

While your fingers are on the glass, don't touch a grounded object, or you'll receive a shock. Why? The plasma streamers are charged to a high voltage, and because the voltage is modulated at a high frequency, it will penetrate the glass and reach your hands. Ordinarily, you won't feel a shock, but touching a grounded object will cause the high voltage to flow through you. The shock is not really dangerous, but it can be unpleasant.

You can experiment more with high-frequency, high-voltage breakdown of ordinary dielectric materials by building the Tesla coil described in Chapter 6. Sparks from the Tesla coil will penetrate 1/4-inch or thicker glass plates.

ADDING GASES TO THE GLOBE

Simply evacuating the air from the globe produces a pleasing blue color plasma. You can generate other colors by adding different gas mixtures. Suitable gases include helium, argon, nitrogen, and carbon dioxide. *Under no circumstances should you use flammable gases* such as hydrogen, fluorine, methane, or propane.

To introduce the gas into the sphere, you need to set up a vacuum pumping system as shown in FIG. 3-12. Most motorized and hand-operating vacuum pumps provide the necessary inlets and outlets already, but if yours doesn't, a simple assortment of valves and tubes will suffice.

Gases such as helium, nitrogen, carbon dioxide, and argon are commonly available at almost any welding supply shop. The gases come pressured in a cylinder; get the smallest cylinder because you don't need much gas. Remember: the gas inside the cylinder is pressured, in many cases up to 2,200 pounds per square inch, so you must use a regulator and pressure gauge. *Never* attempt to fill the plasma globe with gas directly from the cylinder. High-pressure gas could cause the globe to shatter into tiny fragments, seriously injuring you. Wear safety goggles for protection in any case.

First, draw out as much air as you can from the globe. Strive for at least 0.05 to 0.1 torr. Close the vacuum valve, and then crack open the gas regulator to no more than a few tenths of a pound per square inch—as little gas as possible. Now, open the gas valve to let in some gas. Close the gas valve when the globe pressure reaches 0.5 to 2.0 torr.

USING CLEAR LIGHT BULBS

You might not want to go through all the trouble of constructing the plasma globe. An alternative is to use a large, clear decorator light bulb. The bulb

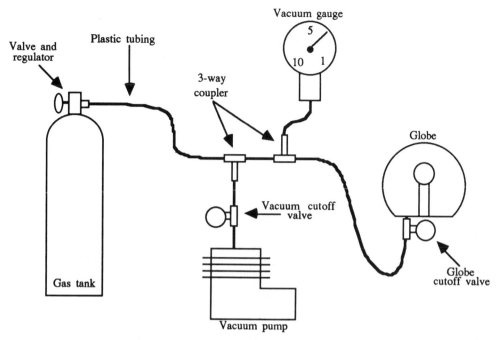

Fig. 3-12. Only a small assortment of tubing, couplers, and valves are needed to introduce nonvolatile gases into the plasma globe.

comes already sealed and evacuated and is ready to be connected to the high-voltage terminal of the power supply. Because the light bulb uses a thin filament instead of a spherical terminal, the plasma streamers are not as pronounced or as thick. Although black-and-white photographs don't do plasma generators justice, you can see the visual effect of the light bulb in FIG. 3-13. The streamers are deep violet in color.

Connect the high-voltage output of the power supply to the middle or rim connection on the light bulb. Because decorator light bulbs come with a standard base, you can use an ordinary socket for rigging up your plasma generator.

The thickness of the plasma streamers is mainly predicated on the frequency of the high-voltage power source. Using the values of R1 and R2 as shown, the frequency is between 22 and 25 kHz. To make the streamers thicker, use a slightly lower value for R2; to make the streamers thinner, use a slightly higher value.

The plasma streamers are best seen when using the globe in a darkened room. To view the globe during the day, place it in an open-ended wooden or plastic box that is painted flat black on the inside.

Fig. 3-13. An ordinary, clear decorator bulb makes an excellent plasma globe. The plasma trails are blue due to the rare atmosphere in the globe.

4

Build a high-tension induction coil

The principles of induction were first discovered by Michael Faraday (1791 – 1867). Out of Faraday's basic work came the induction coil, a device that develops a high voltage inductively from a nearby low voltage.

This chapter details the construction of a high-tension induction coil using easy-to-find and inexpensive parts. The project as described uses an automotive ignition coil, which you can purchase locally for as little as $4 or $5. The remaining components are common electronic store finds.

You'll find plenty of applications for your induction coil. You can use it as the basis of a fence charger to keep rodents or other small animals out of your garden, as an ozone maker for purifying water, as bug and rodent zappers, and much more. These unusual applications are described in this chapter as well.

INDUCTION COIL DESIGN

As shown in FIG. 4-1, the induction coil comprises two coils: a primary winding for the low voltage and a secondary winding for the high voltage. The primary winding consists of a few hundred turns of thick, insulated wire wrapped around a soft iron core. Low voltage—say 12 volts from a car battery— is passed through this winding. The secondary consists of many thousands of turns of thinner wire. The secondary winding is effectively wrapped around the primary.

Although the two windings—primary and secondary—are electrically isolated, the electromagnetic field created in the primary winding and iron core is *inductively* transferred to the secondary winding. Because there are many more windings in the secondary than in the primary, the primary voltage is stepped up by a high ratio, usually 100 or more to 1 (most are 100:1 but many are 250:1 and higher).

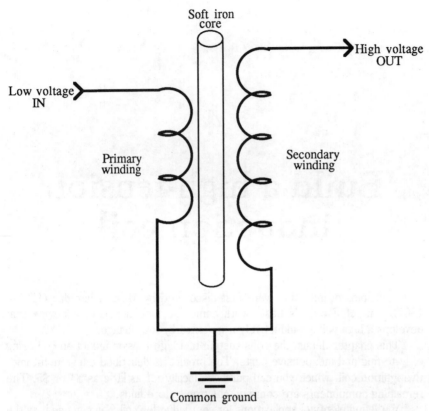

Fig. 4-1. An induction coil consists of just two windings, a primary and secondary, wrapped around a soft iron core. Most coils are shrouded in oil to keep them cool.

The induction coil is the basis of most modern transformer designs except for one important difference: The induction coil is expressly made to generate high voltages (a minimum of 3 kV) from low voltages . Most transformers manufactured today are designed to step moderate voltages down to low voltages—117 Vac house current to six or 12 Vac, for example.

Induction coils are common finds in older automobiles (up to about 1985 for most makes). The coil (see FIG. 4-2) is used to create the high-voltage spark sent to each of the spark plugs. (Nowadays, most cars use electronic ignition, which creates the high voltage by other means.)

Ignition coils typically consist of three terminals:

- A high-voltage anode, usually in the center and recessed deep within the body of the coil. The recessed location is a safety feature.
- A positive battery supply connection. This attaches to the positive rail of the battery.
- A common ground connection. This attaches to the negative rail of the battery, usually through the ignition points of the automobile.

Fig. 4-2. The automotive ignition coil is a common staple of the gadgeteer. They're easy to find and cheap; this one costs about $8 at K-Mart.

The positive and ground connections are generally reversible.

There are plenty of experiments you can conduct with an automobile induction coil. In addition to the experiments detailed later in this chapter, the Tesla coil in Chapter 6 and the Kirlian photography apparatus in Chapter 7 use induction coils to generate high voltages (in fact, all three use the same basic schematic). The project that follows uses an ordinary 12-volt car induction coil to produce up to 40,000 volts!

But first, a word of *caution*: The voltage output from the high-tension induction coil project can *kill you*. Do not expose yourself to its spark. Operate the coil only with the protective guards in place (as described later in the project). When experimenting with the apparatus, follow all safety precautions carefully. Remember, your life depends on it.

BUILDING THE INDUCTION COIL

The high-tension induction coil described here is ideal for both battery or ac operation. The circuitry and induction coil run on 12 volts dc, which can be supplied by a heavy-duty battery (like a car battery) or through a stepdown transfer, bridge diode, and filtering capacitor, as shown in the schematic in FIG. 4-3. A parts list for the induction coil project is provided in TABLE 4-1.

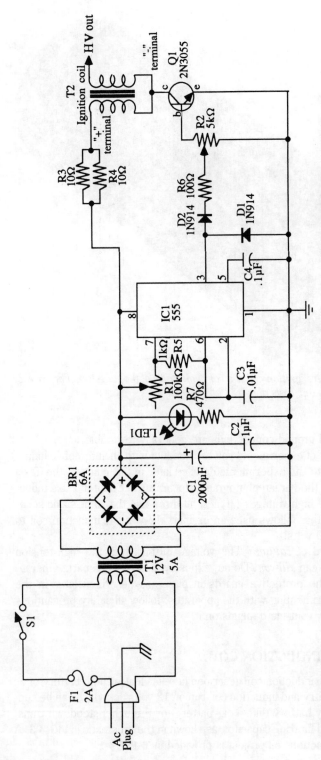

Fig. 4-3. Use this schematic (adapted from Walt Noon's Build a 40,000 Volt Induction Coil) to make your high-tension spark maker.

Table 4-1. Parts List for Induction Coil

R1	100 kΩ potentiometer
R2	5 kΩ potentiometer, 2 watts
R3, R4	10 Ω, 10 watts
R5	1 kΩ, 1 watt
R6	100 Ω, 2 watts
R7	470 Ω
C1	1000 μF electrolytic
C2, C3, C4	.1 μF disc
Q1	2N3055 npn power transistor
D1, D2	1N914 glass diodes
LED1	Light-emitting diode
BR1	Bridge rectifier, 4 amps
T1	120-to-12-volt step-down transformer, 5 amps
T2	Auto ignition coil, 12 volts
S1	SPST switch, rated 10 amps or better at 120 Vac
Misc	Ac plug and line cord, fuse holder, fuse (10 amps at 120 Vac), enclosure, terminal ball, knobs for potentiometers

The circuit uses an ordinary 555 timer IC to provide the pulses required for high-voltage generation. You can build the circuit on perf board or a portion of a predrilled and etched universal experimenter's solder board. Transistor Q1, a 2N3055, must be mounted on a heatsink.

Enclose the circuit and ignition coil in a metal or plastic (plastic is preferred) hobby box. A sample parts layout is shown in FIG. 4-4. The top of the ignition coil can poke through a large hole drilled through the top of the box. Attach a

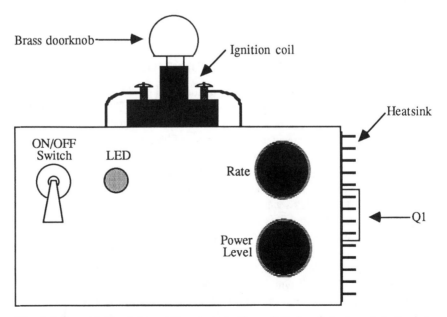

Fig. 4-4. Suggested parts layout for your induction coil device. Cut a large hole through the top of the project box to let the ignition coil poke through.

two-inch-diameter brass ball or brass doorknob handle to a length of heavy-duty spark plug wire. Stick the opposite end of the wire into the recessed high-voltage terminal in the coil.

Note that while the circuit runs under low voltage, it still produces in excess of 20 kV at the high-voltage output, so it should *not* be considered fundamentally safer than the ac-operated version of the induction coil.

ADJUSTING THE COIL OUTPUT

You can control the output of the induction coil by varying R1 and R2. R1 changes the output frequency of the 555, which affects the high-voltage output of the induction coil. Automotive induction coils operate best at fairly low frequencies—about 250 Hz and lower, so you will find that the faster the 555 oscillates, the less output you get from the coil. Conversely, at very low frequencies, the output of the coil is proportionally reduced. A setting between these two extremes provides the best results.

Potentiometer R2 varies the power supplied to the base of transistor Q1. A minimum setting provides little output from the 555 to reach the transistor, hence a lower overall voltage applied to the ignition coil. A maximum setting passes more output from the 555, which drives the ignition coil harder, thus producing more voltage.

You can visually test the output level of the coil by watching the spark. The longer and thicker the spark, the higher the voltage. TABLE 1-1 (in Chapter 1) provides a ready reference for calculating the approximate voltage based on spark length. Be sure to take into account the diameter of the output terminal. Note also that a needlepoint output terminal yields longer sparks for a given voltage.

A READY-MADE HIGH-TENSION INDUCTION COIL

In the projects described above, the ignition coil produces high voltages by introducing a chopped, low voltage to the ground terminal. The chopping is performed by the 555 timer IC, operating in astable multivibrator mode. Some ignition coils include their own voltage chopper, so a 555 timer and all its associated circuitry is not needed. Unfortunately, these self-chopping coils can be hard to get and are often more expensive than standard ignition coils.

The most famous self-chopping coil is the Model T spark coil. This coil, shown in FIG. 4-5, includes a vibrator to chop the voltage to an internal coil. The unit shown in the photograph is a modern reproduction, available through J.C. Whitney or Vintage Auto Parts (see Appendix A for addresses) and some antique auto parts stores. I paid too much for mine (I'm embarrassed to say how much), but if you look carefully, you might be able to find one for about $20.

Connect the Model T ignition coil as shown in FIG. 4-6. The coil is designed to run with 6 to 12 Vdc. You can operate the coil for long periods of time under six volts, but only intermittently under 12 volts. Either case, the coil needs a

Fig. 4-5. A replica of the old Model-T ignition coil. The device includes its own induction coil, condenser, points, and vibrator, and when connected to a 6-volt dc source, is a complete, self-contained spark maker.

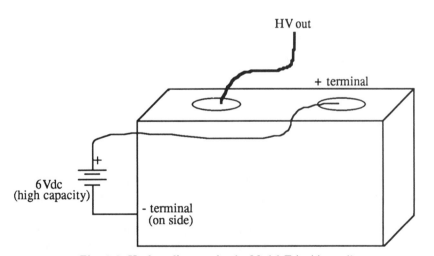

Fig. 4-6. Hookup diagram for the Model-T ignition coil.

heavy-duty car or motorcycle battery to operate. The coil has its own points, which you might need to adjust to obtain a high-voltage output.

In addition to the Model T coil, a 12-volt spark coil made by Delco-Remy (and available through Vintage Auto Parts) can be used in the same manner.

EXPERIMENTING WITH INDUCTION COILS

Now that you've completed your high-tension induction coil, what do you use it for? Some ideas:

- Ozone water purifier.
- Fence charger.
- Bug and rodent zapper.
- Spark platform.

Parts lists for these projects are provided in TABLE 4-2.

Table 4-2. Parts List for Induction Coil Projects

Ozone Water Purifier
Plastic enclosure for induction coil
Aquarium pump
Water tank (plastic preferred)
Nylon tubing for ozone gas

Fence Charger
Wire, 12 to 20 gauge, for perimeter of fence
Ceramic or plastic standoffs for each fence post
Ground rod for ground connection

Insect Zapper
Printed circuit board with etched spiral pattern; nominal size is 6-by-6 inches

Rodent Zapper
Two metal plates (approximate size is 6-by-9 inches)
Two plastic standoffs
Plastic bowl
Food that rodents like to eat

Ozone water purifier

You can create a water purification system using the coil to produce ozone. Construct an ozone concentrator as shown in the simplified diagram in FIG. 4-7. (NOTE: you can also use the Jacob's ladder or plasma sphere power supplies, described in Chapters 2 and 3, respectively, to create the arc for ozone production.)

Place the induction coil in a closed plastic jar. Arrange the ground terminals close to the output sphere so that you obtain the longest, loudest spark possible.

Stretch a length of Teflon aquarium tube to the ozone pump (an ordinary aquarium pump will suffice as long as it has inlet and outlet ports). The output of the pump is connected to another length of Teflon tubing and into the bottom of a small water tank. Use a check valve in the output tubing to prevent the water from the tank from seeping into the pump.

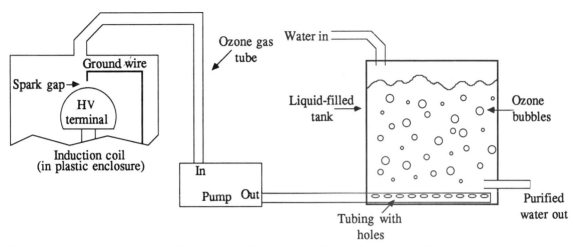

Fig. 4-7. Ozone produced by the discharge from the ignition coil can be used to purify water, using the apparatus shown here. Caution: Ozone is poisonous and extremely corrosive.

In operation, ozone gas percolates from the bottom of the tank, through the water, thus purifying it. You can add to the purification procedure by adding filtration to the water inlet. The filter (paper and carbon works best) removes large particulate matter.

Note that ozone is corrosive, so the aquarium pump will break down after a while. The plastic and rubber parts inside will harden after a while, and simply cease to function. If you're serious about building a permanent water purification system, invest in a lab-grade pump designed to handle corrosive gases.

Fence charger

Used at a low- or medium-voltage level, the induction coil can be effectively used as an electric fence charger. You should use the coil as a charger *only* when humans (adults or kids) are sure not to touch the fence. *Never* electrify a fence where unsuspecting persons might touch it. You could be in deep legal trouble even if the person is unharmed. Appropriate fences for electrification are horse corrals, animal pens, and garden perimeters. You can even put a charge on metal trash cans to help keep pesky raccoons away from your garbage.

Figure 4-8 shows how to electrify a heavy metal wire that stretches across the top of a fence. Number 6 or number 8 aluminum grounding wire works best. Note that the wire *must* be electrically insulated from the ground. Use plastic or ceramic standoffs to insulate the wire from the body of the fence, even if the fence is made of nonconducting wood (some wood does conduct, especially when wet). You can electrify an entire metal fence, but it must be completely insulated from the ground. This is generally a tough order to fill, so most people don't do it.

Connect one end of the charged wire to the high-voltage output of the coil and the other end to a good earth ground. You can obtain a suitable ground by

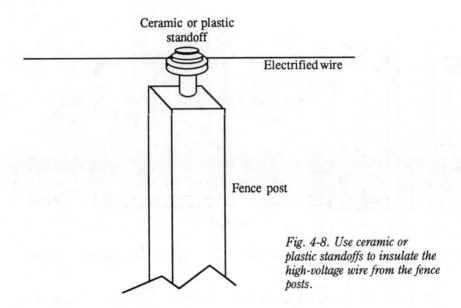

Fig. 4-8. *Use ceramic or plastic standoffs to insulate the high-voltage wire from the fence posts.*

driving a one foot (or longer) metal stake in the ground, and attaching the chassis ground connection to it. If the fence isn't operating up to par, check this ground connection first.

You can also use a cold-water pipe (metal of course) for a ground connection. *Under no circumstances should you use a gas pipe or hot-water pipe for establishing a ground connection.*

Bug and rodent zapper

The larger the bug or animal, the more current its body will conduct and the nastier the shock will be. However, past a point, large enough animals can sustain even a heavy dose of electricity. That makes the induction coil an effective device for the elimination of bugs and rodents, but not for larger animals like cats, dogs, and raccoons.

The idea behind bug and rodent zappers is that the high voltage isn't enough to cause a corona between the output terminals, but it *is* enough to stun or kill the body of the critter that comes across the terminals.

Flying bugs can be readily controlled using a commercially available bug zapper, but crawling bugs need a setup like that shown in FIG. 4-9. Here, a large-area printed circuit board is etched so that a cockroach, potato bug, or other insect comes into contact with the output terminals, which alternate in a lattice. Instead of a big printed circuit board, you can apply self-adhesive copper foil to a 1/8-inch or thicker plastic base. (The copper foil need not have a conductive adhesive.)

Larger rodents can be controlled using the two-plate approach shown in FIG. 4-10. Here, two metal plates or metalized screens are placed on insulators

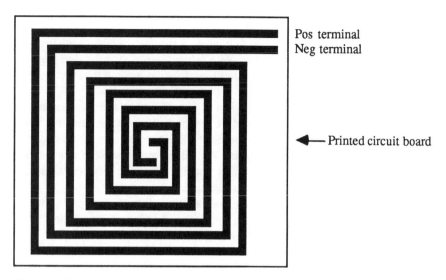

Pos terminal
Neg terminal

Printed circuit board

Fig. 4-9. Etch a set of spiraling traces on a printed circuit board for use in a bug zapper. Experiment with the spacing of the traces to prevent arcing (usually no less than 1/8 inch apart).

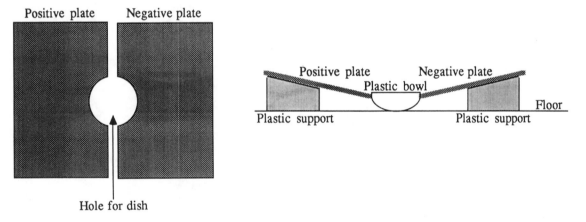

Positive plate   Negative plate

Hole for dish

Positive plate   Negative plate
Plastic bowl
Floor
Plastic support   Plastic support

Fig. 4-10. For zapping (but probably not killing) small rodents, place two metal plates on plastic supports as shown. The critter receives a shock when it touches both plates at once.

on either side of a good bowl. When the rodent touches both plates, it completes the circuit and receives a shock. Most rodents won't be killed, but they won't make it a habit of hanging around your yard or house.

Note that very small bugs (ants, fleas, etc.), are not effected by the zapper. Not only are their bodies too small to stretch between the terminals, but their small size limits the amount of current passing through them. As an experiment, place an ant on the spherical metal terminal of the induction coil and turn the device on. The ant will continue to walk around happily.

Spark platform

A purely fun but educational project involves the use of high voltage from an induction coil and ordinary iron filings. Attach two 1/2-inch strips of thin brass to a 5-by-7 piece of wood or plastic. Place the strips no further than two inches apart.

Clip the high-voltage output (anode and ground) from the coil to the ends of the strips. Be sure the electrical contact is secure. Place some fine iron filings in a pepper shaker and gently shake some of the filings to make a light carpet of black dust between the two metal strips (iron filings are available at many hobby stores in the chemistry department, at school lab supply outlets, and a number of mail-order firms, including Edmund Scientific).

Dim the room lights and turn the coil on. Hundreds of tiny sparks should move between the metal strips and the iron filings. Add more filings if the sparks are weak, but don't lay down too thick of a coat.

GOING FURTHER

If you enjoyed working with the induction coil described in this chapter, you might be interested in building other, more powerful versions. For starters, consult Walt Noon's excellent booklet, *How to Build a 40,000 Volt Induction Coil* (which served as the inspiration for the induction coil in this chapter) for more information on induction coil design and modification. The booklet is available from Lindsay Publications. Also see *High-Voltage Devices II*, published by Consumertronics (see Appendix B for more information on these publications).

Allegro Systems (see Appendix A) publishes plans for an extremely high-voltage induction coil—90 kV to 125 kV output—using some rather novel design approaches. Be aware that this device is not a toy and is too powerful for use in ozone production, bug zapping, and fence charging. Useful applications for the device include an ion source in accelerators, high-voltage power supply for nitrogen superradiant gas discharge lasers, and lightning simulation.

5

Build a high-voltage
Van de Graaff generator

Over 50 years ago, Robert J. Van de Graaff developed a unique contraption that generated potentials in excess of a million volts. It was Van de Graaff's aim to use his device, now universally known as a Van de Graaff generator, to create voluminous amounts of ions for electron acceleration.

The early particle accelerators used in nuclear research during the late 1930s and early 1940s were largely based on Van de Graaff's amazing device.

The Van de Graaff generator is a common staple of schools and industrial labs. Sizes vary and so do the voltage outputs. A typical table model produces 300,000 to 500,000 volts yet stands only a few feet high and draws little more current from the wall socket than a 100-watt light bulb.

Building such a Van de Graaff generator is not complex. With the exception of one component, all the parts are easily available and inexpensive. In less than a full day, you can construct a powerful Van de Graaff generator. In this chapter you'll learn how to construct a moderately-sized Van de Graaff that's suitable for all sorts of scientific experiments.

ABOUT THE VAN DE GRAAFF GENERATOR

A complete Van de Graaff generator is shown in FIG. 5-1. It is composed of a metal base, an insulating tube, and a spherical high-voltage terminal. Inside is a high-voltage ion spray source (more about that later), a motor, a wide rubber band, and metal brushes. Astonishingly enough, this mish-mash of flea-market junk combines to create a device that literally creates miniature bolts of lightning.

The high-voltage output of the Van de Graaff generator is essentially harmless. Though the voltage is extremely high (about 250,000 volts in our prototype), it is pure static electricity and barely measures over a few microamperes.

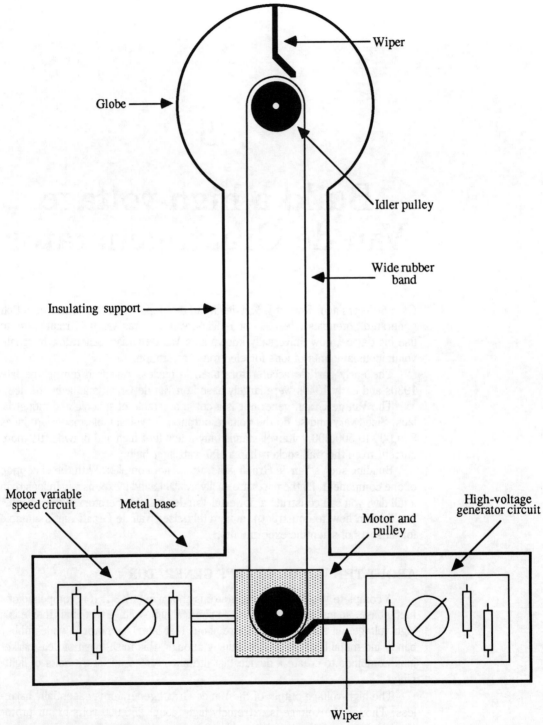

Fig. 5-1. An overview of the construction of a Van de Graaff generator.

A common Van de Graaff parlor trick is to touch the high-voltage output terminal, which charges the entire body with hundreds of thousands of volts of static electricity. So charged, your hair stands on end like a mad scientist who's just perfected his latest diabolic creation. Turn the Van de Graaff generator off, and your hair slowly settles back down.

An overview of static electricity

Before going much further into the Van de Graaff generator, let's pause for a moment and review the nature and character of static electricity. Static electricity is an electrical charge built up in an insulating body. The electricity will remain until discharged. A good example of charged and discharged static electricity is when you walk across a carpet and touch the metal knob of a door. You feel a shock as the static, built up in your body, is discharged on the door knob.

Static electricity is different from dynamic electricity. In the latter, the electrical charge flows through a conductor and quickly dissipates. Dynamic electricity is the kind generated by the induction coils and flyback transformers discussed in the previous chapter. Static electricity is the kind built up in clouds (resulting in lightning) or a cat's fur on a warm, dry day.

The study of static electricity—or electrostatics—goes back thousands of years to the times of ancient Egypt. One of the first controlled experiments with electrostatics involved rubbing a piece of animal fur against a block of amber (amber is a brownish-yellow fossil resin used to make jewelry). The action of rubbing the two materials together caused them to be electrically charged with respect to one another. The animal fur became positively charged, and the amber negatively charged. The fur could then attract other more negatively charged objects (such as silk), and the amber attracted more positively charged objects (like glass and human hair).

TABLE 5-1 represents the *triboelectric series*. The triboelectric series ranks many common materials by their ability to build and retain a positive or negative charge. Materials at the top of the list tend to build a positive charge (a lack in electrons) and materials at the bottom of the list tend to build a negative charge (an abundance of electrons).

For the greatest charge differential, rub together two materials that are as far apart on the table as possible. That is, rubbing plastic kitchen wrap against glass creates a very high charge differential, but rubbing wood and cotton (which are right next to one another) does not.

A practical experiment

An easy electrostatic experiment you can conduct requires:

- A short length of glass rod or a test tube
- A piece of polyester thread
- A swath of wool
- A piece of soft plastic wrap (like the kind used to protect clothes returned from the cleaners)

Positive Polarity (+)
Asbestos
Rabbit's fur
Glass
Mica
Nylon
Wool
Cat's fur
Silk
Paper
Cotton
Wood
Lucite
Sealing wax
Amber
Polystyrene
Polyethylene
Rubber balloon
Sulfur
Celluloid
Hard rubber
Vinylite
Saran Wrap

Table 5-1. The Triboelectric Series

Negative Polarity (–)
Materials high on the list become positively charged (give up electrons) when rubbed with materials lower on the list. The further apart the materials are in the list, the higher the charge will be. Surface conditions, humidity, and other factors can affect the charge potentials.

Vigorously rub the thread on the wool and the plastic on the glass. Note that these materials—polyester/wool, and plastic/glass—are somewhat far apart in the triboelectric series table. Observe the following:

- Unlike charges attract. The negative charge on the thread and the positive charge on the glass cause the two objects to attract each other.
- Like charges repel. The negative charge on the thread is repelled by the negative charge on the plastic.

How Van de Graaff generators work

The Van de Graaff generator produces an abundance of either positive ions or negative ions at its high-voltage terminal (you can control the polarity with the ion sprayer, discussed later). With the high-voltage terminal charged, it repels or attracts materials according to the charge on the materials. In many cases, the Van de Graaff generator attracts almost any material because of the high-voltage potential contained in the terminal. For example, if the terminal is positively charged, just about any material it encounters will have a more negative charge, so the two are attracted to one another.

CONSTRUCTING THE VAN DE GRAAFF GENERATOR

Refer to TABLE 5-2 for a parts list for the general components of the Van de Graaff generator. The Van de Graaff generator uses both mechanics and electronics to produce high-voltage static. Here's how the generator works: At the

Table 5-2. General Parts List for Van de Graaff Generator

1	Output sphere (3 inches in diameter or more)
1	Ac motor
1	Idler pulley (consisting of $3/4$-inch by $1/2$-inch PVC pipe, two $1/8$-inch ID by $4/8$-inch OD ball bearings, and $15/8$-inch-long $1/8$-inch-diameter steel shaft)
1	Motor pulley ($3/4$-by-$1/2$-inch PVC pipe)
1	Insulating support, 13-by-$11/4$-inch PVC pipe
1	Wide rubber band (26 inches in prototype)
2	Ion wipers
Misc.	Enclosure box (nominal size: 8 by 8 by 4 inches), ac cord, on/off switch. Optional: motor variable speed control, high-voltage ion generator, L brackets, nuts and bolts for construction.

core of the Van de Graaff is a wide rubber band. This rubber band acts as sort of conveyer belt for positive and negative ions. The band is kept moving by a small electric motor.

At the bottom of the Van de Graaff machine is an ion sprayer, a high-voltage power supply that literally sprays ions onto the moving rubber band. The high-voltage output terminal of the sprayer is a comb consisting of tiny metal bristles. The ions launch themselves from the tips of each bristle onto the rubber band. The ions are then transported up the rubber band to the spherical output terminal of the Van de Graaff generator. There, metal wipers pick up the ions from the rubber band and distribute them around the metal terminal. Figure 5-2 shows how the rubber band is charged and discharged as it conveys ions up and down length of the Van de Graaff machine.

High-voltage power supplies can create either positive or negative ions. For example, if the power supply is rated at +15kV, its high-voltage terminal creates positive ions. If the power supply is rated at −15kV, it creates negative ions. The Van de Graaff generator can produce either positive or negative ions, depending on the type of power supply you use. You can even build a power supply that lets you switch between positive or negative output.

The hardest part about building the Van de Graaff generator is finding a suitable spherical output terminal. My prototype used a $43/4$-inch-diameter spun aluminum ball originally designed for use in a magic trick. The two halves of the ball easily separate (the seam is not soldered), making it particularly easy to work with. However, the ball is somewhat small: the larger the ball, the higher the potential voltage you can collect.

Finding suitable metal spheres isn't easy, but here are some suggestions:

- Use a magician's sphere, as I did. They are available at most magic shops. Try to get the biggest metal sphere you can.
- Sandwich together two 6- or 8-inch stainless steel salad bowls. This is not the best approach because the ions tend to shoot out from the rims of the bowls.
- Use the brass ball from a flagpole. I haven't tried this but it looks like it might work.

- Spin your own metal sphere. If you have access to a lathe, you can spin two half spheres to make your own metal ball. Metal spinning requires few tools, but it does demand special skills and safety precautions. If you're interested, most books on metal working include a chapter on metal spinning. Since most wood lathes have a maximum capacity of 12 inches, your Van de Graaff output terminal can be up to a foot in diameter, which is ideal.

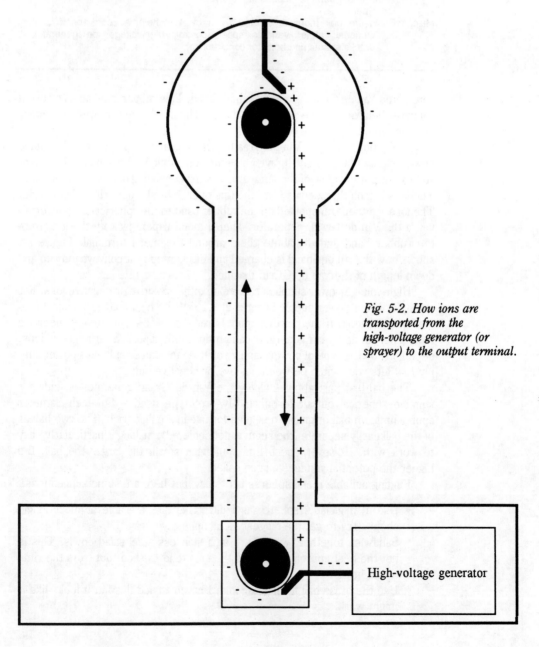

Fig. 5-2. How ions are transported from the high-voltage generator (or sprayer) to the output terminal.

You'll want to adjust the dimensions provided for the prototype to account for the size of the output terminal you use. The dimensions I provide are for a $4^3/4$-inch-diameter terminal; increase the diameter and length of the support insulator if your terminal is smaller or larger. Remember though that the length of the support insulator is greatly dictated by the length of rubber band you use. The prototype uses a 26-inch rubber band (unstretched).

Considering the addition of motor pulleys and idlers plus a 20-percent increase in size due to stretch, the center column of the prototype Van de Graaff generator was cut to 13 inches. The rubber band used in the prototype is $3/4$ inch wide and is the replacement belt for the ready-made Van de Graaff generator sold by Edmund Scientific.

Start construction by cutting a piece of $1^1/4$-inch PVC plastic pipe to 13 inches. File the ends to remove the burrs. If the metal sphere you are using does not already have a hole, you must cut one out to match the outside diameter of the pipe—in this case, a tad larger than $1^5/8$ inches. I found it's easier to start the hole by drilling with a $1/4$-inch bit and then use snips or a metal nibbler to enlarge it to the proper size. Avoid making the hole too big because that will detract from the appearance of the generator and the stability of the terminal.

The base of the Van de Graaff generator is constructed from a 8-by-8-by-4 plastic or metal box (exact size is unimportant as long as all the parts fit inside). If you use a plastic box, you must add a piece of 8-by-8-inch 20 or 22 gauge sheet metal to the top to act as the ground terminal.

Follow the construction plans shown in FIG. 5-3 for mounting the small ac motor. Note that the motor mount is adjustable. Be sure to attach the nylon pulley squarely and securely. For best results, add a small crown to the pulley to keep the rubber band centered. Check the squarness of the motor mount. The pulley must be at right angles to the sides of the base.

Figure 5-4 shows a circuit you can use for altering the speed of the ac motor (a parts list is provided in TABLE 5-3). The circuit uses an SCR to vary the speed of the motor from about 60 percent of normal to 100 percent. The prototype

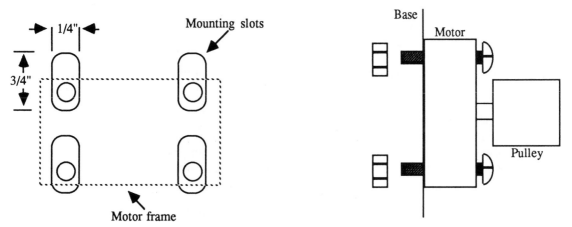

Fig. 5-3. Cut or drill $1/2$- to $3/4$-inch slots for the motor mount so that you can adjust the positioning of the motor.

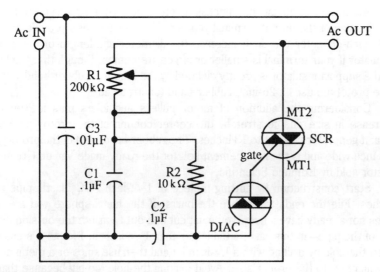

Fig. 5-4. An SCR-controlled ac motor control circuit. This is a full-wave circuit and is best used when the load remains constant.

Table 5-3. Parts List for Ac Motor Speed Control

R1	200 kΩ potentiometer
R2	10 kΩ
C1 – C3	0.1 μF
SCR1	SCR, rated at 200 V at 20 amps or better
DIAC1	DIAC, rated at 200 V at 20 amps or better

All resistors are 5 to 10 percent tolerance. All capacitors are 10 to 20 percent tolerance, and rated at 200 volts or more.

used a ready-made surplus motor speed controller. They are commonly available from surplus mail order outfits (see Appendix A) for less than $5.

The prototype Van de Graaff generator uses a modular high-voltage power supply for the ion sprayer, shown in FIG. 5-5. This supply originally was intended for use in a black-and-white television set. I purchased mine for $5 from Meredith Instruments (again, see Appendix A), but because they are surplus items, there's no telling if they'll still have them by the time you read this or attempt to purchase one. In any case, such self-contained power supplies, which generate up to +15kV from a 12- to 22-volt dc supply, are commonly available from most other mail-order companies.

If you can't locate a suitable modular supply, you can make your own. The flyback transformer power supply designed for the plasma sphere generator in Chapter 3 makes a suitable ion sprayer. You can add a voltage doubler or tripler (obtainable from a discarded TV or from a surplus parts outlet) to increase the output voltage.

Figure 5-6 shows a voltage multiplier you can build from scratch. A parts list is given in TABLE 5-4. The benefit of this multiplier is that you can change the

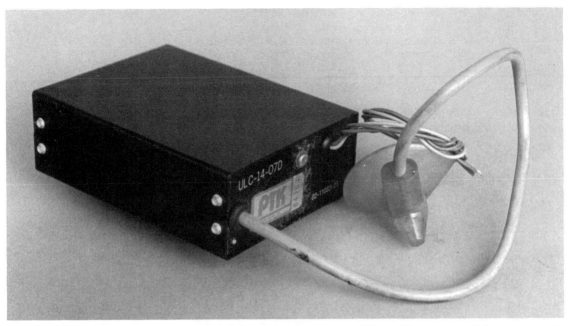

Fig. 5-5. A modular high-voltage power supply salvaged from a portable TV set. If you use a modular power supply, be sure the output is dc; ac won't work (for reasons related to the ion charge on the output terminal of the generator).

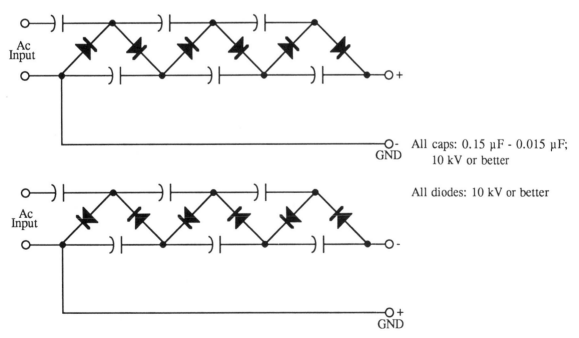

All caps: 0.15 µF - 0.015 µF; 10 kV or better

All diodes: 10 kV or better

Fig. 5-6. This high-voltage tripler can be oriented to provide positive supply (with respect to ground) or negative supply.

Table 5-4. Parts List for Voltage Multiplier

D1 – D6	High-voltage diodes, rated at 10kV or higher
C1 – C6	High-voltage disc capacitors, 0.15 μF, rated at 10kV or higher

output polarity just by inverting the inputs, as shown. A few caveats are in order before you attempt to build the circuit:

- The components should be mounted on perf board. Use care to prevent arcing from nearby components. Coat the circuit and all components with corona dope to prevent arcing.
- High-voltage capacitors and diodes are not that easy to find. Meredith Instruments often carries both, but you'll want to make sure you have a source of the components before attempting to build the voltage multiplier. Don't use capacitors or diodes that are underrated. They could literally blow up.
- The capacitors in the multiplier retain a static charge even after you remove power. You can receive a bad shock if you touch the output terminals. Relieve the charge by temporarily shorting the terminals.

Because the modular power supply operates on 22 to 24 volts dc, you must add a suitable ac stepdown transformer, bridge diode, and filtering capacitor, as shown in the schematic in FIG. 5-7 (parts list in TABLE 5-5). The high-voltage power supply doesn't need a well-regulated voltage, but a variable voltage IC regulator was added anyway to change the output of the power supply. The regulator was attached to a heatsink to help prolong its life.

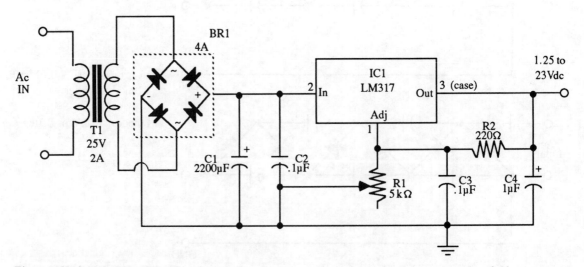

Fig. 5-7. Variable voltage (from about 1.25 to 23 volts dc) can be obtained with the ac-operated variable power supply. Regulator IC1 must be on a heatsink.

Table 5.5 Parts List Variable Input Voltage for High-Voltage Supply

R1	5 kΩ potentiometer
R2	220 Ω
C1	2200 µF capacitor, polarized electrolytic
C2, C3	0.1 µF capacitor, disc
C3	1 µF capacitor, polarized electrolytic
IC1	LM317 adjustable voltage regulator
BR1	4-amp bridge rectifier
T1	Transformer, 117 Vac primary; 25 Vac secondary (rated at 2 amps)

All resistors are 5 to 10 percent tolerance, 1/4 watt, unless otherwise noted. All capacitors are 10 to 20 percent tolerance, rated at 35 volts or more, unless otherwise noted.

You can add the same variable-voltage supply when using the flyback transformer and voltage multiplier discussed earlier, except use a 117-to-24-volt transformer and a 12-volt variable IC regulator.

Complete the electronics package by wiring the switches, control pots, fuse, and power cord as illustrated in FIG. 5-8. The switches and pots should be located as low in the base as possible. This allows you to control the Van de Graaff generator without undo risk of getting shocks from the upper terminal.

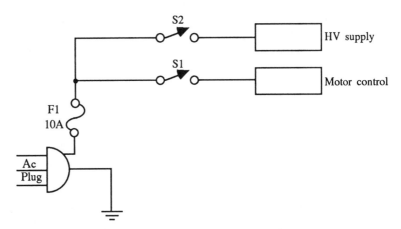

Fig. 5-8. Connect a line cord, fuse, and two ON/OFF switches to power and control your Van de Graaff generator.

With the base finished, attach the plastic insulator pipe using 1/2-inch-by-1-inch L-brackets. Make sure the pipe is secured firmly and is perpendicular to the top of the base. Construct the idler pulley for the output terminal as shown in FIG. 5-9. As with the motor pulley, the idler pulley should be made with a slight crown to self-center the rubber band.

Cut four grooves as shown in the plastic insulator pipe. Two of the grooves are for the pulley; the other two are for the output terminal wiper. Thread the rubber band through the plastic support and wrap it around the motor pulley.

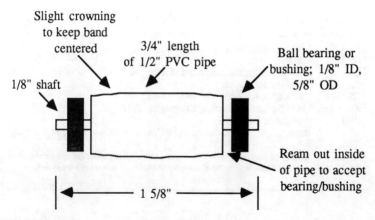

Fig. 5-9. Construction diagram for the idler pulley, which rests in a cut groove on the top of the insulating support. Carefully grind or file a "crown" on the outside of the plastic pipe to prevent the belt from slipping off.

Slip the idler pulley under the band, insert the axle through the pulley, and snap it in place over the notices in the support column.

Refer to FIG. 5-10. Construct the output wiper as shown and attach it to the support column. Adjust the bristles of the output wiper so they are close to the rubber band but not actually touching it. Be sure to place the bristles on the trailing side of the rubber band (that is, the side of rubber band that is going away from the top of the generator). Place the spherical output terminal carefully over the top of the support column until the metal touches the arch of the wiper.

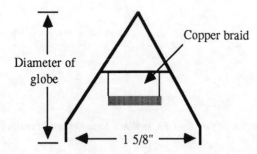

Fig. 5-10. The output wiper is a piece of metal bent in an A-frame shape. A length of copper braid is brazed or soldered to the frame (the braid should just barely touch the rubber band). The height of the frame should equal the diameter of the globe.

Test the tautness of the rubber band at the motor in the base of the generator. The band should be tight enough that it won't "walk" off the idlers but not so tight that it shows signs of gross stretching. If the band seems too tight or loose, unbolt the motor retaining nuts and slip the motor up or down as required. Retighten the nuts when the band is at the proper tension.

As a final step, adjust the wipers from the high-voltage sprayer so they just touch the rubber band. The bristles should point in the direction of rubber band travel.

USING OTHER BELT MATERIALS

The ion conveyer belt used in the prototype Van de Graaff generator is elastic rubber, designed for the Van de Graaff machine sold by Edmund Scientific. As of this writing, replacement belts for this generator cost under $5, and they offer excellent results with the home-built Van de Graaff apparatus. If this belt is no longer available, try one of the science lab mail-order companies listed in Appendix A. They often stock replacement Van de Graaff belts.

You can use materials other than rubber. Heavy paper makes a good belt, especially if you're building a large generator. Paper is easy to glue into a continuous belt. Heavy nylon webbing, especially the rubberized type designed for elastic webbing for furniture, also makes an excellent ion conveyer belt, but it is harder to construct into a continuous band.

TESTING AND OPERATING THE VAN DE GRAAFF GENERATOR

Inspect the wiring and construction one last time and then close the base of the generator. Turn all switches to OFF and turn the control dials to their middle positions. Temporarily remove the spherical output terminal. Plug the Van de Graaff generator into a wall socket.

First, turn the main power switch to ON. That should start the motor and the rubber band should move. Turn the power switch to OFF and inspect the rubber band. Is it still taut? Look carefully at the wipers, both in the base and at the top of the insulated support column. They should still be just above the rubber band.

If all looks satisfactory, replace the spherical terminal and turn the power switch back on. Turn the motor speed control (if present) to vary the speed of the belt motor. You'll hear the motor speed up and slow down. If you don't, the speed control circuit is not wired correctly.

On a dry day, some static will collect in the output terminal even with the high-voltage sprayer off. Run the Van de Graaff generator for a few minutes. Assemble a testing stick from a three-foot length of PVC pipe and a brass coupler. Holding the stick as far back as possible, move the brass end toward the output terminal of the Van de Graaff generator. You might hear or see a faint spark. Don't worry if you don't; the operation of the Van de Graaff is very susceptible to the relative humidity. The more humid the weather, the less static that builds up on the rubber band.

Now turn on the high-voltage ion sprayer. Wait a minute and try the test again. You should get a longer, louder spark. Try adjusting the input voltage to the sprayer (affects its output). Depending on the design of the high-voltage power supply used, the usable range of the control should be about half to full rotation of the knob.

If you fail to get an appreciable spark even with the sprayer turned on and running full blast, there may be something wrong in the construction of the generator. Turn the power switch to OFF and inspect the innards. Inspect the wipers to make sure they are making close contact with the rubber band. Look for any indication that the output terminal is grounded to the base. Check with a sensitive volt-ohmmeter. It should read infinite ohms.

An absent or weak spark may be an indication of poor electrostatic conduction between the output wiper and the spherical terminal. Check to make sure the metal parts are contacting. Clean the arch of the wiper and the inside of the sphere with alcohol. If necessary, file the wiper and sphere to expose new and untarnished bare metal.

With the Van de Graaff working properly, on a fairly dry day (relative humidity less than 25 percent), you should get sparks that reach out three to four inches. With even lower humidity (like a dry winter's night), the sparks could even flash down from the top terminal to the metal plate or chassis below.

VAN DE GRAAFF EXPERIMENTS

The Van de Graaff generator makes for an ideal science fair project. With it, you can explore the world of static electricity and test the electrostatic properties of any number of materials. To indulge in some classic Van de Graaff parlor tricks, here are several you can try. Refer to TABLE 5-6 for a parts list.

Table 5-6. Parts List for Van de Graaff Generator Experiments

1	Fluorescent tube, 18 or 24 inches
1	Egg grate or other insulating pad
1	Metal leaf electroscope consisting of glass flask, cork or rubber stopper, 1-inch-diameter metal ball, $1/16$-inch metal rod, silver or aluminum foil

Fluorescent wand Obtain a short (18- to 24-inch) fluorescent tube. Wearing a pair of protective goggles and heavy rubber gloves, grab the tube at one end *by the glass* (not the metal terminals). Touch the other end to the output terminal. At each spark the tube should ignite and glow. The tube actually needn't touch the Van de Graaff sphere. The spark should reach out for the glass long before contact is made.

Electric hair Stand on a plastic egg or milk crate. The crate acts to insulate you from the ground. With the Van de Graaff generator switched off, place your hand firmly on the output terminal. Don't touch anything else. Have *someone else* turn the generator on. Within seconds, your hair will start to stand on end. With your free hand, comb your fingers through your hair and it will stand to attention even more. After the demonstration, have *someone else* turn the Van de Graaff generator off, but *keep your hand on the terminal*. Wait a few moments for the excess ions to dissipate, step off the crate, and then remove your hand from the terminal. Beware! You'll receive a shock if you remove your hand from the terminal while the generator is on. The shock won't hurt you, but you'll remember it for a while.

Electroscope Build a metal leaf electroscope as shown in FIG. 5-11. The electroscope consists of a glass flask, a cork stopper, a metal stem, and a piece

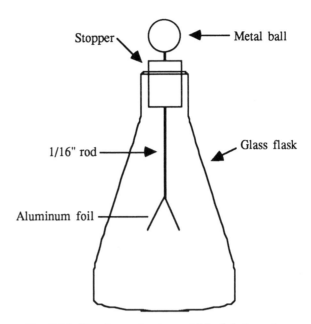

Stopper

Metal ball

1/16" rod

Glass flask

Aluminum foil

Fig. 5-11. How to construct a metal-leaf electroscope.

of aluminum foil. To use the electroscope, place the metal stem that sticks out of the cork near the output terminal of the Van de Graaff generator. If an electrostatic charge is present, the pieces of the aluminum foil will separate due to like charge repulsion. Be careful that your hands don't get too close to the terminal sphere or you could get a dose of some of the electrostatic charge.

6

The power of
the Tesla coil

Thomas Edison reigns as the most famous inventor in world history. Perhaps unfairly, Edison greatly outshadows another noteworthy American inventor, Nikola Tesla, who is responsible for many of the household appliances we now take for granted. Tesla invented the polyphase ac motor (used on every ac motor-equipped appliance in your house) and engineered the giant hydroelectric generators at Niagara Falls. These two advancements helped scuttle Edison's antiquated dc power system, which he designed for operating his new incandescent electrical light.

Tesla was also at hand to develop radio broadcasting, long before Marconi transmitted his first radio signals across the Atlantic. But Tesla's vision was to transmit electrical power, not radio signals. Marconi concentrated on sending voice and music through the air and so became known as the father of modern radio, despite Tesla's earlier experiments.

One of Nikola Tesla's most important contributions to science is the Tesla coil, invented in 1892. The Tesla coil is a special induction coil that takes low-voltage, high-frequency signals and transforms them to extremely high voltages. Because the potentials are at radio frequencies, the sparks created by the device propagate much further through the air than static electricity.

While the Tesla coil is primarily used as a laboratory instrument, it has many beneficial applications. A form of the Tesla coil is used in television sets to create high-frequency high voltage for operating the cathode ray tube. A similar form of Tesla coil is used in the colorful plasma sphere's so popular today. And an all-electronic version of the Tesla coil finds common use in automotive electronic ignition circuitry.

This chapter details the Tesla coil: how it works and how you can build one of your own. You'll also learn several educational experiments you can conduct with your Tesla coil as well as how to treat your coil with the care and respect it deserves.

THE BASICS OF THE TESLA COIL

Tesla coils were born in the nonelectronics age. Sophisticated integrated circuits, transistors, and other electronic components are not required to build a Tesla coil, although—as you'll witness later in this chapter—modern versions of the device can incorporate electronic circuitry.

The basic construction of the Tesla coil is shown in FIG. 6-1. In operation, incoming power—say 60 Hz, 117 volts from an ordinary wall socket—is applied to a soft-iron core transformer or an oil-filled induction coil. The output of the induction coil is now a few thousand volts.

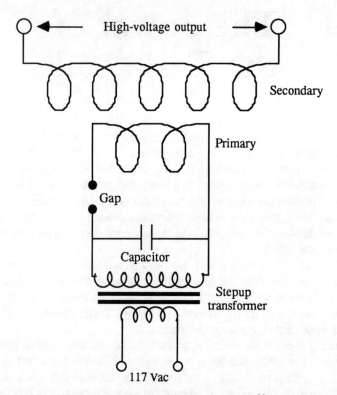

Fig. 6-1. The basic Tesla coil, operating from 117 Vac current.

The potential is then stored in a capacitor, which slowly fills with energy over time. A nearby spark gap allows the capacitor to discharge at regular intervals. The spacing of the spark gap determines the amount of voltage stored in the capacitor. The longer the spark gap, the more voltage is required to cross the contacts of the gap, so the higher the potential in the capacitor. When the voltage level exceeds the air-dielectric capacity of the spark gap, the energy is released from the capacitor and is applied to the heavy primary windings of the Tesla coil. This winding consists of no more than 10 to 20 turns of heavy wire.

Concentric in the primary is the secondary Tesla coil winding. This secondary is inductively coupled to the primary—there is no electrical connection (except perhaps for a common ground) between the primary and secondary.

At first glance, the Tesla coil looks like the induction coil, first devised by Faraday. But closer examination shows that:

- The primary and secondary coils are wound around air, not the soft iron core used in conventional induction coils.
- Thanks to the capacitor, the primary and secondary coil are at resonance, and the increase in voltage between the two coils is much higher than you might expect.

The capacitor and primary are matched as closely as possible so that the output from the secondary winding is the highest frequency possible. Tesla coil frequencies of several tens or even hundreds of megahertz are common.

The output of the secondary is high-frequency high voltage. The voltage and frequency depend on many parameters including the resonance of the circuit, the capacitance of the main capacitor, the spark gap distance, the size of the secondary, and so forth. A typical, well-built Tesla coil operating from 117 volts ac and using a secondary coil measuring 18 inches long can yield an output of 125,000 volts or more.

The length of a 125,000-volt spark of static electricity is normally about 7 inches long, but because the Tesla coil operates under radio frequencies, the length of the spark increases greatly—to about 12 inches. The tabletop Tesla coil described later in this chapter produces 150,000 to 225,000 volts yet creates sparks up to 14 inches long.

A WORD OF CAUTION

Though Tesla coils generate extremely high voltages, getting shocked by a moderately sized Tesla apparatus is not generally dangerous—though the experience can smart. The secret is in the high frequencies. The higher the frequency, the less the voltage that enters the body. Tesla often demonstrated the safe operation of his coils by standing or sitting amidst the lightning bolt sparks. He was not hurt because the electricity skimmed over his skin, and didn't enter his body. While this is fine in theory, other variables could reduce the high-frequency-safety factor of working with Tesla coils. Your coils might not be precisely tuned, or they might not operate at very high frequencies. So not only will the sparks be shorter, but the chance of a heavier shock due to low-frequency potentials will be greater.

Persons in less-than-excellent health might find their heart skips a few beats after they receive a quarter-million-volt jolt from a Tesla coil. Hence, you won't want to get near the high-voltage Tesla coil secondary if you have heart problems or wear a pacemaker. And unless you are an experienced Tesla coil builder and add extra safeguards and circuitry, the 60-cycle component of ac-operated coils can pass through to the secondary coil. That 60-cycle component can be very dangerous, especially when it is magnified by the induction of the coil.

When experimenting with Tesla coils, especially the larger designs, be extra certain that you avoid touching the output terminals. Properly ground all electrical components. Certain parts of the Tesla coil (most specifically the primary coil, spark gap, and capacitor) can deliver a *Lethal* dose of electricity, so be sure these are well shielded.

The open spark gap produces ultraviolet light, which can be extremely damaging to the eyes. Avoid looking at the spark gap, and if possible, enclose the spark gap in an opaque enclosure. The spark gap needs air to operate properly, so make sure the enclosure has holes for air to enter and escape. In addition, the sparking action produces toxic ozone. Use the Tesla coil only in a well-ventilated room.

BUILD YOUR OWN MINIATURE TESLA COIL

Figure 6-2 shows a miniature Tesla coil you can build, using parts indicated in TABLE 6-1. The circuit is shown in FIG. 6-3. Actually, this isn't a true Tesla coil but a high-turns-ratio stepup transformer. No capacitor or spark gap is used, and the circuit is not tuned to resonance. But, it's a fun circuit to build, creates exciting sparks, and is the basis of several other high-voltage projects in this book.

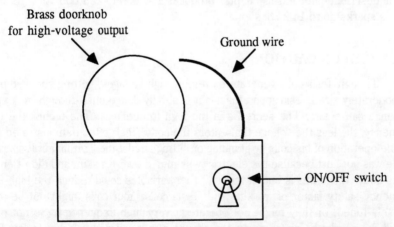

Fig. 6-2. Place the flyback Tesla coil in a small plastic enclosure and orient the ground wire close to the high-voltage output terminal (a brass doorknob makes a good terminal).

The circuit is the same as that found in Chapter 3 for the plasma sphere generator and uses a surplus television flyback transformer. Read that chapter for construction details, especially the instructions for rewinding the flyback transformer.

Table 6-1. Parts List for Flyback Tesla Coil

1	High-voltage output circuit (see circuit in Chapter 3 and parts list in TABLE 3-1)
1	Brass doorknob for output terminal
Misc.	Enclosure box, on/off switch, fuse and fuse holder, 14 to 16 gauge wire for ground wire, ac cord

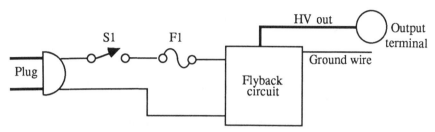

Fig. 6-3. Basic hookup diagram for the flyback Tesla coil; use the flyback circuit in Chapter 3 to complete the project.

Build the circuit in an enclosure. Be sure to mount the two transistors on a suitable heatsink. The output of the flyback transformer is a two-inch brass ball or brass doorknob. The ground terminal is bent so that the sparks leap from the ball and to the wire terminal.

BUILD THE TABLETOP TESLA COIL

The tabletop Tesla coil generates 150,000 to 225,000 volts and when properly tuned produces sparks of about 10 to 14 inches in length. The coil allows for a great deal of experimentation. As you gain experience in building Tesla coils, you can try new techniques and approaches to increase the voltage and spark length. The overall schematic of the tabletop Tesla coil is shown in FIG. 6-4, and a parts list is provided in TABLE 6-2. It consists of five discrete blocks:

- Driving circuit
- Capacitor
- Spark gap
- Primary coil
- Secondary coil

Driving Circuit

The heart of the tabletop Tesla coil is the driving circuit, shown in FIG. 6-5. This circuit was adopted from Walt Noon's booklet on induction coil design, *How to Build a 40,000 Volt Induction Coil* (Lindsay Publications, 1989). Although Noon's circuit is designed as a high-voltage spark maker, it makes a perfect driving circuit for a Tesla coil. This same circuit is used in Chapter 4 for a high-tension induction coil.

Table 6-2. Parts List for Tabletop Tesla Coil

Driving Circuit

R1	100 kΩ potentiometer
R2	5 kΩ potentiometer, 2 watts
R3, R4	10 Ω, 10 watts
R5	1Ω, 1 watt
R6	100 Ω, 2 watts
R7	470 Ω
C1	1000 μF electrolytic
C2, C3, C4	0.1 μF disc
C5	0.0015 μF doorknob capacitor rated at 15 kV or higher (see text)
Q1	2N3055 npn power transistor
D1, D2	1N914 glass diodes
LED1	Light-emitting diode
BR1	Bridge rectifier, 6 amps
T1	Step-down transformer, 120 Vac primary to 12 Vac secondary, at 5 amps
T2	Auto ignition coil, 12 volts
S1	SPST switch, rated at 10 amps or better at 120 Vac
Misc.	Ac plug and line cord, fuse holder, fuse (10 amps at 120 Vac)

Spark Gap

4	5-way heavy-duty plastic binding posts
4	1-inch lengths of 12-gauge copper wire

Primary Winding

4	8-by-$1/2$-inch hardwood dowels
12 ft.	8-gauge aluminum grounding wire

Secondary Winding

1	$14^7/8$-by-$3^1/2$-inch PVC plastic pipe
2	$3^1/2$-inch PVC end caps
1	$1/4$-inch 20 by 3-inch bolt
1	$1/4$-inch 20 by 1-inch bolt
1	High-voltage ceramic insulator
Misc.	Approximately 1000 feet of 28 AWG enamel-coated magnet wire (about 1080 turns), $1/4$-inch 20 nuts and washers

Misc. Parts

1	10-by-24-inch acrylic plastic base ($1/4$ inch or thicker)
1	Enclosure for driving circuit
6	Metal or plastic feet for base

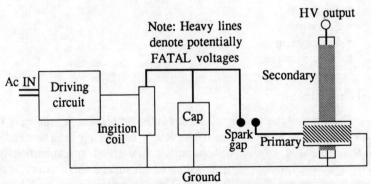

Fig. 6-4. Block diagram of the tabletop Tesla coil.

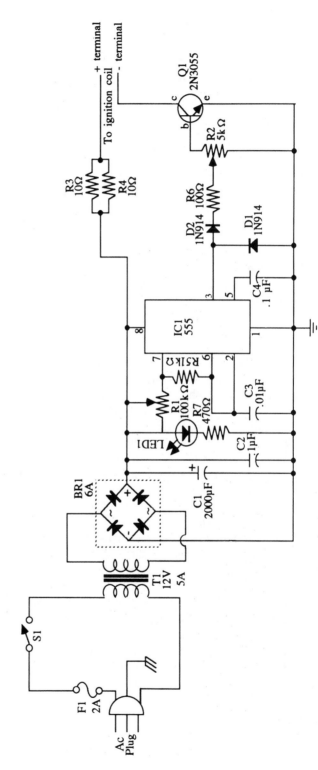

Fig. 6-5. The driving circuit schematic provides the short duration pulses to the ignition coil; the coil in turn steps up these pulses to high voltage.

The driving circuit operates from 117-volt house current that is first stepped down to about 12 volts with a 117- to-12-volt power transformer. A bridge diode network and series of capacitors rectify and filter the line voltage for use by the 555 timer IC. The IC operates as a free-running astable multivibrator. Frequency is set by R1.

The pulsed output of the 555 is chopped and applied to the base of power transistor Q1 (through power level potentiometer R2). The transistor drives an ordinary 12-volt automotive ignition coil. Such coils are available at K-Mart, most automotive parts stores, and auto junkyards. Price is between $6 and $20.

Transistor Q1 must be mounted on an aluminum heatsink using heat transfer paste and a suitable insulator. Resistors R3 and R4, which limit the current to the ignition coil, should be attached to the heatsink as well. Fashion a set of clamps using brass strips (available at most hobby stores) and secure the resistors.

You can build the driver circuit using most any circuit-making technique you wish. The best approaches are printed circuit board, pretinned perf board, and universal solder board. Wiring and lead length are not critical, but you should exercise the usual care to avoid possible shorts between adjacent components.

Capacitor

Discharge capacitors for Tesla coils must be rated at a high voltage—enough so that the voltage from the driving circuit (including induction coil) doesn't blow the capacitor apart. The capacitance value needn't be high. In most Tesla coil designs, the capacitor is between 0.002 and 0.2 microfarads.

The capacitor called for in this circuit has a rating of 15 kV and a capacitance of 0.0015 microfarads (or 1,500 picofarads). While you could make your own high-voltage capacitor for use in the tabletop Tesla coil, a cheaper and more compact alternative is to use a "doorknob" capacitors salvaged from old television sets. Doorknob capacitors are used in the high-voltage rectifier circuit used to drive the picture tube. A particularly good source for doorknob capacitors is mail-order surplus, particularly Fair Radio Sales and John Meshna & Associates (consult Appendix A for addresses and other sources). You can also locate doorknobs at local electronics surplus outfits and from TV repair shops. Doorknob caps are seldom over 500 pF, so if you add three in parallel, you get the 0.0015 μF capacitance required for the Tesla coil circuit, as shown in FIG. 6-6. A deviation of a few hundred picofarads one way or another won't matter, so don't worry if your caps are rated at 600 pF or 400 pF each.

Bear in mind that doorknobs aren't the only kind of caps you can use. As long as the voltage rating is at least 15 kV, you can use at most any capacitor. Combine the caps in parallel if you want higher values. Suitable high-voltage capacitors are often used in transmitter circuits, gyro-compass circuits, and photocopiers. Again, surplus is an excellent source for these items.

Fig. 6-6. Three high-voltage capacitors ganged in parallel. Each cap is 500 pF, so the total capacitance is 1500 pF. These caps are a modern version of the doorknob variety.

SPARK GAP

The spark gap is one of the most important elements of the Tesla coil, yet its design is often ignored. The spark gap works by bridging the air-dielectric between the ground and high-voltage terminals, thus producing an instantaneous high-voltage arc. During the life of the spark, the air is heated and ionized, creating a near vacuum. Another spark can't be produced until more fresh, cool air is introduced to the spark area.

A good spark gap design for Tesla coils consists of two gaps, each with its own adjustment mechanism, as shown in FIG. 6-7. The series spark gap provides better efficiency and the gap terminals don't overheat so easily. Five-way binding posts make great adjustment mechanisms. Unscrew the binding post, slide the metal rod in or out as desired, and tighten the post.

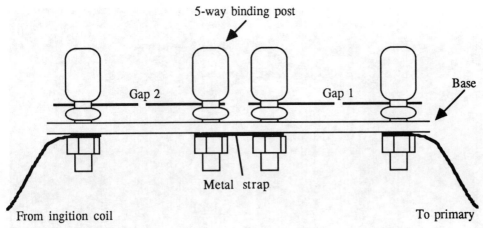

Fig. 6-7. Construction diagram for the series spark gap. Use five-way binding posts and 1/8-inch-diameter welding wire or coat-hanger wire.

The more spark gaps you add, the better the Tesla coil works. You might want to increase the number of gaps from two or four. Construct the extra gaps in the same manner as shown in the figure. You can use a variety of metal for the spark gaps. I prefer welding rod or coat hanger rod, ground polished flat. The rod easily slips through the hole in the larger five-way binding posts.

Primary coil

The primary coil is constructed of 8-gauge aluminum grounding wire, available at Radio Shack. The wire must be wrapped around a coil form, built from four pieces of 1/2-inch doweling. Cut a 3/16-inch deep groove at 1/2-inch intervals in the dowel, as shown in FIG. 6-8. The grooves should be cut using a wide-kerf combination saw blade (radial arm saws work the best, but you can also use a circular or table saw to cut the grooves).

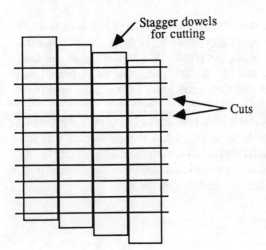

Fig. 6-8. Stagger the dowels as shown for cutting.

Position the dowels on the Tesla coil base as shown in FIG. 6-9. Leave about 6 inches of extra wire in the center of form and carefully wrap the wire counterclockwise around the form. Be sure the wire doesn't slip out of the grooves. When all the grooves are full, cut off the excess wire and wrap it around one of the dowels. The wire should wrap around the base about seven or eight times. Be absolutely sure that each loop of the wire does not touch the loop above or below. Wires touching or too close together will cause a spark and impair the operation of the coil.

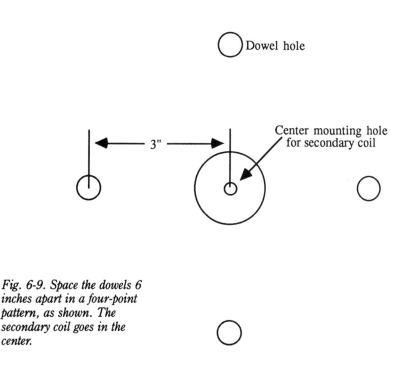

Fig. 6-9. Space the dowels 6 inches apart in a four-point pattern, as shown. The secondary coil goes in the center.

As an alternative to using dowels as the coil form, wind the aluminum wire around an 8-inch diameter Tupperware or other lightweight rubber container. Keep the wire in place with a layer of duct tape or electrical tape. This approach isn't as pretty, but it seems to work well.

Secondary coil

The secondary coil consists of approximately 1080 turns of 28 AWG enamel-coated wire wrapped around a $14^7/8$-inch-long piece of $3^1/2$-inch-diameter PVC pipe. The wire is close-wound so that the strands touch one another as they spiral up along the pipe, but they never overlap. You can build yourself a coil-winding jig using a drill motor and a bushing, but you might find it easier to control the winding process if you do it by hand. Although 1080 turns of thin wire might seem like a tough chore, it should only take an hour or so.

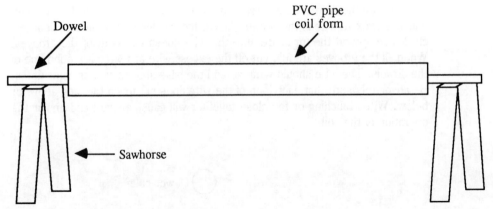

Fig. 6-10. A homemade winding jig makes the task of constructing the secondary coil a lot easier. Have a second person turn the dowel or attach a crank to the end for an even easier job.

Begin by spraying the plastic with a light coat of varnish. Allow the varnish to become tacky. Place the pipe inside a 1/2-inch-diameter wooden dowel and wrap the ends of the pipe to the dowel with heavy duct tape. This secures the pipe to the dowel for easier handling. Straddle the dowel between two saw horses, as shown in FIG. 6-10.

Standing in front of the pipe, position the enameled wire 1 inch from the left edge and secure it with electrical tape. Have someone help you unspool the wire as you wind the coil, or else arrange some sort of jig that lets you unspool the wire on demand. The big trick is to keep constant back tension on the wire so that it doesn't uncoil quickly and form a rat's nest. Have patience and it'll work out for you.

Turn the dowel counterclockwise while you carefully wind the wire onto the pipe. The job is easier if you can concentrate on the winding while someone else slowly turns the dowel. Each turn of wire should be flush against the proceeding one, with no overlap. Continue all the way to 1 inch from the right end of the pipe. Cut off the wire and secure it in place with electrical tape.

Spray two or three coats of varnish on the secondary winding. This helps the wires from uncoiling and also provides extra protection against high-voltage burning though the enameled wire. Let the coil dry between each coat. You can apply brush-on varnish if you wish. A finished coil is shown in FIG. 6-11.

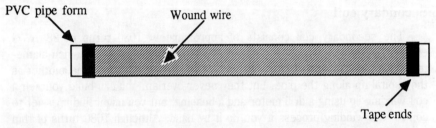

Fig. 6-11. Tape the ends with electrical tape to prevent the winding from coming loose.

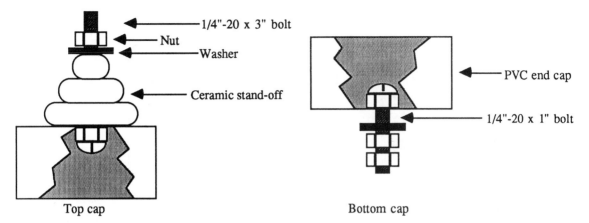

Fig. 6-12. Use PVC end plugs to secure the mounting hardware and output terminals for the secondary coil.

While waiting for the coil to dry, drill a 1/4-inch hole in the exact center of two 31/2-inch PVC end plugs. Assemble mounting hardware as shown in FIG. 6-12 for the plugs. When the coil is dry, carefully unwind 6 to 8 inches of wire from each end of the coil and retape the ends to prevent unraveling.

With a utility knife, scrape one or two inches of enamel off the ends of the wire. Attach the ends to the plug hardware. Slip the plug over the ends of the pipe until they are snug. Set the secondary coil aside.

Assembling the Tesla coil

Cut a piece of 1/4-inch acrylic plastic to 10 by 24 inches. Mount the component parts on the base as shown in FIG. 6-13. Layout is not critical, so exact positioning is not required. However, be sure to provide at least an inch clearance between the capacitor, spark gap, and primary coil. The layout shown in the figure provides quite a deal of clearance; you might want to make your coil more compact if space is a premium for you. I decided to make the base overly large to accommodate future expansion and experimentation. Add feet at the corners and sides of the base using short 1/4-inch 20 bolts and nuts.

Attach the common ground as shown in FIG. 6-14. Make sure all points—primary coil, secondary coil, and one side of the capacitor bank, are connected firmly.

Setting up the coil

The Tesla coil is ''tuned'' by attaching the spark gap alligator clip to different points on the primary coil. For the time being, place the clip in the middle of the coil. Be sure the clip doesn't touch adjacent turns of the coil.

Next, adjust the two spark gaps so that each gap is about 1/8 inch wide. Double-check your wiring, and then plug the coil into the wall socket. Stand back and flick the switch on. If the spark gaps are adjusted properly, you should

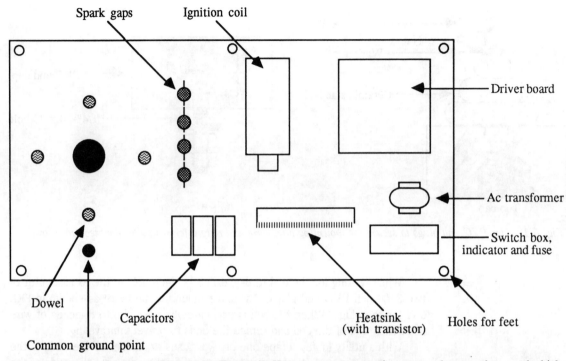

Spark gaps Ignition coil

Driver board

Ac transformer

Switch box,
indicator and fuse

Dowel

Capacitors

Heatsink
(with transistor)

Holes for feet

Common ground point

Fig. 6-13. Suggested layout for the tabletop Tesla coil. You can use most any layout you choose as long as the high-voltage components are not spaced too close together.

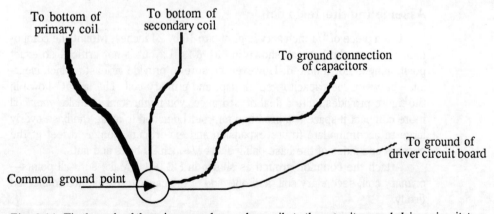

To bottom of
primary coil

To bottom of
secondary coil

To ground connection
of capacitors

To ground of
driver circuit board

Common ground point

Fig. 6-14. Tie the ends of the primary and secondary coils to the capacitors and driver circuit to a common ground point.

hear a crackle and see a white/blue arc from them. If you don't, loosen the binding posts, and with a pair of *heavily insulated* pliers, slowly move the gap points closer together. For best operation, the gap points should be as far apart as possible and still allow a fairly constant arc. *Be sure to keep your hands away from the capacitors, spark gap, and primary coil when making the adjustments.* Potentially *lethal* voltages exist at these points.

Switch the Tesla coil off and connect a grounding wire from the common ground point. The wire should be long enough to reach the top terminal of the secondary. Turn the coil back on, and holding the ground wire as far from the end as possible, move it towards the secondary terminal. Note the spark you get. If you don't get a spark or it isn't at least a few inches long, turn the coil off and move the alligator clip to another part of the winding. Move up or down one loop of the winding at a time. Try again and note the spark length again.

Continue adjusting the placement of the alligator clip until the spark is long as possible. You might need to make occasional changes to the spacing of the spark gap as you change the tuning of the coil.

The prototype used a sharpened piece of $1/4$-inch 20 threaded bolt for the pinpoint output terminal. The bolt is attached to the end of the terminal using a $1/4$-inch 20 coupler.

In the prototype, I was able to draw sparks about 10 to 12 inches long, but it took quite a bit of tinkering with clipping the spark gap wire to the primary and making fine adjustments to R1 and R2. Some of these sparks were pretty weak and best seen in the dark. At the very least, your coil should produce some very brilliant and healthy 6-to-8-inch sparks.

ROTARY SPARK GAP

Tesla coil builders have found that of all the components in the system, the spark gap holds the greatest promise of success or failure. Though series spark gaps are an improvement over a single spark gap, a rotary spark gap is even better. The rotary spark gap is composed of up to several dozen spark points spinning on a disc. Each time one of the points passes the gap terminals, a spark is created. Because each point is used just once per revolution, they aren't subject to the adverse affects of overheating. And the rotary action of the gap wafts fresh, cool air by the spark terminals.

You can construct a rotary spark gap using an inexpensive dc motor, a plastic disc, and $8/32$ hardware. A construction diagram is provided in FIG. 6-15. Cut a piece of $1/8$-inch acrylic plastic to a 4-inch diameter circle. Carefully mark a series of 12 equidistant holes—clock style—$1/2$ inch from the edge of the disc. Insert $6/32$-by-$3/4$-inch flat-headed machine bolts into these holes and tighten with lock washers and nuts as shown in FIG. 6-16. A parts list for the rotary spark gap is provided in TABLE 6-3.

Drill a hole for the motor shaft in the center of the disc. The center hole should be just slightly smaller than the motor shaft so that you can press-fit the disc over the shaft. Be absolutely sure the hole is in the dead center of the disc and that the disc is mounted squarely on the motor shaft. Apply a small amount of heat to the center of the disc to melt the plastic around the shaft.

Mount the motor and construct the spark terminals as shown in FIG. 6-17. As with the series spark gap described earlier in this chapter, you can adjust the spacing between the terminals and rotary points with five-way binding points and short pieces of welding wire or coat-hanger wire.

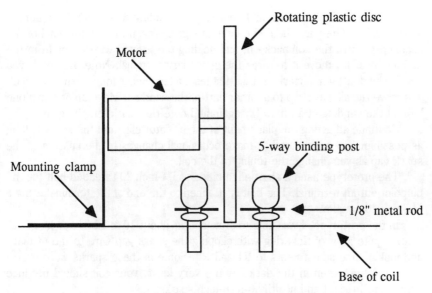

Fig. 6-15. Overall view of the rotary spark gap.

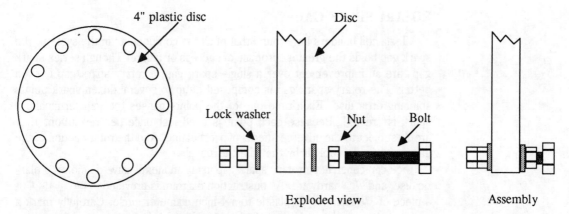

Fig. 6-16. Construction plans for the rotary spark gap disc.

Table 6-3. Parts List for Rotary Spark Gap

1	4-inch-diameter plastic disc
12	$6/32$-by-$3/4$-inch flat-headed machine bolts
36	$6/32$ nuts
24	$6/32$ flat washers
2	Heavy-duty plastic 5-way binding posts
1	Ac or dc motor
Misc.	Bracket and mounting hardware for motor, 12 to 14 gauge wire for 5-way binding posts

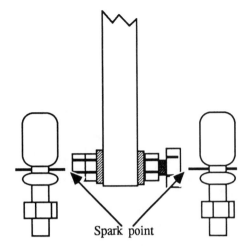

Fig. 6-17. Place the disc (with spark points) between two five-way binding posts; adjust the spacing between the wires and points for best spark.

Spark point

The dc motor called for in the parts list operates on 12 volts dc, but you can vary the voltage to control the rotation speed of the disc. Use a 2-watt potentiometer, wired as shown in FIG. 6-18, to control the motor speed. Although this is not the most high-tech motor control method, it's the easiest to implement.

Fig. 6-18. A simple speed control circuit for a low-voltage rotary spark gap motor. It's not efficient, but it works.

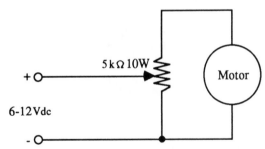

$5\,k\Omega\,10W$

Motor

$+$

6-12Vdc

$-$

ADDING A BALL TERMINAL

The sharpened output terminal on the secondary coil disperses most of the high-voltage energy immediately after the coil generates it. But using a larger surface area at the output terminal provides more electricity and higher output. Figure 6-19 shows a $4^{3}/4$-inch-diameter spun-aluminum (plated brass) ball attached to the output terminal of the prototype Tesla coil. The ball was obtained from a magician's supply store and is the same one used in Chapter 5 (the high-voltage Van de Graaff generator).

You might notice that you need to retune the primary coil when adding the ball terminal. I found I had to select one of the windings further up on the primary.

Fig. 6-19. A brass or aluminum ball makes an ideal output terminal for a Tesla coil.

EXPERIMENTS WITH TESLA COILS

Once you've constructed your Tesla coil, now what? Try these experiments:

Fluorescent tube Place a fluorescent tube close to the output terminal of the coil. Watch the tube light up even though it's several inches from the coil. This is the ''power transmission'' concept Tesla had in mind when he designed his original coils. Figure 6-20 shows a circular fluorescent tube lit up like an angel's halo.

Simplified plasma sphere Place a 4- or 5-inch-diameter clear accent bulb on the pointed output terminal. Inside the bulb you'll see purplish-blue lightning glow from the metal contacts to the glass envelope. This is a simplified version of the plasma sphere described in Chapter 3.

Penetrate insulators Place a heavy, clear glass on top of the pointed output terminal. Place the ground wire on the other side of the glass and watch

Fig. 6-20. The radiated high voltage from a Tesla coil can directly light a fluorescent lamp.

the sparks pierce through the glass! Even though glass is a good insulator, the high frequencies produced by the Tesla coil penetrate right through it. You can also observe this effect by drawing the output wire along the edges of the ceramic insulator on the top of the secondary. The sparks fly through that, as well. Bundle together a few dozen strips of metal-coated tinsel and attach them to the output terminal (point or ball). Watch the tinsel start to stand on end and sparks emit from the ends. The ball of tinsel acts as a giant emitter of electrons, and the strands separate from one another because of the repulsion of like charges. Place an ordinary AM radio a few feet from the Tesla coil. Turn on the coil and listen to the pops and clips you hear through the radio; it's radio interference caused by the discharge of both the spark gaps and the output of the Tesla coil. During my experiments, the prototype coil kept interfering with my cordless phone, making it ring when no one was calling.

This chapter presented only one design for a Tesla coil—and a rather rudimentary one at that—and a few of the experiments you can conduct with one. There are dozens of other Tesla coil designs, many of them producing in excess of a million volts. Check Appendix B for a list of books and magazine articles that provide additional information on Tesla coil construction and use. You might also be interested in joining the Tesla Coil Builder's Association. Their address is provided in Appendix A.

7

Experiments with
Kirlian photography

In 1939, Seymon Kirlian (pronounced *keer-lee-an*) began his work developing a new concept of electrophotography. Kirlian found that by subjecting animate and inanimate objects to a high-voltage, high-frequency pulse, that an *aura* or halo would form around the object and could be recorded on film. The illumination of the halo was sufficient to make the exposure, so no lights were required.

Though Seymon Kirlian lived and died in relative obscurity, Kirlian photography is the subject these days of heated debate. The pro-Kirlians insist that the unique photographic method can capture moods of people, pinpoint illnesses that have yet to manifest themselves in other forms, and graphically depict an individual's psychic aura. The anti-Kirlians, mostly hard-line scientists, insist that there's nothing to the method other than the natural corona discharge of high-voltage, high-frequency potentials.

It is not the purpose of this chapter to debate the meaning of the Kirlian effect but to provide step-by-step instructions on how to build your own Kirlian apparatus so that you can experiment with it in your own garage. Instead of reading about the debate, you can experience it. However, Kirlian photography is *not* for those with heart problems or a pacemaker.

BUILD THE KIRLIAN CAMERA

The Kirlian camera is a simple device, consisting of a high-voltage power source, a metal plate, and a piece of film. You can construct a camera for use with 4-by-5 sheet of film or use ordinary 35 mm film. This chapter shows how to use a 35 mm camera because the film is easier to use and can be readily processed. A parts list for the 35 mm-based Kirlian photography outfit is provided in TABLE 7-1.

Regardless of the camera you use, the high-voltage power source is the same. The Kirlian camera needs 20 kV to 30 kV to work properly, and that

Table 7-1. Parts List for Kirlian Camera

1	35 mm SLR camera, with lens removed
1	12-gauge or thinner brass sheet cut to approximate size of pressure plate on rear of camera
Misc.	Length of insulated wire for high-voltage lead (should not be over 24 inches); high-voltage, high-frequency source (such as Tesla coil in Chapter 6)

voltage can be obtained from the same power supply used as the driver circuit in the high-tension induction coil project in Chapter 4. Refer to that chapter for a schematic diagram and construction plans.

You might already have an old 35 mm camera lying around the house that you no longer use—your trusted 35 mm SLR from your college days got dumped in Uncle Harry's pool and hasn't worked since, but you're reluctant to toss it. These old, discarded cameras can be revived and used to make a Kirlian photographic apparatus.

If you don't happen to have a 35 mm camera that you can donate to the cause (converting it to Kirlian use ruins it for anything else), you can purchase a cheap, used camera for under $30. The only requirement of the camera is that its film advance mechanism works. Its other parts, like the shutter or flash, can be inoperable.

To convert the camera for Kirlian photography, remove the lens and shutter. If the camera has a focal-plane shutter, remove it completely, but be careful not to jam any component parts. If the camera has an iris shutter, the opening through the front to the film will be too small, so you'll need to cut it out with a jeweler's saw or Dremel tool. Cut out the lens if it is permanently mounted to the camera. Therefore, the best cameras for use in Kirlian photography are 35 mm SLRs with removable lenses and cloth focal-plane shutters.

Open the back of the camera and examine the pressure plate there. The plate might be plastic or metal, depending on the quality of the camera. Measure the plate and cut a piece of thin brass sheeting (available at hobby stores) to size. As shown in FIG. 7-1, glue it to the pressure plate and file down the edges so that the film won't catch or scratch on the metal.

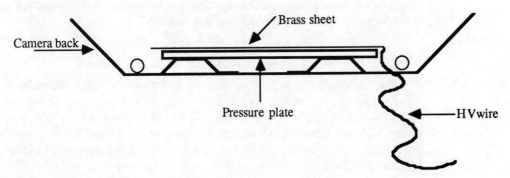

Fig. 7-1. Side view of a 35 mm SLR camera back, showing film rollers, pressure plate, and a brass sheet you add for Kirlian photography.

Next, solder a 2-foot length of wire from one corner of the brass plate. Drill a small hole in the back of the camera behind the pressure plate and feed the wire through it. The hole should be just large enough to feed the wire through. Connect the other end of the wire to the high-voltage output of the power supply.

The camera is now ready to use. Your first text exposures will be the aura around your own fingertips. The exposures must be made in complete darkness. A photographic darkroom is not absolutely necessary, but it comes in handy. If you don't have a darkroom at your disposal, you can make Kirlian photos in a dark, closed closet at night.

USING THE KIRLIAN CAMERA

Remember that the Kirlian camera doesn't have a shutter, so you must use it in complete darkness. Otherwise, you'll fog the film and spoil your Kirlian exposures.

Begin by loading a roll of inexpensive color film into the camera. Generic-brand film (24 exposure) is fine, but the speed of the film should be at least 100 ISO. At this point, don't use black-and-white film because it costs more than color film to have developed and printed. You can graduate to better films—such as slide or high contrast—after you gain some experience.

Advance the film to the first exposure. Don't worry that the lights are still on; the rest of the roll is still in its cassette and is protected from the light. Place the camera on its back on a table, and with the lights still on, turn on the power supply. Rotate the power level knob to its minimum setting. Pass your finger through the opening in the front of the camera and touch the film, as shown in FIG. 7-2. You may already feel a slight tingle—that's okay. The film acts as an insulator and prevents you from getting a more serious shock from the power supply.

Making sure you don't touch any grounded metal objects, slowly rotate the power level knob on the power supply until you feel a slight tingle. You can leave the setting at minimum if you feel a mild shock even before you turn the knob. This is the desired setting for photography. Although you can expose the photograph using higher settings, it might be unpleasant for you to touch the film for long periods of time. This is why experimenting with Kirlian photography is not for those with heart problems or pacemakers.

Cover the pilot light on the power supply with a piece of black electrical tape. Turn the room lights off and wait 5 minutes or so to let your eyes grow accustomed to the dark. If you see any light, cover it up with a towel or blanket. For best results, the room must be *completely dark*.

To make an exposure, advance the film two frames. With the power supply on, place your finger on the film. You'll see a blue arc around your finger and feel a slight tingle; this is normal. Time the exposure so that your finger touches the film for about 8 to 10 seconds. Advance the film another two frames (always skip a frame because the halos can cross into adjacent pictures) and make another exposure, this time with a different finger.

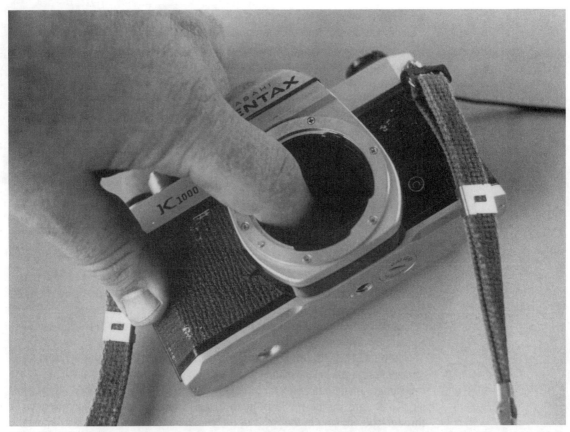

Fig. 7-2. A 35 mm SLR camera with the lens removed is ideal for Kirlian photography because it allows you to place your finger or other objects to be photographed through the shutter opening onto the film underneath.

When you are finished, turn off the power supply, and with the lights still off, rewind the film all the way back into its cassette. Once the film is in its light-proof cassette, you can turn the room lights back on.

Photographing other objects

Kirlian photographs aren't limited just to your fingertips. Try leaves, coins, hair, pieces of leather, and any other small materials you can place directly on the film. You'll need to press the objects against the film during the exposure and select a lower power setting. Or, you can rig up a separate grounding wire from the object to the ground lug on the power supply. You will want to use this technique if you want to increase the power level of the power supply to bring out extra halos in inanimate objects.

One of the most famous Kirlian photographic effects is the "phantom leaf." The effect is rare, occurring in only once or twice in a few hundred exposures.

Here's how to do it. Pick a leaf from a tree and clip a small part off one edge. Immediately take a photograph of it with your Kirlian camera. The phantom effect shows not the outline of the leaf as you cut it, but the whole leaf with the missing segment in place.

Recommended films

Color films bring out the unusual halos and discharges of Kirlian photographs. You can use either print or slide film, but I've had the best results with Kodak Ektachrome 200 ISO. The film seems to be the best for bringing out the deep blues of the electrical discharge and its heavy base (as compared to print film) is a better insulator against the high-voltage potential. Another benefit: Ektachrome can be processed at home with chemistry you can purchase at most any photographic shop (E-6). If you don't care to develop it yourself, most custom photo labs can process Ektachrome in one day or less.

If you need black-and-white photographs, a good film stock to use is Kodak Kodalith. The film yields high contrast images that reproduce well in magazines, newsletters, and books. You can readily develop Kodalith film at home using a single two-step process, or take it to a custom photo lab.

In a hurry? Then try Polaroid's instant slide film. This film is developed in less than five minutes using a portable, manually operated processor. Polaroid offers a variety of film stocks: color, regular contrast black and white, high contrast black and white, and medical imaging. The film is expensive, but the cost does include processing and the results are almost instantaneous.

Exposure tips

Exposure time varies depending on the object being photographed, the amount of electricity provided by the power supply, and the sensitivity of the film. If the exposures are too weak, use a faster film, increase the exposure time, increase the voltage setting (last resort), or apply more pressure on the film. If the exposures are too strong, use a slower film, decrease the exposure time, decrease the voltage setting (but not much below the "tingle" range), or apply less pressure on the film.

You might find that you can decrease the exposure by as much as 50 percent with each doubling of film speed. For example, when going from ISO 100 to ISO 200 film, reduce the exposure by about one-half.

Until you get used to the variables involved in getting a good picture, you might want to "bracket" your exposures by precisely timing the duration your fingertips are pressed against the film. Start with five or six seconds, then make successive exposures and increase the time two or three seconds for each. Record the bracketing times, and when you get the film back from the processing lab, compare it against your notes. Select the best exposure time from the group.

Using larger film stock

The 35 mm roll film allows a usable frame of $1^3/_8$ by $^{15}/_{16}$ inches. The relatively small size of the 35 mm frame limits the size of the objects you can photograph. If you want to photograph larger objects, you must use larger film. One setup you can use is shown in FIG. 7-3. The plans are for using 4-by-5-inch cut sheet film, but you can build a frame for any size film you want. Construct the frame using a 4-by-5-inch metal "speed easel," available at most any photographic shop. Cement a piece of thin, copper-clad board or a layer of copper foil onto the easel. For each photograph, sandwich the specimen between two pieces of thick, transparent plastic sheeting cut to 4 by 5 inches. The specimen is grounded to the common ground point of the power supply. Place the film over the copper cladding and then lay the specimen and plastic over it.

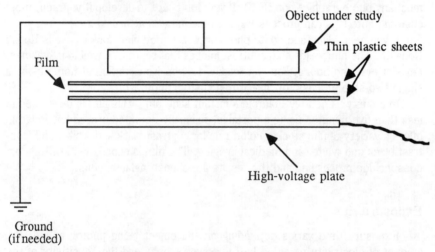

Fig. 7-3. Alternative setup for Kirlian photography, useful when taking large pictures on sheet film. For an interesting effect, try different types of photographic materials, including darkroom papers.

WHAT'S GOING ON HERE?

Once you make a few Kirlian photographs, the inevitable question is: what's going on here? What exactly is the mechanism that produces the weird arcs and halos, and does the size of the discharge have anything to do with moods, diseases, past histories, or any of the other paranormal theories people have attached to the Kirlian process?

You'd do well to obtain a copy of the Spring 1986 issue of the *Skeptical Inquirer* where authors Arleen J. Watkins and William S. Bickel attempt to explain in scientific terms many of the effects regularly observed in Kirlian photography. The article describes numerous parameters, controls, and conditions you might want to follow when conducting your Kirlian experiments so that you can more readily rationalize your results. In addition, a number of photographic trade journals, such as *Functional Photography*, have attempted to tackle the tough question of Kirlian photography.

Kirlian photographs do seem to show a strong aura surrounding certain living tissue (the aura is present around inanimate or dead objects, too, but it's not as great). One test often repeated in Kirlian experiments is to photograph a leaf several times over a period of a few days. As the leaf dies, its aura shrinks.

Is that psychically significant? Perhaps.

Or it could be the result of decreased moisture in the leaf or a microscopic change in the surface of the leaf. Both might have an effect on the discharge of high-frequency high voltage. Why? When Nikola Tesla first described high-frequency, high-voltage potentials, he spoke of how the electricity didn't enter the body but skipped over the skin and quickly dissipated.

This and other effects must be taken into account when debating the verity of Kirlian photography. Also consider *galvanic skin response* (the same as in lie detector tests), which can affect the moisture, resistivity, and even saline content of sweat.

If you plan on making Kirlian photography a hobby, you should keep a log book and record the subjects you've photographed, the films you've used, and the exposure times. Assemble a photobook of your favorite Kirlian photographs. You might even want to enter one or two in a regional photo contest. Kirlian photographs make great abstract art.

8

Projects in piezoelectricity

Who says all great discoveries have already been made?

A new form of electricity was discovered by Pierre and Jacques Curie just a little more than a century ago when the two scientists placed a weight on a certain crystal. The strain on the crystal produced an odd form of electricity—significant amounts of it, in fact. The Curie brothers coined this new electricity piezoelectricity. *Piezo* is derived from the Greek word meaning *press*.

Later, the Curies discovered that the piezoelectric crystals used in their experiments underwent a physical transformation when voltage was applied to them. They also found that the piezoelectric phenomenon is a two-way street: press the crystals and out comes a voltage; apply a voltage to the crystals and they respond by flexing and contracting.

While most people have heard of piezoelectricity, few realize how much of our modern technology revolves around the Curie brothers' discovery. For example, television signals are precisely timed, thanks to a quartz crystal (a piezoelectric element) vibrating under a set voltage potential. Electronic watches beep at hourly intervals thanks to a micro-miniature piezoelectric speaker. Dot-matrix printers tap out characters at high speed thanks to tiny pins actuated by piezoelectric materials.

This chapter details the basic science behind piezoelectricity and investigates some of the more interesting experiments you can conduct with piezoelectric materials. Much of the chapter centers on a unique piezoelectric material called PVDF, a flexible piezo film.

WHAT IS PIEZOELECTRICITY?

All piezoelectric materials share a common molecular structure where all the movable electric dipoles (positive and negative ions) are oriented in one specific direction. Piezoelectricity occurs naturally in crystals that are highly symmetrical—quartz, Rochelle salt crystals, and tourmaline, for example. The alignment of electric dipoles in a crystal structure is similar to the alignment of magnetic dipoles in a magnetic material.

When the piezoelectric material is placed under an electric current, the physical distance between the dipoles changes. This causes the material to contract in one dimension (or axis) and expand in the other. Conversely, placing the piezoelectric material under pressure (in a vise, for example), compresses the dipoles in one or more axes, causing the material to release an electric charge.

While natural crystals comprised the first piezoelectric materials used, synthetic materials have been developed that greatly demonstrate the piezo effect. A common human-made piezoelectric material is ferroelectric zirconium titanate ceramic, often found in piezo buzzers used in smoke alarms, wrist watches, and security systems. The zirconium titanate is evenly deposited on a metal disc. Electrical signals, applied through wires bonded to the surfaces of the disc and ceramic, cause the piezo material to vibrate at high frequencies (usually 4 kHz and above).

Piezo activity is not confined to brittle ceramics. PVDF, or polyvinylidene fluoride (used to make high-temperature PVDF plastic water pipes) is a semicrystalline polymer that lends itself to unusual piezoelectric applications. The plastic is pressed into thin, clear sheets and is given precise piezo properties during manufacture by stretching the sheets, exposing them to intense electrical fields, and/or other techniques.

PVDF piezo film is currently in use in many commercial products including noninductive guitar pickups, microphones, and even solid-state fans for computers and other electrical equipment. One PVDF film you can obtain and experiment with is Kynar, available directly from Pennwalt, the manufacturer (see Appendix A for the address and ordering info). See the section "Experimenting with Kynar piezo film" later in this chapter for more information on using Kynar.

Whether you are experimenting with ceramic or flexible PVDF film, it's important to understand a few basic concepts about piezoelectric materials.

- Piezoelectric materials are voltage sensitive. The higher the voltage, the more the piezoelectric material changes. Apply 1 volt to a ceramic disc and crystal movement is slight, but if you apply 100 volts, the movement is much greater.
- Piezoelectric materials act as capacitors. Apply a set voltage to the material and it will change shape. That shape will remain as long as the voltage is present. Remove the voltage, and the material returns to its original state.
- Piezoelectric materials are bipolar. Apply a positive voltage and the material expands in one axis. Apply a negative voltage and the material contracts in that axis.

EXPERIMENTING WITH CERAMIC DISCS

Piezoelectric ceramic discs are available at Radio Shack and many surplus electronics outlets (see Appendix A) including Edmund Scientific, All Electronics, and Jerryco. The cost is $1 or less per disc. I recently purchased a set of 10 one-inch-diameter discs for $3. That supply will last me a long time.

The most basic application of the piezo ceramic disc is to use it as a buzzer. This requires you to apply an ac voltage to the disc or at the least a voltage that changes between ground and the positive supply.

A simple 555 timer IC, connected as an astable multivibrator, is all that's necessary to create a piezo buzzer. The output of the 555 causes the disc to vibrate, emitting a characteristic buzzing sound. Figure 8-1 shows a circuit that allows you to change the output frequency of the 555, therefore changing the pitch of the vibrating disc. A parts list for the buzzer circuit is provided in TABLE 8-1.

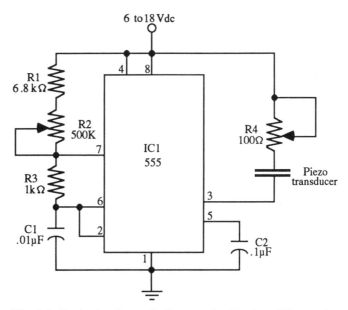

Fig. 8-1. Basic piezo buzzer built around a 555 timer IC operating in the astable mode.

Table 8-1. Parts List for
Piezo Disc 555 Buzzer

R1	6.8 kΩ
R2	500 kΩ potentiometer
R3	1 kΩ
R4	100 kΩ potentiometer
C1	0.01 μF disc
C2	0.01 μF disc
IC1	LM555 timer IC
Misc.	Piezo transducer

All resistors are 5 to 10 percent tolerance, 1/4 watt. All capacitors are 10 to 20 percent tolerance, rated at 35 volts or more.

The greater the supply voltage, the louder the buzz (because your ear is more sensitive to sounds in the 2 to 4 kHz range, these will seem the loudest, even if the supply voltage does not change). To generate powerful sounds, use the highest supply voltage you can—up to 18 Vdc with a 555 timer. Alternatively, you can attach a 2N2222 signal transistor, as shown in FIG. 8-2 (parts list in TABLE 8-2), to operate the disc with supply voltages greater than 18 Vdc. The collector-emitter voltage of the 2N2222 is rated to about 30 Vdc, so don't exceed this or the transistor could be damaged.

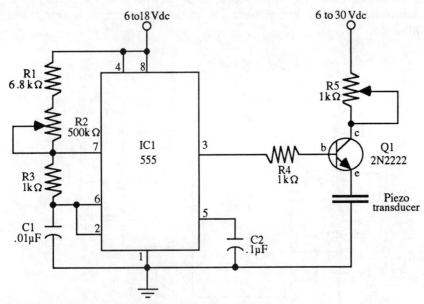

Fig. 8-2. *Adding a common 2N2222 transistor lets you increase the operating potential to about 30 volts.*

Table 8-2. Parts List for
Piezo Disc Transistor Booster

R1	6.8 kΩ
R2	500 kΩ
R3	1 kΩ
R4	100 kΩ
R5	1 kΩ potentiometer
C1	0.01 μF disc
C2	0.01 μF disc
IC1	LM555 timer IC
Q1	2N2222 transistor
Misc.	Piezo transducer

All resistors are 5 to 10 percent tolerance,
1/4 watt. All capacitors are 10 to 20 percent tolerance, rated at 35 volts or more.

Note that some ready-to-go piezo buzzers have an astable multivibrator circuit already built into them. These work merely by plugging them into a low-voltage dc source. The frequency of the buzz is internally set by the multivibrator circuit. The sound gets louder (and a little higher pitched) when you increase the supply voltage.

Caution: Ceramic piezoelectric discs can produce *extremely high* sound pressure levels. Do not place the disc too close to your ears when experimenting, or your eardrums could be exposed to sudden and painful sound levels. You could experience temporary or permanent deafness if exposed to continuous high volumes of sound.

If the sounds are overwhelming, place wax or cotton in your ears. In his September 1986 column on experimenting with piezoelectric buzzers, author Forrest Mims suggests you wear special-purpose protective earplugs, such as the COM-FIT from North Consumer Products (P.O. Box 7500, Cerritos, CA 90701). The suggestion makes good sense: the earplugs allow low-level sounds to pass but block those sound levels higher than about 25 to 40dB.

Using a ceramic disc as a sound source is only half the story. Follow the construction diagram shown in FIG. 8-3 and the parts list in TABLE 8-3. Tape a ceramic disc to a table, metal side down. Solder the ends to an ordinary neon indicator lamp. For safety, tape the exposed terminals of the neon lamp to the table, too. This helps prevent accidental contact of the leads to your skin. That could lead to an uncomfortable shock.

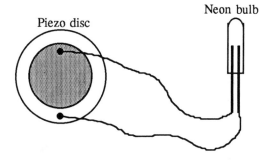

Fig. 8-3. Connect a neon bulb to a piezo disc as shown for demonstrating high-voltage output.

| Table 8-3. Parts List for Piezo Disc and Neon Light Flasher | 1 Piezo disc |
| | 1 Neon lamp, NE-2 or equivalent |

Tap on the disc and watch the neon lamp glow briefly. The harder you tap, the brighter the glow. Since the neon gas in the lamp doesn't ionize until the potential is about 90 Vdc, this experiment shows you how much electricity can be created by a thin carpet of piezoelectric ceramic bonded to a 1-inch disc.

While it is possible to permanently damage the ceramic material by striking it too hard, you can try producing very high voltages by hitting the disc with the blunt end of a plastic screwdriver. Wear protective goggles in case chips of the ceramic fly off under impact.

Using ceramic piezo discs to produce voltages in excess of 20 to 30kV is not uncommon. Piezo strikers—used primarily in outdoor gas barbecues—produce about 20,000 volts merely at the press of a button. You can purchase piezo strikers at surplus suppliers (Edmund, Jerryco, among others) or through the parts department of most department stores like Sears and Montgomery-Wards.

The striker, as shown in FIG. 8-4, consists of a plunger, some heavy-duty springs, and a ceramic piezo disc. Pressing the button causes the plunger to snap against the disc. This produces a brief 20,000-volt (or so) pulse between the output terminals. You can see this pulse as a thin blue spark. Solder a length of wire to the end terminal and bring it to within about 1/2-inch to the side terminal, as shown in FIG. 8-5. Press the button and watch the sparks fly!

A source for miniature piezoelectric strikers is the "electric" cigarette lighter, like the Scripto Electra. Despite its small size, the piezo striker produces a short pulse of 12 to 15kV. If you salvage the striker from the lighter, be sure to deplete the butane gas first (simply keep the lighter on for a minute or two until the gas is gone). Even after the gas has departed, the lighter poses a fire and explosion hazard. Be very careful when removing the striker. Pry the striker out with a small jeweler's screwdriver. Avoid the use of saws.

Another experiment you can try with piezoelectric discs is to connect two of them in a closed-loop, as shown in FIG. 8-6. Tap on one disc, and you'll hear it on the other. The communications link is two-way. Each disc is a sender and

Fig. 8-4. A self-contained high-voltage piezoelectric striker, commonly used in propane barbecue grills.

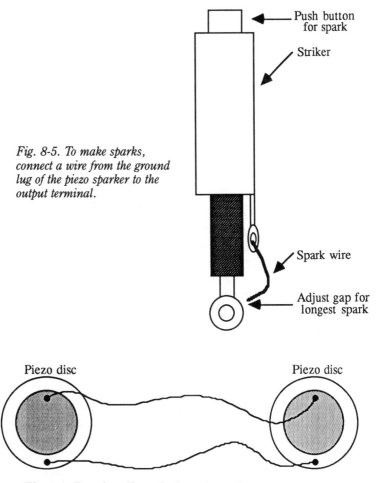

Fig. 8-5. To make sparks, connect a wire from the ground lug of the piezo sparker to the output terminal.

Push button for spark

Striker

Spark wire

Adjust gap for longest spark

Piezo disc

Piezo disc

Fig. 8-6. Two piezo discs wired together to form a transceiver.

receiver, and no intermediate battery or electronic circuit is required. If you and a friend know Morse code, you can send messages over fairly long distances just with two inexpensive ceramic piezoelectric discs. It beats wiring two cups together with string.

EXPERIMENTING WITH KYNAR PIEZO FILM

Samples of Kynar piezoelectric film are available in $1^1/_2$-by-$^1/_2$-inch wafers. The wafers, which are about the same thickness as the paper in this book, have two connection points, as illustrated in FIG. 8-7. Like ceramic discs, these two connection points are used to activate the film with an electrical signal or to relay pressure on the film as an electrical impulse.

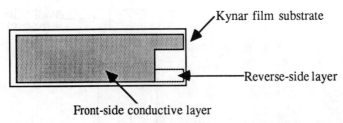

Fig. 8-7. Typical orientation of Kynar film, showing conductive layer and connecting points on the end.

Basic experiments

You can perform basic experiments with the film using just an ocsilloscope (preferred) or a high-impedance digital voltmeter. Connect the leads of the scope or meter to the tabs on the end of the film (the connection will be sloppy; later on we'll discuss ways to apply leads to Kynar film). Place the film on a table and tap on it. You'll see a fast voltage spike on the scope or an instantaneous rise in voltage on the meter. If the meter isn't auto-ranging and you are using the meter at a low setting, chances are that the voltage spike will exceed the selected range.

Next, tape the film (using double-sided tape) to the tabletop. Let it stand there a few moments, and then breathe on it. Watch the increase in output. The output will settle back down as the film returns to room temperature.

A third simple experiment shows the Kynar film as a contact microphone. Tape a piece of sample film to the bottom of a styrofoam cup (again, use double-sided tape, and be sure the contact is complete). Attach alligator clip leads to the contact points on the film. Avoid shorting. Attach the other end of the alligator leads to the microphone input jack of a tape recorder or PA amplifier. First, talk into the piece of film and listen to the results. Tap on the film and rub it with your fingers. You'll hear the sound amplified on the recorder or PA system. If the film is securely mounted to the cup, you can create sounds by picking up and rubbing the cup.

With the tape recorder, record a few moments of tapping, scraping, and rubbing. Rewind the tape, connect the alligator leads to the speaker output instead of the microphone input, and turn the volume up. Replay the tape and you'll hear the sounds (albeit rather subdued) over the Kynar film.

Attaching leads

Unlike piezoelectric ceramic discs, Kynar film doesn't usually come with preattached leads (although you can order samples with leads attached, but the cost is high). There are a variety of ways to attach leads to Kynar film.

Obviously, soldering the leads onto the film contact areas is out of the question. Acceptable methods include applying conductive ink or paint, self-adhesive copper-foil tape, small metal hardware, even miniature rivets. In all instances, use small-gauge wire—22 AWG or smaller. I have had good results using 28 AWG and 30 AWG solid wire-wrapping wire.

Conductive ink or paint   Conductive ink, such as GC Electronics' Nickel-Print paint, bonds thin wire leads directly to the contact points on Kynar film. Apply a small globule of paint to the contact point, then slide the end of the wire in place. Wait several minutes for the paint to set before handling. Apply a strip of electrical tape to provide physical strength.

Self-adhesive copper-foil tape   Copper-foil tape designed for repairing printed circuit boards can be used to attach wires to Kynar film. The tape uses a conductive adhesive and is quick and simple to apply. As with conductive inks and paints, apply a strip of electrical tape to the joint to give it physical strength.

Metal hardware   Use small $2/56$ or $4/40$ nuts, washers, and bolts (available at hobby stores) to mechanically attach leads to the Kynar. Poke a small hole in the film, slip the bolt through, add the washer, and wrap the end of a wire around the bolt. Tighten with the nut.

Miniature rivets   Homemade jewelry often uses miniature brass or stainless steel rivets. You can obtain the rivets and the proper riveting tool from many hobby and jewelry-making stores. To use, pierce the film to make a small hole, wrap the end of the wire around the rivet post, and squeeze the riveting tool (you might need to use metal washers to keep the wire in place).

Using Kynar film as a buzzer

Use the buzzer circuit shown in FIG. 8-1 near the beginning of this chapter to drive a small piece of Kynar film. You'll get best results if you tape the film to a flat or slightly curved surface. Little sound is emitted when the film flutters by itself in midair.

Using Kynar film as a microphone

Although Kynar film can be used alone as a microphone element, best results are achieved when used with a voltage-follower/buffer, as shown in FIG. 8-8 (a parts list is provided in TABLE 8-4). The 2N3819 transistor is a commonly available FET. The output can be applied directly to the microphone input of an amplifier or recorder.

You can make your own microphone as shown in FIG. 8-9. The plastic cap is from a plastic film canister lid. Cut a $3/8$-inch hole in the cap. Sandwich a piece of Kynar film between two slices of lightweight foam (sheets of batting work well, too), and use double-sided tape to adhere the foam and Kynar to the bottom of the canister lid. If you want, you can add a small 6-volt battery and FET circuit to the microphone element and stuff it all into the rest of the film canister. Drill a hole in the canister for the output leads.

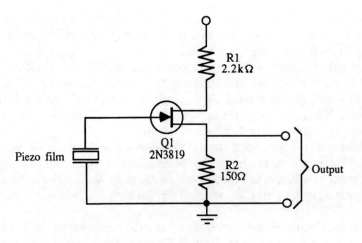

Fig. 8-8. Basic follower/buffer circuit for piezoelectric applications.

Table 8-4. Parts List for Piezo Disc Follower/Buffer Circuit

R1	2.2 kΩ
R2	150 Ω
Q1	2N3819 FET transistor
Misc.	Piezo disc

All resistors are 5 to 10 percent tolerance, 1/4 watt.

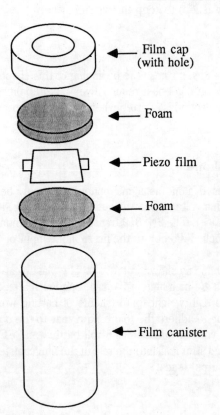

← Film cap (with hole)

← Foam

← Piezo film

← Foam

← Film canister

Fig. 8-9. Use a small piece of piezo film sandwiched between foam discs to make a film canister microphone. The FET preamp electronics can also be stashed in the canister.

Using Kynar as a mechanical transducer

Figure 8-10 shows a circuit you can build that indicates each time a piece of Kynar film is struck. See TABLE 8-5 for a parts list. Tapping the film produces a voltage output, which is visually indicated when the LED flashes. The 4066 IC is an analog switch. When a voltage is applied to pin 3, the connection between pins 2 and 7 is completed, and that finishes the electrical circuit to light the LED.

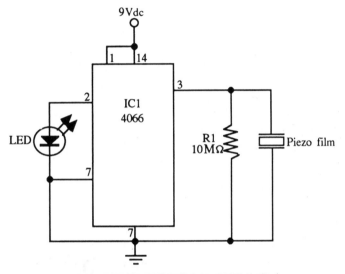

Fig. 8-10. Schematic for strike/vibration circuit.

Table 8-5. Parts List for Piezo Disc Strike/Vibration Detector		
R1	10 MΩ	
LED1	Light-emitting diode	
IC1	4066 CMOS analog switch	
Misc.	Piezo disc	

All resistors are 5 to 10 percent tolerance, 1/4 watt.

Though the voltage output of Kynar film is greater than that of most mechanical transducers, a stepup transformer can be used to boost the voltage to aid in the detection of slight mechanical movements.

First try a 117-to-6.3-volt power transformer. Attach the film leads to the 6.3-volt (secondary) terminals and your oscilloscope or meter to the 117-volt (primary) terminals, as depicted in FIG. 8-11. The transformer provides an 18.5:1 stepup ratio so that impacts or vibrations that normally cause a 500-milli-volt output directly from the Kynar film now produce a 9.25-volt output. Try miniature dc-dc converter and audio transformers to reduce the size of the transducer.

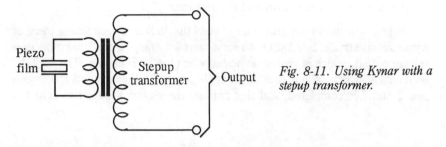

Fig. 8-11. Using Kynar with a stepup transformer.

Kynar speakers

Any piece of Kynar can be used as a speaker, but Pennwalt makes special speaker film, as shown in FIG. 8-12. For best results, use the amplifier circuit shown in FIG. 8-13 (parts list in TABLE 8-6). The circuit is adapted from Pennwalt's application notes for the piezo speaker film. Connect the input of the amplifier to the earphone jack of a portable cassette player or stereo system. Adjust the volume levels of the cassette player or stereo system, and R1 on the speaker amplifier, to a comfortable listening level.

Lay the film on a flat surface and listen to the sound. Although the film doesn't appear to move much, it's actually creating a hefty volume of sound. Gently cup the film and you'll note an increase in volume. Curving the film gives a pumping action to the piezo film, which is more efficient in coupling the mechanical movement to the air.

Fig. 8-12. Kynar speaker film.

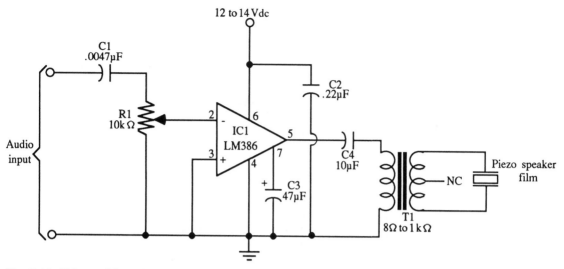

Fig. 8-13. This amplifier drives the Kynar speaker film. Use with a line-level audio source.

Table 8-6. Parts List for Piezo Speaker

R1	10 kΩ potentiometer
C1	0.0047 μF disc
C2, C4	0.22 μF disc
C3	47 μF polarized electrolytic
IC1	LM386 integrated amplifier IC
T1	8 kΩ-to-1 kΩ impedance-matching transformer
Misc.	Piezo speaker film

All resistors are 5 to 10 percent tolerance, 1/4 watt. All capacitors are 10 to 20 percent tolerance, rated at 35 volts or more.

ADDITIONAL EXPERIMENTS

There are a number of other experiments you can conduct with piezo discs or Kynar piezo film.

60-cycle induction

Piezo material is susceptible to 60-cycle induction caused by nearby power cords and appliances. Connect a piezo disc or Kynar film to the simple op amp amplifier in FIG. 8-14 (parts list in TABLE 8-7), and to an oscilloscope. Touch the disc or film; you'll see a dramatic increase in 60-cycle hum on the scope display. You can use this effect to your advantage—as a touch switch, for example. Normally, though, it's an undesired by-product.

You can largely cancel the 60-cycle interference by bonding two discs or pieces of Kynar film together or by folding one piece in half. This acts to cancel out the 60-cycle hum. Figure 8-15 shows a piece of Kynar film folded in half.

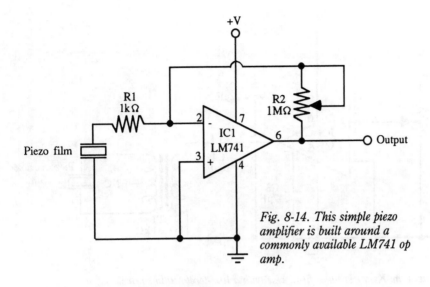

Fig. 8-14. This simple piezo amplifier is built around a commonly available LM741 op amp.

R1	1 kΩ	
R2	1 MΩ potentiometer	
IC1	LM741 op amp	
Misc.	Piezo film	

Table 8-7. Parts List for Piezo Audio Amplifier

All resistors are 5 to 10 percent tolerance, 1/4 watt.

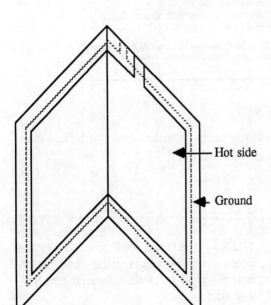

Hot side

Ground

Fig. 8-15. Fold a piece of Kynar film in half to cancel the effects of nearby 60-cycle induction.

This so-called *common-mode rejection* works because each piece of film contributes a 60-cycle sine wave. Because the two films are back-to-back, the sine waves act to cancel each other out.

Demonstrate capacitance of piezo film

As mentioned earlier in this chapter, piezo discs and film act as capacitors. You can readily demonstrate the capacitance effect with Kynar piezo film. Tape a small piece of film to a credit card using double-sided tape. Next, connect the leads to a high-impedance digital voltmeter (1 to 10 MΩ input required). Bend the card and watch the reading on the meter. Keep the card bent and watch the reading fall back to zero. Unbend the card and watch the reading; it will have an opposite sign as the voltage when the card was first bent.

Kynar's makers, Pennwalt, likens this action to a sponge filled with water. When you push on the sponge, the water is forced out. As you hold the sponge down, nothing happens. But when you release the sponge, the water is drawn back in.

To demonstrate the dielectric capacity of the film, bend the card again, but this time disconnect one of the leads to the voltmeter while you hold the card bent. Release the card and wait 10 seconds. Reconnect the lead: the voltage is still there but quickly dissipates.

Instrument pickups

Kynar film makes an excellent pickup for acoustic instruments including piano, guitar (electric or otherwise), violin, flute, saxophone, or even drums. The advantage of the piezo material is that it picks up the sound by vibration, not induction. Mount the film to the instrument using double-sided tape. The film should be positioned close to the sound board, port, or other orifice where the tones actually come out.

9

The science of lasers

Lasers rank as one of the most important innovations of the century. Unlike many other technological advances, lasers of one type or another find heavy use in industry. You'll find lasers in supermarket scanning systems, compact audio disc players, sophisticated surveying equipment, artistic light shows, holographic photography, and much more. The growing application of lasers means that laser systems, particularly the helium-neon gas variety, are readily available in the surplus market.

In this chapter, you'll learn the basic operation of helium-neon lasers (solid-state diode lasers are detailed in Chapter 11). Use the basic information presented here to begin your laser experiments; Chapters 10 through 17 offer a number of useful and exciting laser projects you'll want to try. However, be certain to read through the last section in this chapter on working with lasers and their power supplies.

THE BASICS OF LASER OPERATION

Albert Einstein was responsible for first proposing the idea of the laser in about 1916. Einstein knew that light was a series of particles (now known as *photons*) traveling in a continuous wave. These photons could be collected, using an apparatus not yet developed, and focused into a narrow beam.

To be useful, all the photons would be emitted from the apparatus at specific intervals. As important, much of the light energy would be concentrated in a specific wavelength—or color—making the light even more intense and powerful.

Photons can be created via a variety of means including ionizing gas within a sealed tube, burning some organic material, or heating a filament in a light bulb. In all cases, the atoms that make up the light source change from their usual stable or ground state to a higher, excited state by the introduction of some form of energy, typically electricity. The atom can't stay at the excited state for long. As depicted in FIG. 9-1, when the atom drops back to the ground state, it gives off a photon of light.

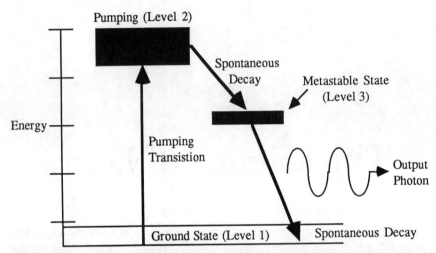

Fig. 9-1. Energy levels of a typical laser. After initial pumping by an electrical source of flash tube, the atoms in the lasing medium spontaneously decay to a metastable state; they then decay once more and output photons.

The release of photons by natural methods results in spontaneous emission. The photons leave the source in a random and unpredictable manner, and once a photon is emitted, it marks the end of the energy transfer cycle. The number of excited atoms is relatively low, so the great majority of photons leave the source without meeting another excited atom.

Einstein was most interested in what would happen if a photon hit an atom that happened to be at the excited, high-energy state. He reasoned that the atom would release a photon of light that would be an identical twin to the first. If enough atoms could be excited, the chance of photons hitting them would be increased. That would lead to a chain reaction where photons would hit excited atoms and make new photons—the process continuing until the original energy source was removed. Einstein had a name for this phenomenon and called it *stimulated emission of radiation.*

Raising atoms to a high energy state is referred to as *pumping.* In the common neon light, for example, the neon atoms are pumped to their high-energy state by means of a high-voltage charge applied to a pair of electrodes. The gas within the tube ionizes, emitting photons. If the electrical charge is high enough, a majority of the neon atoms will be pumped to the high-energy state. A so-called "population inversion" occurs when there are more high-energy atoms than low energy ones. A laser cannot work unless this population inversion is present.

Protons scatter all over the place, and left on their own, they simply escape the tube and do not strike many excited atoms. But assume a pair of mirrors are

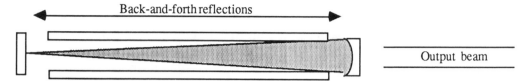

Back-and-forth reflections

Output beam

Fig. 9-2. The facing mirrors in a laser tube form an optical resonator, which amplifies the light levels.

mounted on either end of the tube and that some photons bounce back and forth between the two mirrors.

At each bounce, the photons collide with more atoms. If many of these atoms are in their excited state, they too release photons. Remember, these new photons are twins of the original and share many of its characteristics including wavelength, frequency, polarity, and phase. The process of bouncing from one mirror to the next, each time photons striking atoms in the path, constitutes *light amplification*.

In theory, if both mirrors are completely reflective, the photons would bounce back and forth indefinitely. Rub a little of the reflective coating off one mirror, however, and it passes some light. Now, a beam of photons passes through the partially reflective mirror, like that in FIG. 9-2, after the light has been sufficiently amplified. In addition, because the mirror is partially reflective, it holds back some of the light energy. This reserve continues the chain reaction inside the tube.

The combination of light amplification and stimulated emission of radiation makes the laser operate. As you probably already know, the word *laser* is an acronym for its theory of operation—*l*ight *a*mplification of *s*timulated *e*mission of *r*adiation.

ANATOMY OF A HELIUM-NEON LASER TUBE

The helium-neon (He-Ne) tube—as shown in FIG. 9-3—is the staple of the laser experimenter. He-Ne tubes are in plentiful supply, especially in the surplus market. The laser emits a bright, deep red glow that can be seen for miles around. Although the power output of He-Ne tubes is relatively small compared to other laser systems, it is perfectly suited for many homebrew and school experiments.

Fig. 9-3. A commercially manufactured helium-neon laser.

The helium-neon laser is a glass vessel filled with 10 parts helium and one part neon and is pressurized to about 1 torr (exact gas pressure and ratios vary between laser manufacturers). Electrodes placed at the ends of the tube provide a means to ionize the gas, thereby exciting the helium and neon atoms. Mirrors mounted at either end form an optical resonator, or Fabry-Perot resonator. In most He-Ne tubes, one mirror is totally reflective and the other is partially reflective. The partially reflective mirror is the output of the tube. A schematic diagram of the components of the typical laser tube is shown in FIG. 9-4.

Modern He-Ne lasers are composed of few parts, all fused together during manufacturing. Only the very old He-Ne tubes, or those used for special laboratory experiments, use external mirrors. The all-in-one design of the typical He-Ne tube means they cost less to manufacture and the mirrors are not as prone to misalignment.

Helium-neon lasers are actually composed of two tubes: an outer plasma tube that contains the gas and a shorter and smaller inner bore or *capillary* where the lasing action takes place. The bore is attached to only one end of the tube. The loose end is the output and faces the partially reflective output mirror. The bore is held concentric by a metal element called the *spider*. The inner diameter of the bore largely determines the diameter width of the beam, usually 2.0 mm to 0.5 mm.

The ends, where the mirrors are mounted, typically serve as the anode (positive) and cathode (negative) terminals. On other lasers, the terminals are mounted on the same end of the tube. A strip of metal or wire extends to the cathode on the other end. The output mirror can be on either the anode or cathode end, but on most tubes, it is the cathode. Many manufacturers prefer this arrangement, claiming it is safer and provides more flexibility.

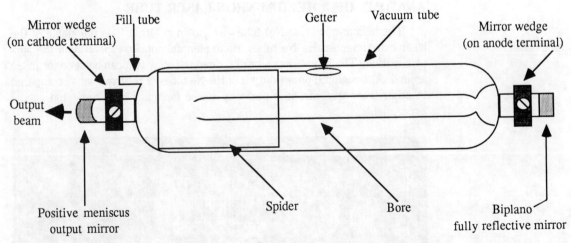

Fig. 9-4. An x-ray view of a typical He-Ne laser tube, with component parts indicated. The arrangement and style of your laser tube might be slightly different.

Metal rings with hex screws are often placed on the mirror mounts as a means to tweak the alignment of the mirrors. Unless you suspect the mirrors are out of alignment, you should *not* attempt to adjust the rings. They have been adjusted at the factory for maximum beam output. Tweaking them unnecessarily can seriously degrade the performance of the laser.

He-Ne lasers are available in two general forms: bare and cylindrical head. Bare tubes are just that—the tube is not shielded by any type of housing and should be placed inside a tube or box during operation for protection. Cylindrical head lasers (or just "laser heads") are housed inside an aluminum sheath, as shown in FIG. 9-5. Leads for power come out the back end of the laser. The opposite end might have a hole for the output beam or be equipped with a safety shutter. The shutter prevents accidental exposure to the beam.

POWER OUTPUT

The greatest difference among helium-neon laser tubes is power output. There are some He-Ne tubes designed to produce as little as 0.5 mW of power, while others generate 10 mW or more of light energy. The difference in power output is not always visible to your eye.

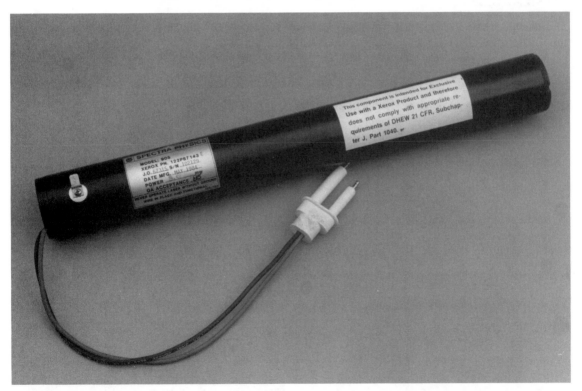

Fig. 9-5. A typical cylindrical laser head, removed from a Xerox laser printer. Laser heads such as these are common finds in the surplus market.

By far, the majority of helium-neon tubes are rated at 1 to 2 mW. This is adequate for most laser experiments and you rarely need more. In fact, lower-power lasers are often easier to work with because the power supply requirements are not as stringent. However, higher power comes in handy if you are engaged in holography (the more power, the faster the exposure), outdoor surveying, and other applications where a bright beam is necessary.

Cylindrical-head lasers, generally designed for use in telecopiers, laser facsimile machines, and supermarket bar-code scanners, are generally engineered for high output. Most tubes in this class are rated at 5 to 8 mW. If you need lots of power, check these out. Just be sure that the power supply delivers sufficient current and voltage to the tube.

Bear in mind that power output varies with the tube age. A surplus tube that produced two milliwatts when new might only generate 1.5 mW after several thousand hours of use. Careless handling also reduces the power output. Every shock or jolt can tweak the mirrors out of alignment, which reduces the power output. You can readily measure the power output of a laser by using a calibrated power meter such as the Metrologic model 45-450.

PHYSICAL SIZE

Helium-neon laser tubes come in a variety of sizes, depending on power output. Most are about 1- to $1^{1}/4$-inches in diameter by 7 to 10 inches long. Some very small ''pee-wee'' tubes are designed for use in hand-held bar code readers and generate less than 0.5 mW, but these are fragile and hard to use. Higher-powered tubes measure about 1.5 inches in diameter and 13 inches long. When placed in an aluminum housing, the entire laser head measures 1.75 inches in diameter by 15 inches in length.

BEAM CHARACTERISTICS

The beam from a helium-neon laser doesn't vary much between tubes. Except in special cases, the light has a wavelength of 632.8 nanometers and can measure anywhere between 0.5 to 2 mm in diameter. Note that the diameter is less than the actual side-to-side measurement of the laser beam. Most manufacturers eliminate the outer 13.5 percent of the beam diameter, leaving the bright inside core.

Most He-Ne lasers work in what's called TEM_{00} mode, which provides a nicely rounded beam and a Gaussian transverse irradiance profile. The TEM_{00} mode has the minimum possible diffraction loss, minimum divergence, and can be focused to a very small spot. The Gaussian irradiance profile of a laser operating in TEM_{00} mode is shown in FIG. 9-6. The center 86.5 percent of the beam is the brightest; the irradiance falls off as you approach the edges of the beam.

Even without external optics, the divergence (spreading) of a helium-neon laser is typically not over 1.2 milliradians (mrads). That means that at 100

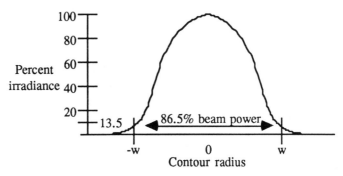

Fig. 9-6. The Gaussian irradiance profile of the He-Ne laser in TEM_{00} mode. The peak of the irradiance profile is the center of the beam.

meters, the beam will spread 120 millimeters, or about 4.7 inches. You can readily calculate beam spread for any distance by multiplying the divergence (in thousandths of a radian) by the distance (in meters). The answer is beam spread in meters.

EXAMPLE: if the divergence is one milliradian (0.001 radian) and the distance is 120 meters, the divergence is 0.12 meters (0.001 times 120), or 12 centimeters. That's a beam less than 5 inches wide, even though the light has traveled a distance longer than a football field. You can reduce divergence even more with the addition of simple optics: pass the beam through a cheap telescope so that it enters the eyepiece and exits the objective, as shown in FIG. 9-7. Assuming the focus is correct, the beam will spread less over longer distances.

A laser beam might appear grainy or spotty when reflected off of a white or lightly colored wall. This effect is called *speckle* and is caused by local interference within your eye as well as at the target surface. Even a smooth, painted surface has many hills and valleys in comparison to the wavelength size of the laser beam. The light experiences constructive and destructive interference as it is reflected off of the uneven surface of the target.

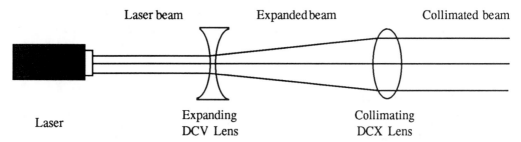

Fig. 9-7. A laser beam collimator is as simple as two lenses, as shown, and is similar to a telescope used in reverse.

HE-NE COLORS

A He-Ne tube generates many different colors. Most colors (plus infrared wavelengths you can't see) are transmitted out of the tube before they are amplified, thus preventing them from turning into laser light.

He-Ne lasers emit a deep red beam at 632.8 nanometers because it is the strongest wavelength produced within the tube. The other colors are weak or might not be sufficiently coherent or monochromatic. Yet there are some special helium-neon lasers that are made to operate at different wavelengths, namely 1.523 micrometers (infrared) and 543.5 nanometers (green). Additional colors can be generated with the current crop of specialty He-Ne tubes including yellow and yellow-green. Some He-Ne tubes are even "tunable" and produce a number of colors. Nonred He-Ne tubes are exceptionally expensive and fairly rare in the surplus market.

POWERING THE TUBE

Most surplus He-Ne tubes are sold without instructions of any type. It's up to you to know how to use it. All He-Ne tubes have an anode (positive terminal) and cathode (negative terminal). In some cases, the terminals are marked with an "A" and "C" (sometimes "K" for the latter) or simply " + " and " – " for anode and cathode, respectively. A few others have a red dot that indicates the anode. Following the polarity specified, connect the power supply so that the positive lead connects to the anode and the negative (or ground) lead connects to the cathode.

By far, most He-Ne tubes have no markings at all and you are left wondering which one is which. Connecting the tube backwards to the power supply can damage the supply, the tube, or both. You can readily identify the cathode of most tubes by looking for the ring-shaped "getter" resting near the inside wall of the tube and (typically) connected to the cathode by way of the metal spider. Another tip-off is the filling line originally used to pump the gases into the tube. The filling line usually (but not always) denotes the cathode end. In fact, the metal line often doubles as the cathode connection.

Lasers enclosed in aluminum housings have leads coming out an end cap. These leads might terminate in a male high-voltage Alden connector (shown in FIG. 9-8), designed for use with a corresponding female connector attached to commercial laser power supplies. The thin prong of the connector denotes the anode.

THE ROLE OF THE BALLAST RESISTOR

He-Ne tubes are current sensitive. They use a ballast resistor to limit the current reaching the laser and to provide stable electrical discharge. The value of the ballast resistor varies depending on the tube and power output, but it's generally between about 60 and 230kΩ. The ballast resistor is placed close to the anode of the tube to minimize anode capacitance and to provide more stable operation.

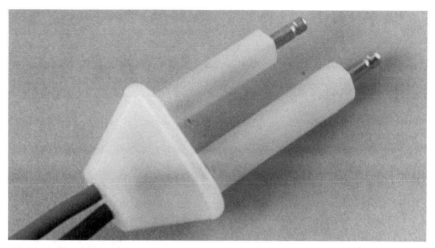

Fig. 9-8. The Alden high-voltage connector is usually found on cylindrical laser heads.

You can safely operate most lasers between the 3.5 to 7mA band gap typically recommended by manufacturers for He-Ne tubes, but the lower the current, the better. The tube will last longer and the power supply will operate more efficiently.

If you have a choice of ballast resistor values, choose the highest one you can that fires the tube and keeps it running. If the tube sputters or blinks on and off, it could be a sign that it's not receiving enough current. Lower the ballast resistor and try again. The benefit of the pulse-modulated power supply described below is that it works with most any laser, regardless of the ballast resistor used. You adjust the current supplied to the laser by turning a control.

You can test the amount of current supplied to the tube by connecting a meter as shown in FIG. 9-9. Dial the meter to mA and be sure to keep your hands away from the meter probes.

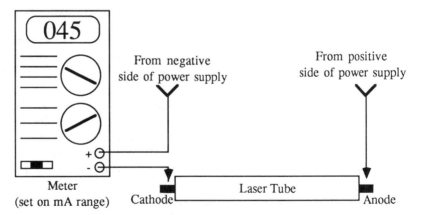

Fig. 9-9. Use this arrangement with an analog or digital voltmeter to test the current consumption of a laser tube. Do not touch the test leads—you could receive a bad shock.

IMPORTANT NOTE: Commercially made power supplies designed for use with cylindrical head lasers are typically engineered *without a ballast resistor*. Rather, the resistor is housed in the end cap of the laser head. If you use the power supply with a bare tube that lacks a ballast resistor, you could permanently ruin the laser. If in doubt, add the ballast resistor; you can always take it out later.

THE PROPERTIES OF LASER LIGHT

The light from a laser is special in many respects.

- Laser light is *monochromatic*. Laser light coming from the output mirror consists of one wavelength, or in some instances, two or more specific wavelengths. The individual wavelengths can be separated using various optical components.
- Laser light is *spatially coherent*. The term spatial coherence means that all the waves coming from the laser are in tandem. That is, the crests and the troughs of the waves that make up the beam are in lock-step.
- Laser light is *temporally coherent*. Temporal coherency is when the waves from the laser (which can be considered as one large wave, thanks to spatial coherency) are emitted in even, accurately spaced intervals. Temporal coherence is similar to the precise clicks of the metronome, timing out the beat of the music.
- Laser light is *collimated*. Because of monochromaticity and coherence, laser light does not spread (diverge) as much as ordinary light. The design of the laser itself, or simple optics, can collimate the laser light into a parallel beam.

The four main properties of laser light combine to produce a shaft of illumination that is many times more brilliant than the light of equal area from the sun. Because of their coherency, monochromaticity, and low beam divergence, lasers are ideally suited for a number of important applications. For example, the monochromatic and coherent light from a laser is necessary to form the intricate swirling patterns of a hologram. Without the laser, optical holograms would be much more difficult to produce.

Coherence plays a leading role in the minimum size of a focused spot. With the right optics, it's possible to focus a laser beam to an area equal to the wavelength of the light. With the typical infrared-emitting laser diode, for instance, the beam can be focused to a tiny spot measuring just 0.8 micrometers wide. Such intricate focusing is the backbone of compact audio discs and laser discs.

Minimum divergence (owing to the coherent nature of laser light) means that the beam can travel a longer distance before spreading out. The average helium-neon laser, without optics, can form a beam spot measuring only a few inches in diameter from a distance several hundred feet away. With additional optics, beam divergence can be reduced, making it possible to transmit sound, pictures, and computer code many miles on a shaft of light. The signal is intercepted by a receiving station in the light path.

BUYING AND TESTING HE-NE TUBES

Apart from size and output power, tubes vary by their construction, reliability, and beam quality. After buying a He-Ne tube, you should always test it; return the tube if it doesn't work or if its quality is inferior.

Should you need a laser for a specific application that requires precision or a great deal of reliability, you might be better off buying a new and certified tube rather than one from surplus. The tube will come with a warranty and certification of power output.

The first step in establishing the quality of the tube is to inspect it visually. If you get a used tube, be on the lookout for scratched, broken, or marred mirrors. After inspection, connect the tube to a suitable power supply, point the laser toward a wall, and apply power. If the laser is working properly, the beam will come out one end only and the beam spot will be solid and well defined.

Occasionally, the totally reflective mirror allows a small amount of light to pass through and you see a weak beam coming out the back end (this is especially true if the mirror is not precisely aligned). Usually, this poses no serious problem unless the coating on the mirror is excessively weak or damaged or if the mirrors are seriously out of alignment.

All lasers exhibit satellite beams—small, low-powered spots caused by internal reflection that appear off to the side of the main spot. In most cases, the main beam and satellites are centered within one another, so you see just one spot. But slight variations and adjustment of the mirrors can cause the satellites to wander off axis. This can be unsightly and if it matters to you, choose a tube that has a solid beam.

Should the tube start but no beam comes out, check to be sure that nothing is blocking the exit mirror. If the beam still isn't visible, the mirrors might be out of alignment and the laser should be returned for a replacement.

If the tube doesn't ignite at all, check the power supply and connections. Try a known good tube if you have one. The tube still doesn't light? The problem could be caused by:

- Bad tube. The tube is "gassed out," has a hairline crack, or is just plain broken.
- Power supply too weak. The tube might require more current or voltage than the levels provided by the power supply.
- Insulating coating or broken connection on terminal. New and stored tubes might have an insulating coating on the terminals. Be sure to clean the terminals thoroughly. A broken lead can be mended by soldering on a new wire.

Some "problems" with laser tubes are really caused by the power supply. In fact, if your laser doesn't work, inspect the power supply first. One common problem is that the tube sputters when you turn it on. This fault is most often caused by a tube that isn't receiving enough current, either because the connections from the power supply are loose or broken, the power supply is not producing enough current for the tube, or the ballast resistor is too high or too low.

Hard-to-start tubes flick on but quickly go out. If the power supply incorporates a trigger transformer, the tube might "click" on and off once every 2 to 3 seconds (correlating to the time delay between each high-voltage trigger pulse). Tubes that haven't been used in a while can be hard to start, so once you get it going, keep it on for a day or two. In most cases, the tube will then start normally. Hard starting could also be caused by age and degassing, two factors you can't fix.

WORKING WITH LASERS

Lasers emit electromagnetic radiation, usually either visible light or infrared. The level of "radiation" is generally quite small in hobby lasers and has about the same effect on external bodily tissues as sunning yourself with the living room lamp.

Skin is fairly resilient, even to exposure to several watts of laser energy. But the eye is much more susceptible to damage, and it is the effects of laser light on the retina that is the greatest concern. Even as little as 20 to 50 milliwatts of focused visible or infrared radiation can cause eye damage.

The longer the exposure to the radiation and the more focused the beam, the greater the chance that the laser will cause a lesion on the surface of the retina. Retinal lesions can heal, but many leave blind spots. Retinal damage when using hobby lasers—those with outputs of less than 5 or 10 milliwatts—is rare, but you should nevertheless avoid looking directly at the beam.

All gas lasers, including the popular helium-neon variety, require high-voltage power supplies (one such do-it-yourself supply is shown in Chapter 12, High-tech laser projects). These power supplies boost the main voltage from 12 volts dc or 117 volts ac to several thousand volts. The typical 12-volt dc laser power supply produces up to 3,000 volts dc, an extraordinary amount when you consider that the electric chair puts out only 2,000 volts!

Some laser experimenters tend to disregard the high voltages. They assume that although the voltage is high, the current level is low. While it is true that the current demand of the typical helium-neon laser is low—between 3 and 7 milliwatts—at the 1.2 to 3kV level of standard laser operating voltages, even a 7-milliwatt jolt can be injurious.

Laser power supplies should be properly shielded and insulated. Avoid operating a power supply in the open, and always cover exposed high-voltage parts. Insulating material such as high-voltage putty (available at TV repair shops and electronic stores) restricts arcing and provides a relatively shockproof layer.

It's easy to forget that several thousand volts are coursing through a wire, and you could inadvertently touch it or brush across it. Even a direct contact is not necessary. A 3kV arc can jump a quarter of an inch or more and, like light-ning, discharge on the nearest object. If that object happens to be your fingers or elbow, you will receive a painful shock.

Admittedly, most shocks from laser power supplies won't kill you, nor will they burn your skin. But don't underestimate the power of low-current high voltages. Your body's protective mechanism automatically reacts to the jolt. If you touch a live wire with your hand, your body quickly contracts to prevent further shock. If the jolt is large enough, you could be knocked backward onto the ground. Should you be holding the laser tube at the time, you could drop and shatter it, endangering yourself even more.

Most laser power supplies use high-voltage capacitors at the output stage. Like all capacitors, these can retain current even after the power supply has been turned off. Although many laser supplies use bleeder resistors that drain the capacitors after power has been removed, not all do. Play it safe and assume that the output of any power supply—plugged in or not—is potentially hot.

When working with the laser, make sure the power supply is off, and then temporarily short the leads of the power supply together or simply touch the positive terminal of the supply to ground, as shown in FIG. 9-10. That will dis-charge any remaining current, making the power supply safe to work on.

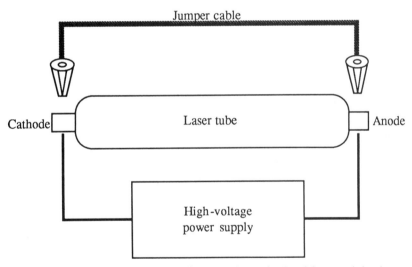

Fig. 9-10. For safety's sake, always discharge the static electricity remaining in a gas laser and its power supply by shorting the terminals with a jumper cable (be sure power is removed first!). If you can't reach both terminals, short the anode to ground.

Likewise, the laser tube behaves like a Leyden jar, which is a type of capac-itor. It, too, can retain a current after power has been removed. Drain the remaining charge by shorting the terminals or leads together.

10

Helium-neon laser pistol

You can readily build your own low-power, handheld laser using a commonly available and affordable helium-neon tube and 12-volt dc power supply. The pistol is made from 2- and 1 1/4-inch schedule PVC plumbing pipe, which you can cut with an ordinary hacksaw.

This chapter details the construction of two laser pistols. One uses commonly available PVC water pipe as the body of the pistol, while the other is designed around a plastic toy "laser" gun. The innards of the toy gun are removed and replaced by real laser components.

BUILD THE PVC LASER PISTOL

The PVC laser pistol is made with white plastic pipe, so the parts are easy to find, easy to work with, and inexpensive. The laser pistol uses a 1 to 2 mW red He-Ne tube, a modular power supply, and an external battery pack.

Defining barrel length

The exact length of the pistol barrel depends on the laser tube and power supply you use. The laser used in the prototype pistol is a common 2 mW tube measuring 1 1/2 inches in diameter by 7 1/2 inches. You can obtain even smaller tubes through some laser dealers, such as Allegro and Meredith Instruments, and make your pistol more compact.

The power supply is one of the smallest commercially made, measuring a scant 7/8 inch in diameter by slightly under 4 inches in length. This particular power supply was purchased through Meredith Instruments (see address in Appendix A) but is also available from Melles Griot and several laser manufacturers.

Cutting the barrel and grip

TABLE 10-1 provides a parts list for the laser pistol. Assuming you use the same or similar tube and power supply, cut a piece of 2-inch schedule 40 PVC to 12¹/₂ inches. Sand or file the cut ends to make them smooth. Drill a ³/₈-inch-diameter hole 4 inches from one end of the tube, as shown in FIG. 10-1. This hole serves as the leadway for the power wires.

Next, cut a length of 1¹/₄-inch schedule 40 PVC to 6¹/₄ inches. Using a wide, round file, shape one end of this piece so that its contour matches the 2-inch PVC. The angle of the smaller length of pipe, which serves as the grip, should be approximately 10 to 15 degrees. The match between the barrel and grip does not need to be exact, but avoid big gaps. You might want to cut the grip a little long so that you can allow yourself extra room to shape the contour. The grip is about the right length if it measures 6 inches top to bottom.

Cut a ¹/₄-inch hole 1¹/₄ inches from the bottom of the grip and another ¹/₄-inch hole 90 degrees to the right, but at a distance of 4³/₄ inches from the bottom (see FIG. 10-2). The lower hole is for the ¹/₄-inch phone jack for the power; the upper hole is for the pushbutton switch. Note that the size of the upper hole depends on the particular switch you use. The switch detailed in the parts list is commonly available at Radio Shack and other electronics outlets. If you use another switch, you should measure the diameter of the shaft and drill a hole accordingly.

Table 10-1. Parts List for Laser Pistol

1	12¹/₂-inch length of 2-inch schedule 40 PVC pipe
1	6¹/₄-inch lengths of 1¹/₄-inch schedule 40 PVC pipe
1	3-inch adjustable hose clamp
2	2-inch test plugs
1	1¹/₄-inch PVC end cap
J1	¹/₄-inch phone jack
S1	SPST momentary switch (normally open)
Misc.	Miniature 12 Vdc He-Ne power supply (see text), laser tube (see text)

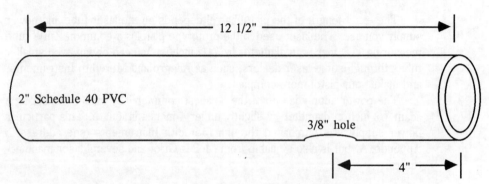

Fig. 10-1. Cutting and drilling guide for the He-Ne laser pistol barrel.

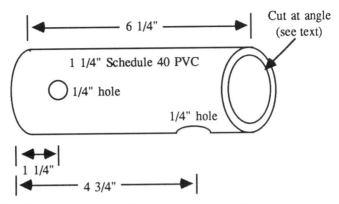

Fig. 10-2. Cutting and drilling guide for the He-Ne laser pistol grip.

Cut two 1/6-inch-wide slits approximately 1/2 inch from the top of the grip. The slits should be opposite one another and at right angles to the top hole (used for the switch). Unthread the loose end of a 12-inch-long (3 1/2-inch-diameter) hose clamp through the slits. Tighten the clamp one or two turns, but not so much that you can't insert the barrel in it. The pistol so far should look like FIG. 10-3.

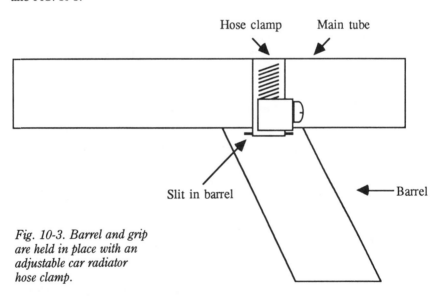

Fig. 10-3. Barrel and grip are held in place with an adjustable car radiator hose clamp.

Wiring the jack and switch

Use 20- or 22-gauge stranded wire to connect the components as shown in FIG. 10-4. The wire lengths are approximate and provide some room for easily fitting the jack, switch, power supply, and tube into the PVC enclosure. After you have soldered the jack and switch, mount them in the grip, feed the wires through the hole in the barrel, and mount the grip to the barrel. Center the hole in the grip and tighten the clamp.

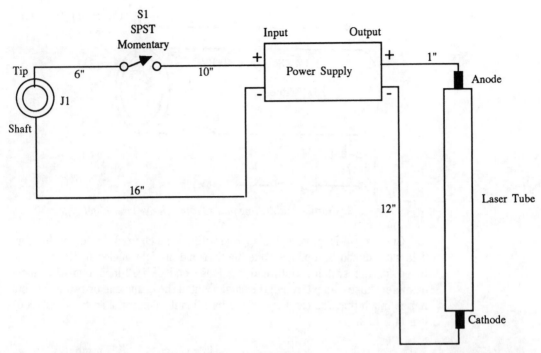

Fig. 10-4. Wire the He-Ne laser pistol as shown in this diagram. Be aware of the high voltages present at the output of the power supply. The diagram assumes the power supply contains its own ballast resistor; be sure to add one if yours doesn't.

Mounting the laser tube and power supply

Attach the power supply leads to the tube. With the power supply and tube used, the anode lead from the supply connects directly to the anode terminal of the tube. This wire must be kept short to minimize current loss and high-voltage arcing. The cathode lead stretches from the supply to the opposite end of the tube. This design works well because the tube emits light from the cathode side, opposite the power supply.

Since not all He-Ne lasers operate this way, you'll want to choose the tube carefully. The tube must emit its light from the cathode side, or the power supply will get in the way. Loosely attach the power supply to the tube by wrapping them together with electrical tape.

The tube should be protected against shock by wrapping it in a thin Styrofoam sheet—the kind used for shipping fragile objects. One or two wraps should be sufficient. Close up the sheet with electrical tape.

Final electrical connection and assembly

Solder the wires from the jack and switch to the power supply. Be sure to observe proper polarity. Slide the tube and power supply into the barrel so that the output end of the laser faces forward. Drill a 3/8-inch hole in the approximate center of a 2-inch knockout "test" plug (available in the ABS plumbing pipe department of most hardware stores). Because the tube might not rest in the exact center of the barrel, determine the proper spot for drilling by first inserting the plug and observing the location of the output window. Mark the spot with a pencil, remove the plug, and drill out the hole.

After drilling, replace the plug and push the tube into the barrel so that the output window is flush with the inside of the plug. If you allow the window to protrude outside the plug, you run a greater risk of chipping or breaking the output mirror. Insert another 2-inch plug in the rear of the barrel.

Test plugs are not routinely available for 1¹/4-inch pipe, so you must use an ordinary end cap for the bottom of the grip. Slip the end cap over the pipe; the fit should be tight. If not, try another end cap.

Building the battery pack

The battery pack for the laser is separate, not only because it allows you to build a smaller gun, but also because it allows you to power the device from a variety of sources.

The main battery pack consists of two 6-volt 8 AH lead-acid batteries contained in a 4³/8-by-7³/4-by-2³/8-inch phenolic or plastic experimenter's box (available at Radio Shack). The batteries are held in place inside the box with heavy double-sided foam tape. The battery pack contains a fuse and power jack, mounted as shown in FIG. 10-5. A parts list for the pack is in TABLE 10-2.

Fig. 10-5. The internal arrangement of the twin 6-volt batteries, fuse, and power jack for the battery pack.

Table 10-2. Parts List for Battery Pack

B1, B2	6 Vdc high-output (4 AH or more) gelled electrolyte or lead-acid rechargeable battery
F1	5-amp fuse
J1	1/4-inch phone jack
1	Project box measuring $4^{3}/_{8}$ by $7^{3}/_{4}$ by $2^{3}/_{8}$ inches
2	$5/_{8}$-inch eyelets
1	Camera or guitar strap

The fuse is absolutely necessary in case of a direct short. Lead-acid batteries of this capacity produce a heavy amount of current that can easily burn through wires and cause a fire. A 5-amp fast-acting bus fuse provides adequate short-circuit protection without burning out during the short-term shorts that can occur when you are plugging in the battery pack cable.

Wire the battery pack as indicated in FIG. 10-6. When wired in series, the two 6-volt batteries produce 12 volts. The high capacity of the batteries means you can operate the laser for at least an hour before it needs recharged. I have operated the prototype pistol for up to 6 hours before needing a recharge.

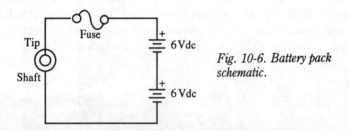

Fig. 10-6. Battery pack schematic.

To make the battery pack conveniently portable, insert two $5/_{8}$-inch eyelet screws near the top of the box. Snap on a wide camera strap and adjust the strap for your shoulder. You can sling the battery pack over your shoulder while holding the pistol in your hands.

Current is delivered from the battery pack to the pistol by means of a 6-to-12-foot coiled guitar extension cord. These are available at Radio Shack and most music stores. Buy the two-wire mono variety; you don't need the three-wire stereo type. Remember to double-check the hookup of the power jacks in the battery pack and pistol. Make sure that the positive terminal connects to the tip of the jack.

Battery recharger

The battery recharger is a surplus battery eliminator/charger pack designed for 12-volt systems. The pack outputs 13 to 18 volts at about 350 mA. Recharge the batteries in the box by unplugging the power cord that stretches between the pack and pistol and plugging in the recharger. At 350 mA, recharging takes from 10 to 14 hours. Because the battery eliminator/recharger is not "intelligent," be sure to remove it after the recharge period. Otherwise, the batteries could be damaged by overcharging.

ADDING ON TO THE LASER PISTOL

There are several modifications you can make to the laser pistol to increase its functionality and versatility, including a power indicator, modulation bypass jack, and single-shot circuit.

Power indicator

A power light provides a visual indication that the laser gun is on. The light is especially helpful if you are using the pistol in daylight, when the beam is hard to see and you want to be sure that the gun is working.

The power light consists of an LED and current-limiting resistor. Drill a hole for the LED in the rear test plug and mount it using all-purpose glue. Connect the LED in parallel with the laser power supply: when you pull the switch, current is delivered to the power supply and LED.

Modulation bypass

The modulation bypass permits you to easily modulate the beam with an analog or digital signal. The bypass consists of a 1/8-inch miniature earphone jack connected between the cathode lead of the power supply and the cathode terminal of the tube. Both LED and modulation bypass add-ons are shown in FIG. 10-7.

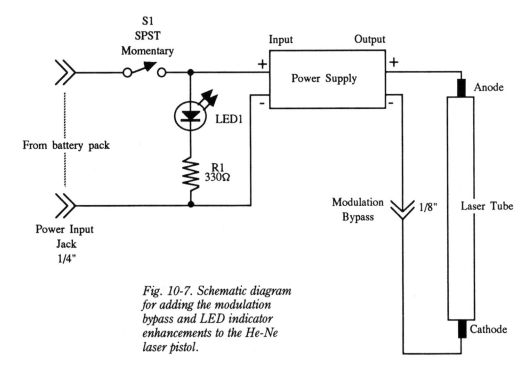

Fig. 10-7. Schematic diagram for adding the modulation bypass and LED indicator enhancements to the He-Ne laser pistol.

To add the bypass to the pistol, drill a hole for the jack in the rear test plug. Apply high-voltage putty around the terminals to the jack, but be sure that none of the putty interferes with the contacts of the jack. The putty helps prevent arcing from high voltages. When not using the bypass jack, insert a shorting plug into the jack. The shorting plug has its internal contacts shorted together.

To use the modulation bypass, remove the shorting plug and insert the leads from a modulation transformer, as detailed in Chapter 15. The transformer can be driven by an audio amplifier so signals can be transmitted via the laser beam. Depending on what you connect to the modulation transformer, you can transmit audio signals or digital data over distances exceeding 1 mile.

Continuous/single-shot circuit

Pressing the switch on the pistol grip produces a steady beam of red light. If you plan on using the pistol for target practice, you'll want to convert the pistol to single-shot operation. (See the subsequent section on how to build an electronic target.)

The circuit in FIG. 10-8 (parts list in TABLE 10-3) shows how to add a switch and simple circuit to provide either continuous or single-shot firing of the pistol. A good place to mount the toggle switch is near the power jack on the grip. Tuck the circuit, including the miniature relay, inside the grip. You can mount the components on a suitable predrilled circuit board or directly solder the leads of the components together. (I prefer the first method; it looks cleaner and lasts longer.)

Note that you must experiment with the value of R3 and the miniature relay. The coil resistance of the relay might require a higher or lower value for R3, so try several values until you come up with one you like. The schematic

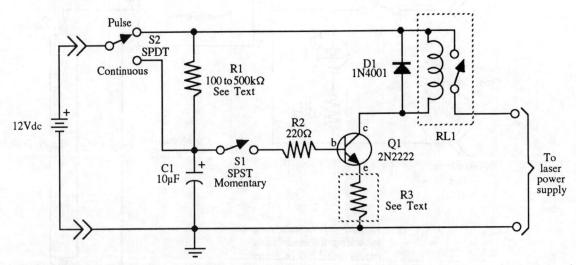

Fig. 10-8. Single-shot schematic add-on for the He-Ne laser pistol. You might need to experiment with the value of R3, depending on the resistance of the coil in the relay.

R1	100 to 500 kΩ
R2	220 Ω
R3	1 kΩ (see text)
C1	10 μF tantalum
Q1	2N2222 transistor
D1	1N4001 diode
RL1	12 Vdc SPST relay
S1	SPST momentary switch (normally open)
S2	DPDT switch

Table 10-3. Parts List for Single-Shot Enhancement

All resistors are 5 to 10 percent tolerance, 1/4 watt. All capacitors are 10 to 20 percent tolerance, rated at 35 volts or more.

indicates the resistor and relay used in the prototype. He-Ne power supplies consume from 250 to over 650 mA of current, often requiring 1 or more amps of current when the tube is first turned on. Some miniature relays, especially reed relays, cannot pass this much current. If you notice that the tube doesn't fire or that it sputters, try another relay.

Aiming sights or scope

If you plan on using the pistol for target practice, you'll want to add front and rear sights along the top of the barrel. Because the beam comes out of the barrel almost $1^1/4$ inches from the top, you are bound to experience parallax problems with any type of sight system. For best results, mount the sights as close to the barrel as possible—the farther away they are from the barrel, the more pronounced the effects of parallax error are.

You can make effective homemade sights by inserting two small, flat or pan head machine screws into the top of the barrel (use flat-blade screws, not hex or Phillips). The screws should be short and should not overly extend inside the barrel. If they do, the tips of the screws could interfere with the laser and power supply.

Adjust the leveling of the sights by turning the screws clockwise or counterclockwise. Position the "slits" of the tops of the screws so that they are parallel to the length of the barrel. By adjusting the height of the screws, you compensate for the differences in parallel between the sight and the laser tube. Note that there is no need to adjust for windage since the light is not affected by the wind, nor do you need to compensate for bullet trajectory because the light beam will continue in a straight path.

Painting

Although painting the laser pistol won't make it work better, it certainly improves its looks. Prior to painting, remove the end caps, tube, switch, and jack or use masking tape to prevent paint from spraying on them. Be particularly wary of paint coming into contact with the output window of the tube and the internal contacts of the power plug.

Fig. 10-9. The completed He-Ne laser pistol, with modulation bypass shorting plug installed.

Spray on a light coat of flat black paint. Testor's hobby paints are a good choice and when used properly won't sag during drying. The paint dries to a touch in 10 to 15 minutes but isn't cured until overnight. You can, of course, paint the pistol any color. A complete, painted pistol is shown in FIG. 10-9.

Black ABS plastic pipe does not need to be painted, saving you from this extra step. However, straight pieces of 1¹/4- and 2-inch black ABS plastic are hard to find. Most plumbing and hardware stores carry only preformed fittings for drains and other wastewater systems.

APPLICATIONS FOR THE LASER PISTOL

Let's face it, the He-Ne laser pistol isn't going to make you Luke Sky-walker, so there is little chance you'll be going around the galaxy disintegrating bad guys. The output of the tube used in the pistol (like all other helium-neon lasers) isn't enough to be felt on skin—it won't burn holes in anything, nor is it capable of any kind of destruction. That leaves rather peaceful applications of the pistol.

Before detailing some of the fun you can have with the laser pistol, I must reiterate the safety requirements. The handheld nature and design of the pistol might prompt you to use it in a laser tag-like game. Don't. The eyes of you and your opponent (man or beast) could be exposed to the laser beam, a definite health hazard. Point the pistol only at inaminate, "blind" objects.

Target practice

You can use any bullseye target to test your shooting skills. As long as the light doesn't reflect in his/her eyes, you can have a friend stand near the target and announce your score at each hit.

Another more professional method is to build the electronic target, which consists of a phototransistor sensor, amplifier, and sound-effects generator. The circuit for the basic target is shown in FIG. 10-10, with a parts list provided in TABLE 10-4. Mount the sensor in the center of a 3-by-3-inch smoked or black piece of plastic. One method of making the target is shown in FIG. 10-11.

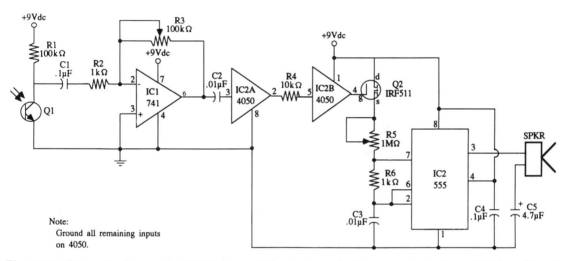

Fig. 10-10. Schematic diagram for light beam amplifier and sound-effects generator (using a 555 timer IC and speaker). The light striking Q1 generates a siren-like sound. Circuit courtesy Forrest Mims III.

Table 10-4. Parts List for Target Practice Add-on

R1	100 kΩ
R2, R6	1 kΩ
R3	100 kΩ potentiometer
R4	10 kΩ
R5	1 MΩ potentiometer
C1, C4	0.1 μF disc
C2, C3	0.01 μF disc
C5	4.7 μF polarized electrolytic
IC1	LM741 op amp
IC2	4050 CMOS buffer IC
IC3	LM555 timer IC
Q1	Infrared phototransistor
Q2	IRF511 power MOSFET transistor
SPKR	8 Ω speaker
Misc.	Heatsink for Q2, 3-inch-square acrylic plastic (1/8 to 1/4 inch thick), red filter

All resistors are 5 to 10 percent tolerance, 1/4 watt. All capacitors are 10 to 20 percent tolerance, rated at 35 volts or more.

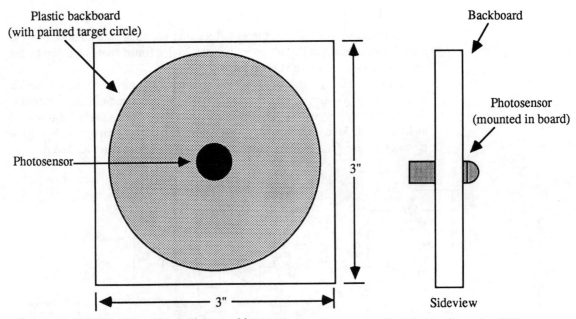

Fig. 10-11. Suggested layout for the light-sensitive target.

Note that the circuit is relatively immune to ambient light. Capacitor C1, connected across the output of the phototransistor and the input of the op amp, makes the circuit sensitive mainly to fast rise and fall pulses of light. However, direct sunlight on the phototransistor could swamp it, making it ''blind'' to the laser light. Firing at the target will have no effect because the sensor cannot ''see'' the laser light. You can reduce the effects of swamping in a number of ways:

- Use the target in shaded areas only, for example in the backyard under a tree or under the eave of the garage.
- Place a red filter over the phototransistor. It will block all but the red wavelengths. The best filter to use is the kind specifically designed for helium-neon lasers—it passes 632.8 nm light but sharply cuts off light at other wavelengths.
- A variable filter can be built using two small polarizers. Place the polarizers in brass tubes and mount the tubes in front of the sensor. Rotate the front filter to keep light from reaching the sensor.

Two fun circuits you can try are a programmable sound generator and a counter. The programmable sound generator replaces the ''gunshot'' noise generator of the original circuit. The programmable sound generator uses the versatile Texas Instruments SN74577 chip (other similar chips are available). As shown in FIG. 10-12, the chip is programmed to make a gunshot sound. TABLE 10-5 indicates the required parts for the sound generator.

A second addition is the counter (see the parts list in TABLE 10-6). Each direct hit on the sensor causes the counter to increment by one. Although the

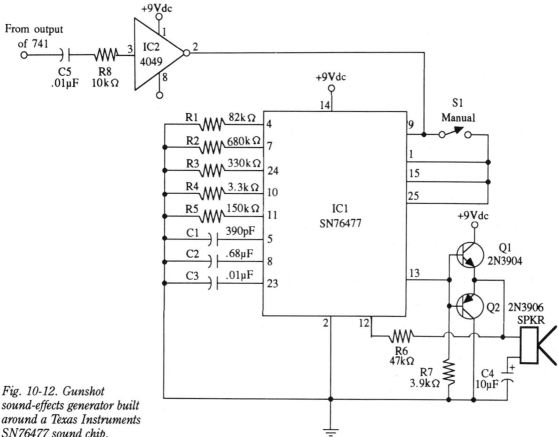

Fig. 10-12. Gunshot sound-effects generator built around a Texas Instruments SN76477 sound chip.

Table 10-5. Parts List for Gunshot Sound Effects Generator

R1	82 kΩ
R2	680 kΩ
R3	330 kΩ
R4	3.3 kΩ
R5	150 kΩ
R6	47 kΩ
R7	3.9 kΩ
R8	10 kΩ
C1	390 pF disc
C2	0.68 μF disc
C3, C5	0.01 μF disc
C4	10 μF polarized electrolytic
IC1	Texas Instruments SN76477 sound effects generator IC
IC2	4049 CMOS hex inverter IC
Q1	2N3904 transistor
Q2	2N3906 transistor
S1	SPST momentary switch (normally open)
SPKR	8 Ω speaker

All resistors are 5 to 10 percent tolerance, 1/4 watt. All capacitors are 10 to 20 percent tolerance, rated at 35 volts or more.

Table 10-6. Parts List for Two-Digit Counter

R1, R2	10 kΩ
R3 – 9	1 kΩ (required when supply is 5 volts or more)
C1, C2	0.01 µF disc
IC1	National Semiconductor 74C926 integrated counter IC
Q1, Q2	2N2222 transistor
Q3, Q4	Infrared phototransistor
LED1, 2	Common-cathode seven-segment LED display
S1	Momentary, normally open SPST switch

All resistors are 5 to 10 percent tolerance, 1/4 watt. All capacitors are 10 to 20 percent tolerance, rated at 35 volts or more.

chip used in the circuit is capable of counting up to 9999 (four digits), it's unlikely you'll be scoring that high in any one session, so only two digits are used for a total count of up to 99. After 99, the counter resets to 00. You reset the counter manually by pressing the RESET switch (S1). A schematic for the counter appears in FIG. 10-13.

You can devise any number of interesting and challenging target practice games. Manually count the number of shots you squeeze off; then compare it to

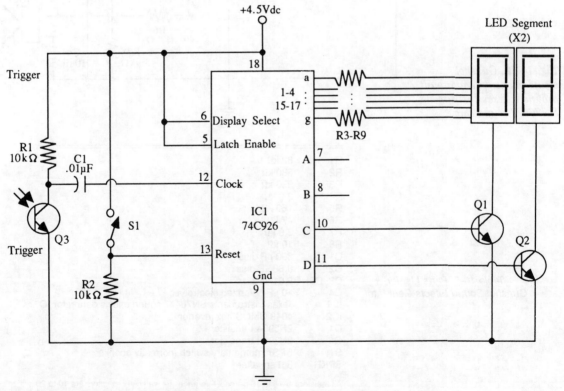

Fig. 10-13. A two-digit counter built around the versatile 74C926 counter chip (this chip includes a counter and segment encoder). Press S1 to reset the counter.

the number on the counter. The ratio of shots versus hits is your shooting average. Try practicing at various distances from the target. First try 5 yards (15 feet), 10 yards, 25 yards, then 35 yards. How does your shooting average compare at each distance?

Handheld pointer

Although the He-Ne pistol is rather large for the task, it can be used effectively as a handheld pointer. Even in a large auditorium with a brightly lit movie, the laser pointer can be easily seen on the screen.

If you don't care for the pinpoint of light, try shaping the beam with optics and a shape mask. Use a double-concave lens to expand the beam from its nominal 0.75 mm to about 10 mm. A biconvex lens *collimates* the beam; that is, makes the rays parallel again. In front of the collimating lens, place a mask of an arrow cut from a piece of thick black plastic, aluminum foil, or photographic film. (Your local offset printer can provide a high-contrast mask of any artwork.) When using aluminum foil, paint both front and back sides to cut down light reflections.

Since the positioning of the optical components depends on the focal lengths of the lens, you should experiment with some lenses and try the system out before building it onto the pistol. When you have determined the proper placement of the lens and mask, mount them in PVC pipe or a paper tube and attach the pipe to the end of the pistol. You can use a 2-inch coupler to attach extensions to the front of the laser.

The ultimate cat toy

Believe it or not, many owners of helium-neon lasers spend countless hours using the bright red beam as an electromagnetic cat toy. If you have a cat that is fairly playful, try this experiment. When the animal is least expecting it, shine the beam on the floor (not into its eyes). Most cats react to the beam by pouncing on it. Of course, they can never get it because it's simply a spot of light on the carpet. Scan the beam at a fairly slow rate and have the cat chase after it.

Dogs don't seem much interested in laser beam spots. Mine just licks the end of the laser and wags his tail.

TOY PISTOL

The "toy" laser pistol project uses a commercially available Laser Tag (or equivalent) pistol as the basis for a real laser gun. While the toy pistol is easier to make because you don't have to fashion parts from PVC pipe, it does require that you use miniature components, which are harder to find and more expensive.

Figure 10-14 shows the insides of a laser tag-type gun—the "Laser Shot," purchased on sale at Radio Shack for about $8. The sound board and infrared

Fig. 10-14. The original (unmodified) "Laser Shot" toy pistol, originally designed for use in an infrared light tag game (similar to laser tag).

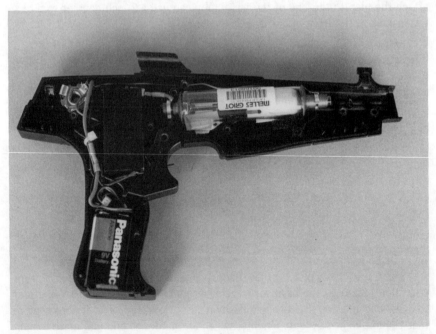

Fig. 10-15. The modified toy pistol now contains all the necessary parts for a real laser gun. Note the miniature high-voltage power supply nestled behind the trigger. That (and the miniature Melles-Griot laser tube) is the secret behind the miniaturization of the pistol.

LED (used to "shoot" the opponent) parts are removed and a real laser tube, miniature power supply, and switch are added, as shown in FIG. 10-15.

The parts just fit. Well, that's not true. The plastic around the trigger had to be trimmed down because the power supply wouldn't fit in the grip. The original plastic trigger of the gun was made immobile, and a miniature momentary pushbutton switch was carefully inserted in the trigger. All components are installed with heavy, double-sided foam tape. There is little room for mounting hardware.

TABLE 10-7 provides a parts list for the toy laser pistol.

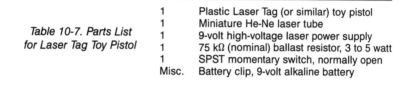

Table 10-7. Parts List for Laser Tag Toy Pistol	1	Plastic Laser Tag (or similar) toy pistol
	1	Miniature He-Ne laser tube
	1	9-volt high-voltage laser power supply
	1	75 kΩ (nominal) ballast resistor, 3 to 5 watt
	1	SPST momentary switch, normally open
	Misc.	Battery clip, 9-volt alkaline battery

Note the miniature power supply. This supply is slightly larger than a 9-volt battery, yet operates from 9 volts dc from a transistor battery. With the miniature "007" laser tube used, an alkaline transistor battery lasts about 3 or 4 minutes before pooping out.

To provide longer play time, a pistol is outfitted with an external power jack. The jack is the normally closed type, as shown in FIG. 10-16, so that when no

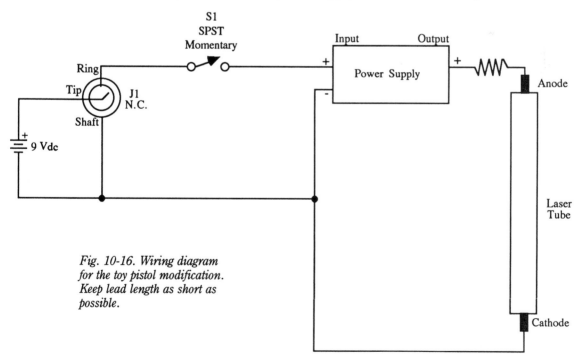

Fig. 10-16. Wiring diagram for the toy pistol modification. Keep lead length as short as possible.

plug is inserted, juice from the battery flows to the power supply. But insert a jack into the plug and the battery-to-power supply connection is broken. The power supply is then operated from the external dc provided through the jack. You can use the same 12-volt battery pack built for the PVC laser pistol, or design a more compact one for belt-clip operation.

11

Working with diode lasers

Imagine stuffing a laser into a size no larger than the dot in the *i*. Such is the semiconductor laser. A close relative to the ordinary light-emitting diode, the semiconductor laser is made in mass quantities from wafers of gallium arsenide or similar crystals.

In quantity, low- to medium-power semiconductor lasers cost from $5 to $35. Such lasers are used in consumer products such as compact audio disc players and laser disc players as well as bar code readers and fiberoptics data links. With the proliferation of these and other devices, the cost of laser semiconductors (or laser diodes) is expected to drop even more. The low cost of semiconductor lasers—typically $10 on the surplus market—makes them ideal for school projects where a tight budget doesn't allow for more expensive gas lasers.

This chapter presents an overview of the diode laser: how it's made, the various types that are available, and how to use them in your experiments. You'll also find numerous circuits for powering diode lasers.

THE INSIDES OF A SEMICONDUCTOR LASER

The basic configuration of the diode laser (sometimes called an injection laser) is shown in FIG. 11-1. The laser is composed of a pn junction, similar to that found in transistors and LEDs. A chunk of this material is cut from a larger silicon wafer, and the ends are cleaved precisely to make the diode chip. Wires are bonded to the top and bottom. When current is applied, light is produced inside the junction. As it stands, the device is an LED—the light is not coherent.

An increase in current causes an increase in light output. The cleaved faces act as partially reflective mirrors, which bounce the emitted light back and forth

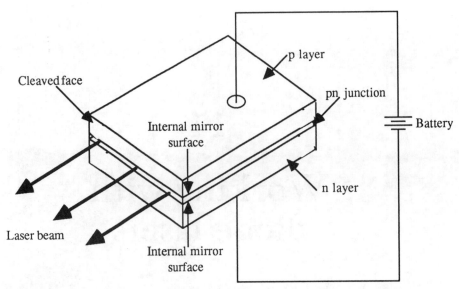

Fig. 11-1. Design of a semiconductor laser chip showing the cleaved face and pn junction.

within the junction. Once amplified, the light exits the chip. This light is temporally and spatially coherent, but because of the design of the diode chip, it is not very directional. The beam of most laser diodes is elliptical with a spread of about 10 by 35 degrees.

The first laser diodes, which were created in 1962 shortly after the introduction of the first ruby and helium-neon lasers, were composed of a single material forming one junction—a *homojunction*. These could be powered only in short pulses because the heat produced within the junction would literally cause the diode to explode. Continuous output could only be achieved by dipping the diode in a cryogenic fluid such as liquid nitrogen (whose temperature is minus 196 degrees C).

As manufacturing techniques improved, additional layers were added in varying thicknesses to produce a heterojunction diode. The simplest heterojunction semiconductor lasers have a gallium arsenide (GaAs) junction topped off by a layer of aluminum gallium arsenide (AlGaAs). These can produce from 3 to 10 watts of optical output when driven by a current of approximately 10 amps. At such high outputs, the diode must be operated in pulsed mode.

Typical specifications for single-heterostructure (sh) lasers call for a pulse duration of less than 200 nanoseconds (200 billionths of a second). Most drive circuits operate the diode laser conservatively with pulse durations under 75 or 100 nanoseconds. Output wavelength is generally between 780 nm and 904 nm.

A double-heterostructure (dh) laser diode is usually made by sandwiching a GaAs junction between two AlGaAs layers. This helps confine the light generated within the chip and allows the diode to operate continuously (called *continuous wave* or *cw*) at room temperature. The wavelength can be altered by varying the amount of aluminum in the AlGaAs material. The output wavelength can be between 680 nm and 900 nm, with 780 nm being the most common and least expensive (680 nm diode lasers are commonly used in compact rifle and pistol laser sights, but they cost over $100, even in quantity).

Power output of a double-heterostructure laser is considerably less than with a single-heterostructure diode. Most dh lasers produce 3 to 5 mW of light, although some high-output varieties can generate up to 500 mW yet can still be operated at room temperature (indeed, some high-cost cw lasers can produce up to 2.6 watts of optical power, but these are rare and very expensive).

POWERING SINGLE-HETEROSTRUCTURE DIODE LASERS

Single-heterostructure lasers are typically driven by applying a high-voltage, short-duration pulse. The duration of the pulse is controlled by an RC network. The power pulse is delivered by a transistor. Care must be exercised to ensure that the pulse duration does not exceed the maximum specified by the manufacturer. Longer pulses cause the laser to overheat, annihilating itself in a violent puff of smoke.

A common method for powering an sh injection diode is shown in FIG. 11-2 (see the parts list in TABLE 11-1). The power supply provides pulses of about 45 amps at a short duration of around 50 ns. The supply provides sufficient drive current to exceed the threshold of the laser (typically about 7 or 8 amps) with some room to spare. The laser might still glow as currents less than threshold, but the light won't be stimulated emission. In other words, the device will not emit laser light but rather behave like an expensive LED.

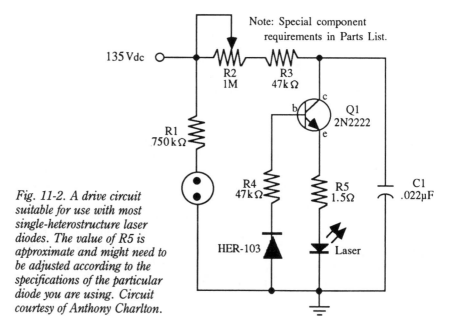

Fig. 11-2. A drive circuit suitable for use with most single-heterostructure laser diodes. The value of R5 is approximate and might need to be adjusted according to the specifications of the particular diode you are using. Circuit courtesy of Anthony Charlton.

Table 11-1. Parts List for Single-Heterostructure Laser Drive

R1	750 kΩ
R2	1 MΩ
R3, R4	47 kΩ
R5	1.5 Ω, 5 watt, carbon composition
C1	0.022 μF disc, 250 V or higher
D1	HER-103
Q1	2N2222 transistor
Misc.	Single-heterostructure diode laser, two 67.5-volt batteries, NE-2 neon lamp

All resistors are 5 to 10 percent tolerance, 1/4 watt, unless otherwise noted. All capacitors are 10 to 20 percent tolerance, rated at 35 volts or more.

The sh laser diode circuit uses a common npn transistor operated in avalanche mode. The batteries are 67.5-volt type (NEDA 217, Eveready number 416), used in older tube-type equipment. You'll have better luck finding the required batteries at an electronic store specializing in communications or ham gear. The price can be steep—up to $10 each depending on the source—so make sure they are fresh before you sign the check.

Quality control in low-cost plastic npn transistors is not great, so not all transistors will work in the circuit. The schematic calls for a 2N2222, but you might need to experiment with several until you find one that oscillates in the circuit.

The diode operates at a wavelength of about 904 nm, which is beyond normal human vision, so don't expect the same bright red beam that's emitted by a helium-neon laser. You can test the operation of the laser by using one of the infrared sensors described in the previous chapters.

POWERING DOUBLE-HETEROSTRUCTURE DIODE LASERS

Double-heterostructure semiconductor lasers can be operated either in pulsed or cw mode. In pulsed mode, the diode is driven by short, high-energy spikes as with an sh laser. Power output might be on the order of several watts, but because the pulses are short in duration, the average power is considerably less. In cw mode, a low-voltage constant current is applied to the laser, which outputs a steady stream of light. Cw lasers and drive circuits are used in compact disc players, where the light emitted by the laser is more coherent than even the beam from the revered He-Ne tube.

Forward drive current for most cw lasers is in the neighborhood of 60 to 80 mA. That's 50 to 200 percent higher than the forward current used to power light-emitting diodes. If a cw laser is provided less current, it might still emit light, but it won't be laser light. The device lases only when the threshold current is exceeded—typically a minimum of 50 to 60 mA. Conversely, if the laser is provided too much current, it generates excessive heat and is soon destroyed.

Monitoring power output

All laser diodes are susceptible to changes in temperature. As the temperature of a semiconductor laser increases, the device becomes less efficient and its light output falls. If the temperature decreases, the laser becomes far more efficient. With the increase in output power, there is a risk of damaging the laser, so most cw drive circuits incorporate a feedback loop to monitor the temperature or output power of the device, and adjust its operating current accordingly.

Sensing temperature change requires an elaborate thermal sensing device and a complicated constant-current reference source. An easier approach is to monitor the light output of the laser. When the output increases, current decreases. Conversely, when the output decreases, current increases.

To facilitate the feedback system, the majority of cw laser diodes now incorporate a built-in photodiode monitor. This photodiode is positioned at the opposite end of the diode chip, as shown in FIG. 11-3, and samples a small portion of the output power. The photodiode is connected to a relatively simple

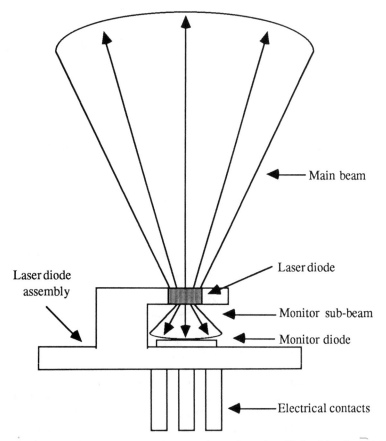

Main beam

Laser diode assembly

Laser diode

Monitor sub-beam

Monitor diode

Electrical contacts

Fig. 11-3. The orientation of the laser and monitor photodiode chips in a typical double-heterostructure semiconductor laser.

comparator or op amp circuit. As the power output of the laser varies, the current (and voltage) of the photodiode monitor changes. The feedback circuit tracks these changes and adjusts the voltage (or current) supplied to the laser. The feedback circuit can be designed around discrete parts or a custom-made IC. Actual driving circuits using both designs are presented in the following chapter. You'll also find a schematic for driving a cw laser in pulsed mode.

Connecting the laser to the drive circuit

The laser and photodiode are almost always ganged together, using one of two approaches. Either the anode of the laser is connected to the cathode of the photodiode, or the anode and cathode are grouped together. That leaves three terminals for connecting the diode to the control circuits. Schematic diagrams for the two approaches are illustrated in FIG. 11-4. A sample terminal layout for the popular Sharp laser diodes (as used in bar code readers and compact disc players) is shown in FIG. 11-5.

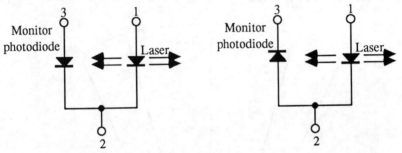

Fig. 11-4. Two ways of internally connecting the laser and monitor photodiode.

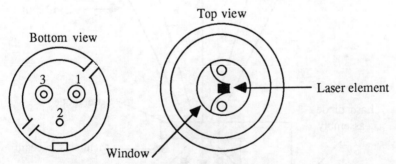

Fig. 11-5. Package outline and terminal configuration for the Sharp LT020 laser diode, often used in compact audio disc players.

There is a danger of damaging a laser diode by improperly connecting the drive circuit. Connecting a 60 to 80mA current source to the photodiode will probably burn it out, which could destroy the entire laser. Moral: follow the hookup diagram carefully. If no diagram came with the laser diode you received, write the seller or manufacturer and ask for a copy of the specifications sheet or application note.

Pulsed dh injection diode supply

The popularity of compact audio discs, as well as many forms of laser bar code scanning, have made double-heterostructure laser diodes plentiful in the surplus market. A number of sources (many of which are listed in Appendix A) offer dh laser diodes for prices ranging from $5 to $15. Depending on the power output of the laser, new units are even affordable. A typical 5 mW laser diode lists for about $25 to $30 in low quantities. Sharp is a major manufacturer of dh laser diodes; write them for literature and a price list.

As discussed in the previous chapter, one of the most attractive features of dh laser diodes is that they work with low-voltage power supplies. A dh laser can easily be run off of a single 9-volt transistor battery. However, dh laser diodes are sensitive to temperature. They become more efficient at lower temperatures and their power output increases.

Unless the temperature is very low (such as when the diode is dunked in liquid nitrogen, as described later in this chapter), the increase in power output can damage the laser. That's why most dh lasers are equipped with a monitor photodiode. The current output of the monitor photodiode is used in a closed-loop feedback circuit to keep the power output of the laser constant.

Although dh lasers are designed for constant wave operation, they can also be used in pulse mode. An astable multivibrator such as a 555 can be used to pulse the laser. Such a circuit is shown in FIG. 11-6 with a parts list in TABLE 11-2. Because the laser is pulsed, the forward current can exceed the maximum allowed for constant wave operation (generally 60 to 80 mA). However, care must be taken to keep the pulses short. Pulses longer than about 50 percent duty cycle (half on; half off) can cause damage to the laser. Duty cycle is not a critical consideration when the current is maintained under 80 mA. The circuit shown in the figure lets you alter the frequency of the astable multivibrator (and therefore the duty cycle).

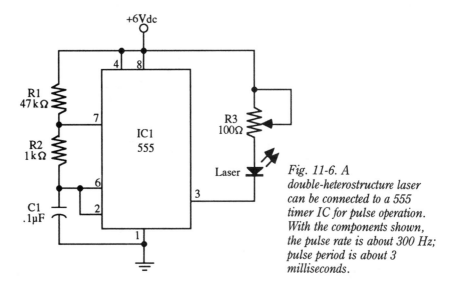

Fig. 11-6. A double-heterostructure laser can be connected to a 555 timer IC for pulse operation. With the components shown, the pulse rate is about 300 Hz; pulse period is about 3 milliseconds.

*Table 11-2. Parts List for Pulsed
Double-Heterostructure Laser Drive*

R1	47 kΩ
R2	1 kΩ
R3	100 kΩ potentiometer
C1	0.1 μF disc
IC1	555 timer IC
Misc.	Double-heterostructure laser diode, heatsink

All resistors are 5 to 10 percent tolerance, $1/4$ watt. All capacitors are 10 to 20 percent tolerance, rated at 35 volts or more.

Constant wave supply with current feedback

One of the best means for operating a double-heterostructure laser diode in constant wave mode is to provide a closed-loop feedback system. The feedback mechanism constantly watches over the output of monitor photodiode and maintains the proper current to the laser diode. One such circuit is shown in FIG. 11-7, designed around the IR3C02 chip, a special-purpose IC manufactured by Sharp. See TABLE 11-3 for a list of required parts. This IC is made for use with

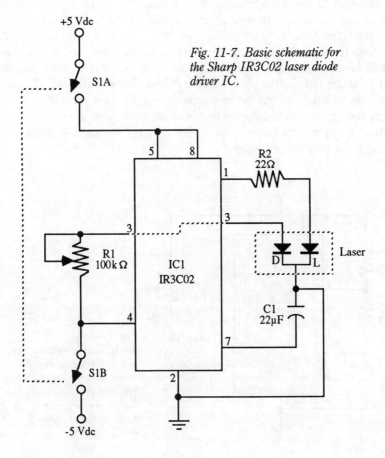

Fig. 11-7. Basic schematic for the Sharp IR3C02 laser diode driver IC.

Table 11-3. Parts List for Sharp IC Laser Drive

R1	100 kΩ
R2	22 Ω
C1	22 μF polarized electrolytic
IC1	Sharp IR3C02 laser diode drive IC
S1	DPDT switch
Misc.	Double-heterostructure laser diode (such as Sharp LT020), heatsink

their extensive line of dh lasers, and while hard to find, it is relatively inexpensive (you can obtain the chip through Sharp's parts service or from a distributor dealing with Sharp components).

Another method using discrete components is shown in FIG. 11-8 (parts list in TABLE 11-4). Here, an op amp acting as a high-gain comparator checks the current from the monitor photodiode. As the current increases, the output of the op amp decreases, and the output of the laser drops. The gain of the circuit—the ratio between the incoming and outgoing current—is determined by the settings of R1, R4, and R5.

The circuit in the schematic was adapted from an application note for a General Electric C86002E laser diode and uses a CA3130 CMOS op amp. You

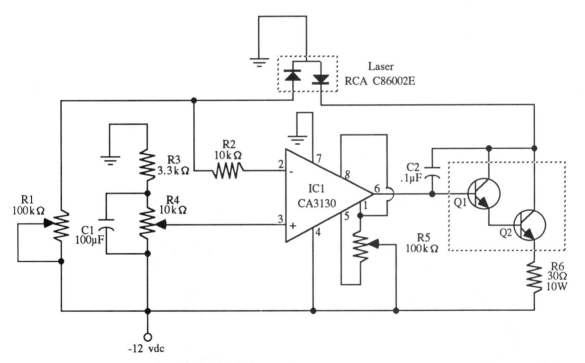

Fig. 11-8. One way to automatically adjust drive current using a discrete op amp. Use the transistors specified or replace with a suitable Darlington power transistor such as the TIP120.

*Table 11-4. Parts List
for Op Amp Laser Drive*

R1, R5	100 kΩ potentiometer
R2	10 kΩ
R3	3.3 kΩ
R4	10 kΩ potentiometer
R6	30 Ω, 10 watt
C1	100 μF polarized electrolytic
C2	0.1 μF disc
IC1	RCA CA 3130 operational amplifier
Q1	2N2101 transistor
Q2	2N3585 transistor
Laser	RCA C86002 (or equivalent laser diode)

All resistors are 5 to 10 percent tolerance, $1/4$ watt, unless otherwise indicated. All capacitors are 10 to 20 percent tolerance, rated at 35 volts or more.

can readily modify the circuit if you use another op amp or laser diode. Both output transistors are available through most large electronics outlets, but if you have trouble locating them, you might be able to substitute them with a single TIP120 Darlington power transistor.

HANDLING AND SAFETY PRECAUTIONS

While the latest semiconductor lasers are hearty, well-made devices, they do require certain handling precautions. And even though they are small, they still emit laser light that can be potentially dangerous to your eyes. Keep these points in mind:

- Always make sure the terminals of a laser diode are connectd properly to the drive circuit (we've covered this already, but it is crucial).
- Never apply more than the maximum forward current (as specified by the manufacturer) or the laser will burn up. Use the pulsed drive described earlier in this chapter if you are not using the laser with a monitor photo-diode feedback circuit.
- Handle laser diodes with the same care you extend to CMOS devices. Wear an antistatic wrist strap while handling the laser, and keep the device in a protective, antistatic bag until ready for use.
- Never connect the probes of a volt-ohmmeter across the terminals of a laser diode (the current from the internal battery of the meter could damage the laser).
- Use only batteries or *well-filtered* ac power supplies. Laser diodes are susceptible to voltage transients and can be ruined when powered by poorly filtered line-operated supplies.
- Take care not to short the terminals of the laser during operation.
- Avoid looking into the window of the laser while it is operating, even if

you can't see any light coming out. This is especially important if you have added focusing or collimating optics.

- Mount the laser diode on a suitable heatsink, preferably larger than 1 inch square. Use silicone heat transfer paste to assure a good thermal contact between the laser and the heatsink. You can buy heatsinks ready made or construct your own. (Some ideas for heatsinks appear in the next section.)

- Insulate the connections between the laser diode and the drive to minimize the chance of a short circuit. Use shielded three-conductor wire to reduce induction from nearby high-frequency sources.

- Laser diodes are subject to the same government regulations as any other laser in its power class. Apply the proper warning stickers and advise others not to stare directly into the laser when it is on.

- Use only a grounded soldering pencil when attaching wires to the laser diode terminals. Limit soldering duration to less than 5 seconds per terminal.

- Unless otherwise specified by the manufacturer, clean the output window of the laser diode with a cotton swab dipped in ethanol. Alternatively, you can use optics-grade lens-cleaning fluid.

MOUNTING AND HEATSINKS

Most laser diodes lack any means by which to mount them in a suitable enclosure. Their compact size does not allow for mounting holes. However, with a bit of ingenuity you can construct mounts that secure the laser in place as well as provide the recommended heatsinking. One approach is to clip the laser in place using a fuse holder. You might have to bend the holder out a bit to accommodate the laser. Mount the clip on a small piece of aluminum or a TO-220 heatsink. Use silicone paste at the junction of all metal pieces to assist in proper heat transfer.

Another method is to drill a hole the same diameter as the laser in an aluminum heatsink. Use copper retaining clips (available at the hobby store) to secure the laser in place. Once again, apply silicone paste to aid in heat transfer.

Some lasers are available on the surplus market already attached to a heatsink and mount (as shown in FIGS. 11-9 and 11-10). In FIG. 11-10, the mount doubles as a rail for collimating and beam-shaping optics. You can use the laser with or without these optics or substitute them with your own.

SOURCES FOR LASER DIODES

Laser diodes are seldom sold at the neighborhood electronics store, and as of this writing, Radio Shack does not carry the device as a replacement or experimenter's item. That leaves buying your laser diodes directly from the manufacturer, through an authorized manufacturer's representative, or through surplus.

Fig. 11-9. Many commercial products build the heatsink for the laser diode into the mounting package.

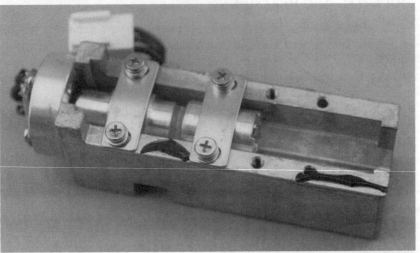

Fig. 11-10. A commercially made "sled" with laser diode (left) and beam-shaping optics installed. This nifty device (which measures about 3 inches long) is an outtake from a Xerox bar code scanner. Even in surplus, the product shows signs of impressive engineering.

Buying direct from the manufacturer or a rep assures you of receiving prime, new goods, but the cost can be high. Average cost for a new 3 to 5 mW laser cw diode is about $30. You can locate local representatives by writing the manufacturer or looking in Yellow Pages under Electronics—Wholesale and Retail.

The same or similar device on the surplus market is about $10 to $15, depending on the power output. Several of the surplus mail-order dealers listed in Appendix B offer sh and dh laser diodes; write them for a current catalog. Many of them, including Allegro Electronic Systems, also provide kit and ready-made drive/power supply circuits. Be aware that at this time, most surplus laser diodes are take-outs, meaning that they were used in some product that was later retired and scrapped. While buying used He-Ne tubes is a chancy affair, the risk of buying preowned laser diodes is minimal. Like all solid-state electronics, the lifespan of a laser diode is extremely long—in excess of 5,000 to 10,000 hours of continuous use.

CRYOGENIC COOLING OF SEMICONDUCTOR LASERS

All semiconductor lasers become super efficient at low temperatures. The first diode lasers could only be operated at the subfreezing temperature of liquid nitrogen, which is minus 197 degrees Centigrade. Only by the mid 1970s had they perfected the room-temperature laser diode. By dipping a low-power (1 to 5 mW) double-heterostructure cw laser in a glass filled with liquid nitrogen, you can increase its operating efficiency by several hundred percent.

Try this experiment. Connect a laser diode as shown in FIG. 11-11 and dip it into a glass Pyrex measuring cup. Connect a meter to the monitor photodiode to register light output. Note the reading from the photodiode on the meter. Now slowly fill the cup with liquid nitrogen. The liquid will bubble violently as it boils. Watch the reading on the meter jump as the diode is cooled.

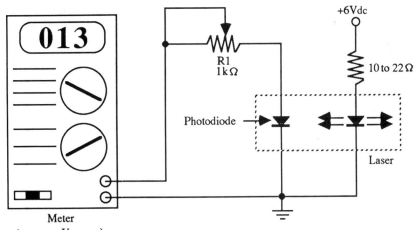

Note:

Use low value resistor only when laser is submerged in liquid nitrogen and cooled to cryogenic temperature!

Use 47Ω to 56Ω resistor (60-80 mA forward current) to test output when laser is at room temperature.

Fig. 11-11. Meter hookup diagram for testing the output of a cryogenically dipped laser diode. Such dipping could ruin the laser, so be sure it's not the only one you have.

In my experiments, the reading on the meter was some 850 times higher at cryogenic temperatures than it was at regular room temperature. However, this reading is misleading because the efficiency of both the laser and the monitor photodiode are increased by the cool temperatures. Use an external phototransistor for a more accurate appraisal of the increase in power from the laser.

This extra light can be readily seen not only because of the higher output of the laser, but because the laser operates at a lower wavelength when cooled. Instead of operating at the threshold of visible light (about 780 nm), the liquid nitrogen brought the operating wavelength down to perhaps 700 nm in the far red region of visible light. Use a lab-grade spectroscope to measure the wavelength of the emitted light.

Handling liquid nitrogen

Though liquid nitrogen is nontoxic and nonflammable, you must exercise care when handling it. Some liquid nitrogen tips and techniques follow.

You can buy liquid nitrogen at welding, hospital, and doctor supply outlets. Price is typically between $2 and $3 per liter. Use a stainless steel or glass Thermos bottle to hold the liquid nitrogen (some outlets won't fill these canisters due to breakage and waste; check first). Drill a hole in the top of the Thermos to allow the nitrogen vapor to escape because without a hole, the Thermos will explode.

The liquid nitrogen will last about a day in a Thermos. If you want to keep it longer, use a Dewar's flask or other approved container designed for handling refrigerated liquid gas.

Wear safety goggles and waterproof welder's gloves when handling liquid nitrogen. Never allow the liquid to touch your skin or you could receive serious frostbite burns. If the liquid touches clothing, immediately grasp the material at a dry spot and pull it away from your body.

While experimenting, fill liquid nitrogen only in Pyrex glass or metal containers. Avoid plastic and regular glass containers because they can shatter on contact with the extremely cold temperatures.

12

High-tech
laser projects

Lasers provide an endless fascination to the electronics gadgeteer. There are literally thousands of useful applications of lasers in everyday life, and thousands more are waiting to be discovered. This chapter provides some useful projects in laser technology. You'll learn how to build your own universal pulse-modulated laser power supply for operating helium-neon tubes (up to about 5 milliwatts output). You'll also learn how to build a handy laser pointer and even an optical timer/tachometer with the pencil-thin beam of a laser.

PULSE-MODULATED DC-OPERATED POWER SUPPLY

While the helium-neon laser tubes themselves are common finds—at costs as low as $20 or $25—the high-voltage power supplies required to run the laser are noticeably in short supply. You can purchase surplus commercially made laser power supplies for $75 to $100, but a less-expensive approach is to make one yourself. Not only do you save money, but you learn more about lasers and how laser power supplies work.

There are numerous approaches to designing a helium-neon laser power supply, but most suffer from a variety of shortcomings. The difficulty is in designing a power supply that's simple and uses readily available parts but also works with a wide variety of helium-neon laser tubes. Too often, the power supply is either extremely complex with hard-to-find specially made parts, or the supply works only with a select few laser tubes.

The universal pulse-width-modulated laser power supply described in this section endorses the simple design approach by using only a few, readily available components. Yet its design allows you the flexibility of using just about any helium-neon laser tube up to a power output of about 5 milliwatts. Two controls on the power supply let you adjust both voltage and current, thus tailoring the output to the particular laser tube you are using.

CAUTION: *All gas lasers, including the popular helium-neon variety, require high-voltage power supplies. These power supplies boost the main voltage from 12 volts dc or 117 volts ac to between 1,200 and 3,000 volts. Some laser experimenters tend to disregard the high voltages, assuming that although the voltage is high, the current level is low. Indeed, while the current demand of the typical helium-neon laser is low, the 1.2 to 3 kV jolt can still harm you.*

Laser power supplies should be properly shielded and insulated, as recommended at the end of this section. Avoid operating a power supply in the open and always cover exposed high-voltage parts.

Most laser power supplies use high-voltage capacitors at the output stage. Like all capacitors, these can retain current even after the power supply has been turned off. When working with the laser, make sure the power supply is off, and then temporarily short the leads of the power supply together or simply touch the positive terminal of the supply to ground. Likewise, the laser tube itself can retain some current after power has been removed. This current should be drained by shorting the terminals or leads together or to ground.

The power supply circuit shown in FIG. 12-1 can be used with He-Ne lasers rated up to 5 milliwatts, depending on the actual power consumption of the tube. The power supply is best constructed using an etched and drilled printed circuit board. A completed power supply, using a printed circuit board available from the source listed in the parts list in TABLE 12-1, is shown in FIG. 12-2.

Theory of operation

The operating conditions of helium-neon laser tubes, regardless of size or power output, vary widely. A new tube starts easily and runs very efficiently.

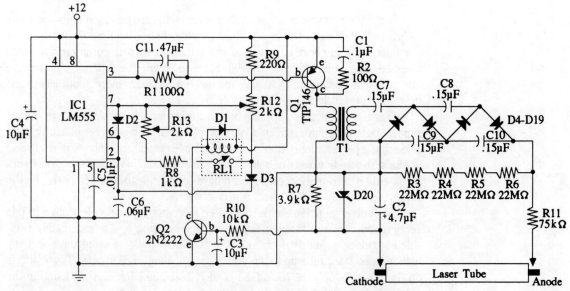

Fig. 12-1. The circuit schematic for the pulse-width-modulated laser power supply. Circuit courtesy Roger Sonntag.

Table 12-1. Parts List for Pulse Width Laser Power Supply

R1,R2	100 Ω
R3 – R6	22 MΩ
R7	3.9 kΩ
R8	1 kΩ
R9	220 Ω
R10	10 kΩ
R11	75 kΩ, 3 to 5 watt
R12,R13	2 kΩ potentiometer
C1	0.1 μF disc
C2	4.7 μF polarized electrolytic
C3,C4	10 μF polarized electrolytic
C5	0.01 μF disc
C6	0.06 μF disc
C7 – C10	0.15 μF disc, 3 kV or more
C11	0.47 μF disc
IC1	555 timer IC
D1 – D3	1N914 diode
D4 – D19	High-voltage diodes (3 kV or more; four diodes in series for each diode symbol in schematic)
Q1	TIP146 transistor (on heatsink)
Q2	2N2222 transistor
RL1	12-volt SPST relay
T1	High-voltage step-up transformer; 9-volt primary, 375-volt secondary
Misc.	Heatsink for Q1, high-dielectric wire for connecting tube to supply

All resistors are 5 to 10 percent tolerance, $1/4$ watt, unless otherwise indicated. All capacitors are 10 to 20 percent tolerance, rated at 35 volts or more, unless otherwise indicated. Parts kit available from General Science & Engineering (see Appendix A).

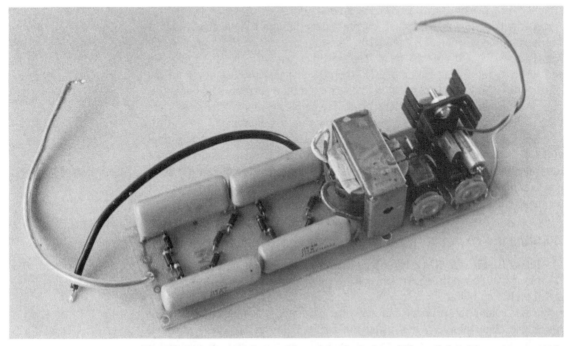

Fig. 12-2. A completed circuit board for the pulse-width-modulated laser power supply.

An older or used tube is harder to start and needs more current to lase continuously. The pulse-modulated laser power supply was designed to accommodate a wide variety of helium-neon tubes—both old and new, up to a maximum laser power output of about 5 milliwatts. Using pulse-width modulation (that is, varying the duty cycle of the square wave), the power supply separately controls the start current as well as the run current of the laser tube.

The power supply has two potentiometers that control the pulse width of the square wave applied to the inverting transformer, T1. In the start mode, R12 varies the pulse width until there is sufficient voltage to start the tube (typically 1.2 to 1.8 kV). As current starts to flow through the tube, R13 is engaged in the circuit. Resistor R13 controls the tube current and is adjusted so that the laser draws the minimum current possible while still allowing the tube to lase.

The power supply operates from a 12 Vdc source delivering a minimum of 750 mA. IC1, an LM555 operating as an astable multivibrator, oscillates at a nominal 16 kHz. Relay RL1 is initially not energized, so R13 and R8 are disconnected from the circuit. The setting of R12 determines the duty cycle, and thus the pulse width, of the square wave coming from the output (pin 3) of IC1. This signal drives the base of power transistor Q1 through current-limiting resistor R1. Transistor Q1, operating as a high-current, low-voltage chopper, delivers a series of square waves to the primary winding of stepup transformer T1.

With a 12-volt square wave at the primary of T1, the output voltage at the secondary is between 800 and 2,000 Vac, depending on the setting of R12. Capacitors C7, C8, C9, and C10 with diodes D4, D5, and D6 form a standard voltage doubler ladder. The output of the voltage doubler, with no tube attached, is approximately 3 to 5 thousand volts dc.

As the tube begins to conduct, current flows through R7, causing a voltage to appear at the junction of R7 and R10. This voltage turns on Q2, a small signal transistor, which activates the relay. The closed contact of relay RL1 brings R8 and R13 into the timing circuit of IC1, thereby changing the duty cycle of the square wave. Now R13 is readjusted to control the tube current. The best position is determined by rotating R13 clockwise until the relay "chatters," and then turning it counterclockwise until the relay remains energized.

Resistors R3, R4, R5, and R6 serve to drain the charge from the capacitors in the voltage doubler ladder as well as the electrostatic charge from the laser tube. Note that the very high resistance of R3 through R6 prohibits them from quickly draining the excess charge, so you should still manually short the output terminals together before working on the laser or power supply.

Building the circuit

Install R1 through R14 onto the circuit board and solder into place. Insert D1 through D9, bend the leads, and solder. Follow by installing C1 through C11 and then IC1 and Q2.

Next, solder the primary and secondary winding leads from T1 to their respective components. When using the ready-made circuit board provided from the source in the parts list, the secondary leads go to the holes marked A and B onto the foil side of the PCB (the leads will protrude through to the component side). The primary winding leads from the transformer are soldered from the component side in holes marked C and D.

Cut the leads of the transistor to a length of $1/8$ inch. On each side of the leads, solder a 20 gauge or larger $21/2$-inch-long wire. Solder the loose end of the wires to the circuit board.

Cut a piece of 20 gauge or heavier sheet metal $23/4$ by $41/4$. Drill holes for mounting to the transformer and board, and then bend the sheet lengthwise to make a mounting bracket. Using $1/4$-inch threaded spacers and $6/32$-by-$1/2$-inch machine bolts and nuts, mount the bracket on the foil side of the board. Then, using $6/32$-by-$11/4$-inch machine bolts and nuts, secure one side of the transformer frame to the outside edge of the bracket.

Using a suitable insulating washer and heat-transfer paste, mount Q1 to the metal bracket. Be sure to add the insulating washer for the mounting bolt. Use a meter to check for a short between the metal tab of Q1 (which serves as an alternate connection to the collector of the transistor) and the metal bracket. Remount Q1 if your meter indicates a short.

Make two high-voltage leads as shown in FIG. 12-3. Cut a length of high-dielectric wire to approximately 8 inches. Strip and tin $1/2$ inch from each end, and slip a 6-inch length of clear neoprene (aquarium) tubing over the wire. Solder one wire into the NEG OUTPUT near R7. Solder a second wire to R11 and R14, which you can mount on the board. On the opposite end of this wire, solder R15 and R16.

If your laser is equipped with flying leads (no connectors or terminals), solder the ends of the output wires to the laser, being sure to observe proper polarity. If the power terminals for your laser are mounted on the ends of the tubes around the mirrors, make suitable clips by coiling a length of lightweight wire around the tube ends.

Before using the power supply, inspect it carefully for solder bridges, loose connections, and improperly installed components. Double-check the orientation of IC1, C2, C3, C4, both Q1 and Q2, and all diodes.

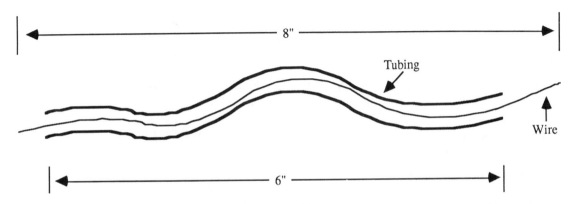

Fig. 12-3. To make a high-voltage lead, encase the insulated wire in a length of aquarium tubing.

Using the power supply

Operating the power supply is straightforward. Secure the power leads to the tube. If necessary, wrap high-voltage putty or electrical tape around the leads to hold them in place, but be sure you don't block the output mirror.

Rotate potentiometers R12 and R13 to their center positions. Apply power and watch the tube. Slowly rotate R12 clockwise until the tube triggers. You will hear the relay click in and possibly a high-pitched whine. Both are normal. If the relay chatters and the tube sputters, keep turning R12. If the tube still won't ignite, rotate R13 slightly counterclockwise.

Once the tube lights and stays on, rotate R13 clockwise so that the tube begins to stutter and the relay clatters. This marks the threshold of the tube. Turn R13 just a little the other way until the tube turns back on and remains steady. Every tube, even those of the same size and with the same output, have slightly different requirements. You will need to readjust R12 and R13 for every tube you own.

Resistors R11, R14, R15, and R16 form the ballast for the laser tube. With the components indicated in the parts list, total resistance is 75 kilohms. You can safely use ballast resistor values of 60 kΩ to 120 kΩ, and adjust for current variations with R13. If the laser doesn't trigger or run after adjusting R12 and R13, try initially reducing the value of the ballast resistor, but avoid going below 60 kΩ. Use a voltmeter, as explained in Chapter 9, to monitor the output current. If the tube begins to flicker after warming up, readjust R13 until the tube is stable.

Most 1 mW tubes draw between 750 mA to 1 amp of current from the 12 Vdc source. You will find you need higher current when operating a laser with greater power output. A typical 5 mW laser draws 2.5 to 3 amps from the 12 Vdc power source. Note that the 12 Vdc power supply must be able to deliver an initial surge of 3 to 5 amps. If your ac-operated power supply cannot handle this requirement, try powering the laser with a 12-volt alkaline lantern battery. Two 6-volt lead-acid or gelled electrolyte batteries wired in series make a good, permanent 12-Vdc power source for the laser power supply.

Enclosing the power supply

Your laser power supply should never be used without placing it in a protective, insulated enclosure. Electronics stores, including Radio Shack, sell project boxes of all shapes and sizes. If you plan on using the supply to power a number of tubes, mount heavy-duty (25-amp) banana jacks to provide easy access to the anode and cathode leads. Keep the jacks separated by at least one inch and apply high-voltage putty around all terminals to them to prevent arcing. Avoid using high-voltage power leads longer than 6 to 9 inches, especially for the anode connection.

BUILD A LASER POINTER

Few sources of light are as strong as a laser beam. A handheld pointer designed around a small He-Ne laser tube and miniature power supply can be used to project a sharp and bright pinpoint of light onto movie and slide screens or the curved surface of a planetarium.

The laser pointer uses a tiny "007" tube, available from several laser surplus sources including Meredith Instruments. The tube measures $4^7/8$ inches long by $7/8$ inches in diameter (that includes the mirrors and mounts; the actual glass tube is only $3^3/8$ inches long). To make the pointer as compact as possible, a miniature high-voltage power supply is used that is powered by two 9-volt batteries hooked up in parallel.

The laser tube fits in a $4^3/4$-inch length of 1-inch schedule 125 PVC (the schedule 125 plastic pipe has thinner walls than standard schedule 40 pipe). The laser tube is fitted into the plastic, with connections to the anode and cathode as shown in FIG. 12-4. Two 1-inch end caps protect the ends of the tube, as shown in FIG. 12-5.

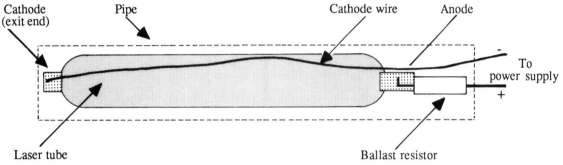

Fig. 12-4. *The handheld portion of the laser pointer consists of a length of schedule 125 PVC pipe stuffed with a laser tube, cathode wire, and ballast resistor.*

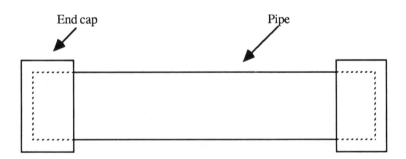

Fig. 12-5. *The pointer "head," showing the pipe and end caps. Drill holes in the end caps for the beam and power wires.*

The power supply, battery, and on/off switch mount in a $3^5/16$-by-$1^3/8$-by-$2^3/16$-inch plastic project box. The exact size of the box is not important, but this box represents the smallest you can get away with when using the components specified in TABLE 12-2. Fit the power supply, batteries and clips, and on/off switch as shown in FIG. 12-6.

Table 12-2. Parts List for Laser Pointer

1	Miniature "007" laser tube, measuring 4$^7/_8$ by $^7/_8$ inches
1	Modular high-voltage laser power supply (9-volt input)
1	75 kΩ, 3-to-5-watt (nominal) ballast resistor
1	4$^3/_4$-inch length of 1-inch schedule 125 PVC plastic pipe
2	1-inch PVC end caps
1	Plastic project case, 3$^5/_{16}$ by 1$^3/_8$ by 2$^3/_{16}$ inches
2	9-volt transistor battery clips
2	9-volt transistor batteries
1	SPST switch

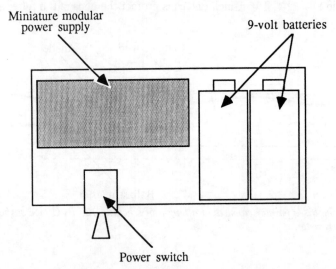

Fig. 12-6. Suggested parts layout for the laser pointer power supply.

The prototype laser pointer is designed with a separate head and power supply section. As a theatre pointer, you'll want to connect the two to make the device more convenient to use. Simply attach the PVC tube to the power supply enclosure—double-sided foam tape works well.

Because the tube and power supply are separate, the pointer can also be used in a number of other applications. You can use it as a laser sight on a competition air rifle. The enclosure can be attached to the wooden stock with screws; attach the tube to suitable mounts.

The parts specified in TABLE 12-2 can also be placed in a larger pipe enclosure to make a "saber"-like pointer. Cut a piece of 1$^1/_2$-inch PVC pipe to 10 inches. Fit the laser components as shown in FIG. 12-7. The laser is placed in the front of the tube, followed by the power supply and then the two batteries. The pair of batteries can't fit in the tube side-by-side, so you must install them one after the other. Place 1$^1/_2$-inch end plugs, with a hole drilled in the front one, to protect and hold the components in the pointer.

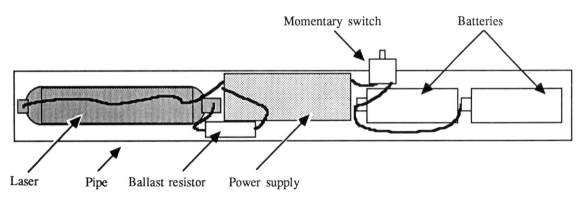

Fig. 12-7. Parts placement for an all-in-one laser pointer with the tube, ballast resistor, power supply, batteries, and on/off switch inserted inside a length of PVC pipe.

LASER TACHOMETER

Pulses of light can be used to count the rotational speed of just about any object when you can't be near that object. A heart of the laser tachometer is the counter circuit, shown in FIG. 12-8A. The remaining parts of the figure show the trigger logic circuit, the crystal-controlled time base, and the phototransistor inputs. A parts list is provided in TABLE 12-3.

To use the tachometer, shine a helium-neon laser at the rotating object you want to time. If you can, paint a white or reflective strip on the object so the beam is adequately reflected. Place a photosensor and amplifier at the point where the reflected beam lands, and connect the amp to the input of the counter circuit.

At each revolution, the light beam is reflected off the reflective strip on the object and directed into the photosensor, amplifier, and counter. The counter resets itself every second, thereby giving you a readout of the rate of the revolution per second. For example, if a car wheel is turning at 100 revolutions per second, the number 200 appears on the display. To compute revolutions per minute (rpm), multiply the result on the readout by 60.

The maximum count using the three digits of the counter is 999. That equates to 59,940 revolutions per minute, faster than even a jet turbine. You shouldn't run into too many things that can outpace the tachometer. If the object you are timing is distant, use the riflescope arrangement presented in the previous section. It allows you to aim the sensor directly at the reflective strip.

A

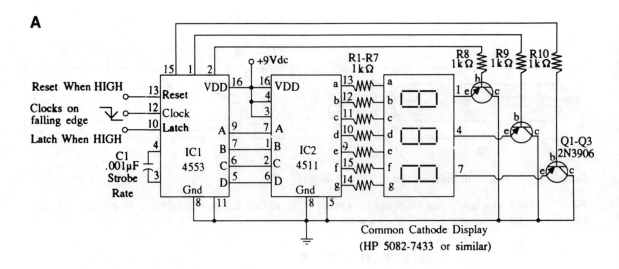

B

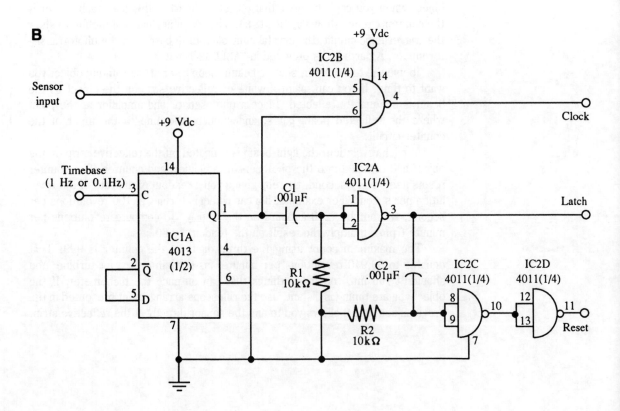

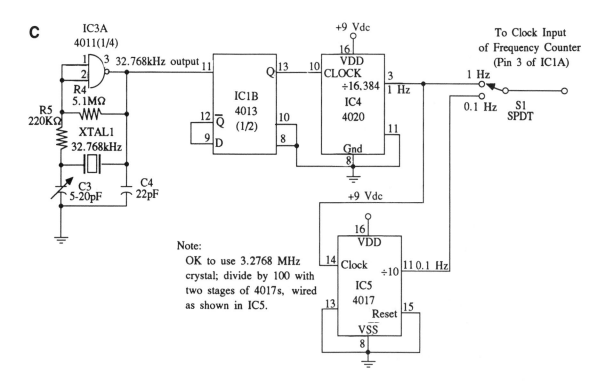

C

D

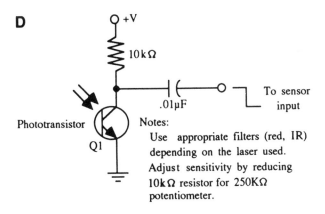

Fig. 12-8. Schematic for the laser light tachometer (actually, the tachometer will work with any kind of light).(A) shows the counter block, (B) is the trigger logic circuit, (C) is the crystal-controlled time base circuit, and (D) shows the phototransistor inputs.

Table 12-3. Parts List for Laser Tachometer

Counter Block (FIG. 12-8A)
R1 – R10   1 kΩ
C1   0.001 μF disc
IC1   4553 CMOS counter IC
IC2   4511 CMOS LED driver IC
Q1 – Q3   2N3906 transistor
LED1 – 3   Common-cathode seven-segment LED display

Trigger Logic Circuit (FIG. 12-8B)
R1,R2   10 kΩ
C1,C2   0.001 μF disc
IC1   4013 CMOS flip-flop IC
IC2   4011 CMOS NAND gate IC

Crystal-Controlled Time Base (FIG. 12-8C)
R1   220 kΩ
R2   5.1 MΩ
C1   5 to 20 pF miniature variable capacitor
C2   22 pF disc
IC1   4013 CMOS flip-flop IC (from trigger logic circuit)
IC2   4020 CMOS counter IC
IC3   4017 CMOS counter IC
IC4   4011 CMOS NAND gate IC
S1   DPDT switch

Phototransistor Inputs (FIG. 12-8D)
R1   10 kΩ
C1   0.01 μF disc
Q1   Infrared phototransistor

All resistors are 5 to 10 percent tolerance, $1/4$ watt. All capacitors are 10 to 20 percent tolerance, rated at 35 volts or more, unless otherwise indicated.

13

Experiments in laser holography

Armed with a helium-neon laser, some assorted optics, and a pack of film, you can create your own three-dimensional laser holograms. Your subjects can be anything that is small enough to illuminate with the laser beam and patient enough to sit still for the exposure. After developing the film, you can display your holograms for others to see and appreciate.

This chapter introduces you to the art and science of making holograms and what you need to set up shop. You'll also learn how to build a holographic table to create and later view transmission and reflection holograms.

A SHORT HISTORY OF HOLOGRAPHY

The idea of holography is older than the laser. Dr. Dennis Gabor, a researcher at the Imperial College in London, conceived and produced the first hologram in 1948. Dr. Gabor's first holograms were crude and difficult to decipher. Part of the problem was the lack of a sufficiently coherent source of light. No matter how complex the setup, the images remained fuzzy and indistinct. The introduction of visible-light lasers in the early sixties provided the final ingredient required to make sharp and clear holograms.

In the 1960s and most of the 1970s, the equipment required to make professional-looking holograms was beyond the reach of most amateur experimenters. Now, however, it's possible to build a workable holography setup for less than a few hundred dollars.

Developing holographic film is similar to processing ordinary black-and-white film and paper. If you've never been in a darkroom before, you might want to pick up a basic book on the subject and read up on the various processes and procedures involved. You can't learn everything about photography and holography in one chapter, so you are urged to expand your knowledge by further reading. A partial list of titles on holography can be found in Appendix B.

WHAT IS A HOLOGRAM?

A hologram is a photographic plate that contains interference patterns that represent the light waves from a reference source as well as from the photographed object itself. The patterns, like those in FIG. 13-1, contain information about the intensity of the light (just as in regular photography) as well as its instantaneous phase and direction. These elements together make the three-dimensional reproduction of the hologram possible.

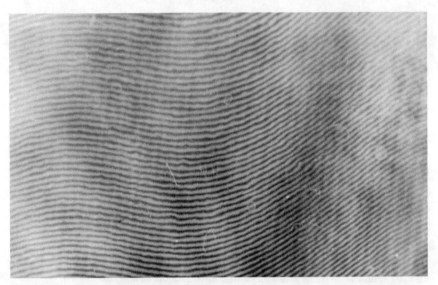

Fig. 13-1. It's not a thumbprint but a close-up of the interference fringes of a hologram captured on high-resolution film.

The interference patterns constitute a series of diffraction gratings, and form what is, in effect, an extremely sophisticated lens. The orientation of the gratings, along with their width and size, determine how the image is reconstructed when viewed in light. In most types of holograms, the image becomes clear only when viewed with the same wavelength of coherent light used to expose the film.

Not all holograms are the same. As you'll see later in this chapter, there are two general forms of holograms: *transmission* and *reflection*. The terminology refers to how the hologram is viewed, not exposed. The typical transmission hologram needs a laser for image reconstruction. The reflection hologram, while made with a laser, can be viewed under ordinary light.

The hologram stores an almost unlimited number of views of a three-dimensional object. Put another way, specific areas on the surface of the film contain different three-dimensional views of the object. You see these views by moving your head up and down or right and left.

This effect is most readily demonstrated by cutting a hologram into small pieces (or simply covering portions of it with a piece of cardboard). Each piece contains the entire image of the photographed object, but at a slightly different angle. As you make your own holograms, you will get a clearer view of how the image actually occupies a three-dimensional space.

WHAT YOU NEED

You need relatively few materials to make a hologram:

- Very stable isolation table (often nothing more than a sandbox on cement pillars)
- Laser—a 1 mW 632.8 nm helium-neon is perfectly suited for holography, though other visible light lasers will do as well; it must operate in TEM_{00} mode
- Beam-expander lens
- Film holder
- Film and darkroom chemicals
- Object to holograph

Other materials are needed for more advanced holographic setups, but this is the basic equipment list. These and other materials are detailed throughout this chapter.

THE ISOLATION TABLE

The fringes produced on the photographic plate that render the holographic image represent the swirling patterns of the laser beam's light waves. Depending on the table or workbench you use, vibrations from people walking nearby or even passing cars and trucks can cause the lightwave fringes to bob.

Since the interference fringes create the holographic image in the first place, you can understand why the table you use to snap the picture must be isolated from external vibrations. If the laser setup moves at all during the exposure—even a few millions of an inch—the hologram will be ruined. Optical tables for scientific research cost in excess of $15,000 and weigh more than your car. But they do a good job at damping vibrations in the lab and preventing them from reaching the laser and optical components.

A professional-grade optical table is probably out of the question, but you can build your own with common yard materials. Your own isolation table uses sand as the heavy damping material. The sand is dumped in a box that sits on cement blocks. The homebrew sandbox table is not as efficient as a commercially made isolation table, but it comes close. Add to your sand table a reasonably vibration-free area (California during an earthquake is out!) and you are on your way to making sharp and clear holograms.

While sandbox isolation tables are inexpensive and easy to build, they have their limitations and disadvantages. Another approach to building a holographic table involves using specially made rubber cushions between a top plate and base. Meredith Instruments (see Appendix A) offers a complete isolation table system using these cushions. As of this writing, the table is a little pricy—about $500 for a complete setup—but it's an ideal system for school or small lab.

Selecting the right size

The size of the table dictates the type of setups you can design as well as the maximum dimensions of the holographed object. The ideal isolation table measures 4 by 8 feet or larger, but a more compact 2-by-2-foot version can suffice for beginners. You will be limited to fairly simple optical arrangements and shooting objects smaller than a few inches square, but the table will be reasonably portable and won't take up half your garage.

The design for the sand table is shown in FIG. 13-2. The table consists of five concrete brick-blocks (the kind used for outside retaining walls), a 2-by-2-foot sheet of $3/4$-inch plywood, carpeting, four small pneumatic inner tubes, and an 8-inch-deep box filled with sand. You can build this table using larger dimensions (such as 4 by 4 or 4 by 8 feet), but it requires considerably more sand. You can figure about one 75- or 100-pound bag of fine sand for each square foot of table area.

Building the table

First, build the box using the materials indicated in TABLE 13-1. Use heavy-duty construction—heavy screws, wood glue, and battens. Make the box as sturdy as possible, strong enough for an adult to sit or stand on. Next, lay out the blocks in a 2-foot-square area inside the house or shop. You'll need a fairly lighttight room to work in, so the day basement is out unless you can *completely* cover the windows. A garage with all cracks and vents closed up is another good area. In addition, be sure that the room is not drafty. Air movement is enough to upset the fringes. Obviously you need a flat, level floor, or the sand table might rock back and forth.

Place the 2-by-2-foot sheet of plywood over the blocks and cover the plywood with one or two layers of soft carpeting (plush or shag works well). Partially inflate the four inner tubes (to about 50 – 60 percent) and place them on the carpet. Don't over-inflate the tubes or fill them with liquid. If possible, position the valves so that you can reach them easily to refill the inner tubes because you might need to jack up the sand box to access the valves. If the tubes are good to begin with, you won't need to perform this duty often.

Carefully position the box over the inner tube. Don't worry about stability at this point: the table will settle down when you add the sand. Be sure to evenly distribute the sand inside the box as you pour out the contents of each bag. There should be little dust if you use high-quality, prewashed and sterilized play sand. Fill the box with as much sand as you can, but avoid overfilling where the sand oozes out over the top edges of the box.

Testing the table

You can test the effectiveness of the sand table by building a makeshift Michelson interferometer, as shown in FIG. 13-3. The optical components are

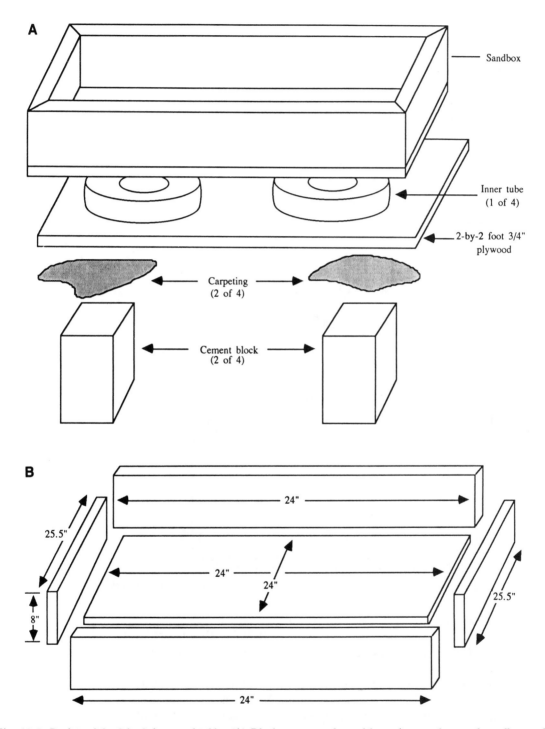

Fig. 13-2. Design of the 2-by-2-foot sand table. (A) Blocks, carpet, plywood base, inner tubes, and sandbox and how they go together. (B) Construction of the sandbox. Use ³/4-inch plywood for the sandbox.

Table 13-1. Parts List for 2-by-2-foot Holography Sand Table

Base and Pedestal

4	Cement building blocks
4	8-inch-square pieces of carpet
1	2-foot-square piece of carpeting
4	10- to 14-inch inner tubes
1	2-by-2-foot sheet of 3/4-inch-thick plywood

Sand Box

1	2-by-2-foot sheet of 3/4-inch-thick plywood
2	8-by-25.5-inch, 3/4-inch-thick plywood
2	8-by-24-inch, 3/4-inch-thick plywood
4	75- or 100-pound bags of washed, sterilized, and filtered sand

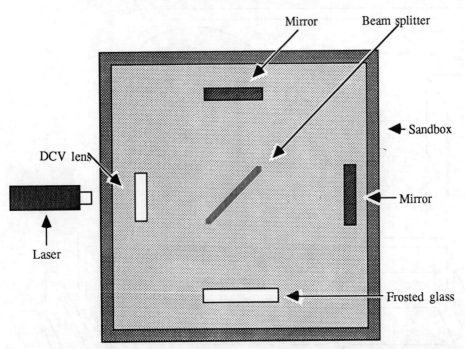

Fig. 13-3. Optics arranged as an interferometer for testing the stability of the sandbox table.

mounted on PVC pipe, as detailed later in this chapter. You can place the laser inside the box, on the top of the sand, or locate it off the table. As long as the beam remains fairly steady, any slight vibration of the laser will not affect the formation of the interference fringes necessary for successful holography.

Once you have the optics aligned and the laser on, wait for the table to settle (5 to 15 minutes). Then look at the frosted glass. You should see a series of circular fringes. If the fringes don't appear, wait a little longer for the table to settle some more. Should the fringes still not materialize, double-check the arrangement of the laser and optics. The paths of the two beams must meet exactly at the frosted glass, and their propagation must be parallel—vertically as well as horizontally.

Determining settling times

Test the isolation capabilities of the table by moving around the room and watching the fringes (you might need someone to stand by the glass and closely watch the fringes). The fringes shouldn't move at all, but if they do, it should be slight. More energetic movement, such as bouncing up and down or shaking the walls, greatly affects the formation of the fringes.

Watch particularly for the effects of vibrations that you can't control, such as other people walking around or cars passing by in front of your house. Problems in adequate isolation might require a more advanced table, more sand, or a change in location. You might also want to wait until after hours, where activity of both people and automobiles is reduced.

Keep in mind that you will have better luck when the sand table is located on a ground floor and on a heavy wood or cement. Floorings in above stories, as well as old, wooden floors used anywhere in a building, pass vibrations easily.

Note the time it takes for the table to settle each time you change the optics or disturb the sand. You need to wait at least this time before you can make a holographic exposure. Also note how long it takes for the table to recover from small shocks and disturbances. Recovery time for most tables should be a matter of seconds.

THE DARKROOM

Part of holography takes place in the darkroom, where the film is processed after exposure. Although it's most convenient to place the isolation table and processing materials in the same room, they can be separate. As with the isolation table, the darkroom must be reasonably lighttight. It should have a source of running water and a drain to wash away unwanted chemicals and processed film. A bathroom is ideal, as long as you have room to move and can work uninterrupted for periods up to 15 to 20 minutes (using the only bathroom in a house full of kids and adults makes holography an unwelcome hobby).

If the bathroom has windows, cover them with opaque black fabric or painted cardboard. If you plan on making many holograms, arrange some sort of easily installed curtain system. Drape some material over the mirror and other shiny objects, like metal towel racks and fixtures. You'll need access to the sink and faucet, so leave these clear.

Only two chemicals are needed for basic holography, though advanced holographers often use a laboratory full of different chemicals. You need to set out two shallow plastic bowls or trays large enough to accommodate the film. If space is a premium inside your bathroom, place a wooden rack over the tub and set the trays on it. The bathtub makes a good location for the trays because the processing chemicals can stain clothing, walls, floors, and other porous materials. If any drips into the tub, promptly wash it away.

OPTICS, FILM, FILM HOLDERS, AND CHEMICALS

Here's a rundown of the basic materials you need to complete your holographic setup.

Optics

Basic holography requires the use of a plano-concave or biconcave lens and one or more front surface mirrors. The lens can be small—on the order of 4 to 10 mm in diameter. Focal length is not a major consideration but should be fairly short. Mount the lens in 1- or 1¼-inch PVC pipe, painted black, as shown in FIG. 13-4.

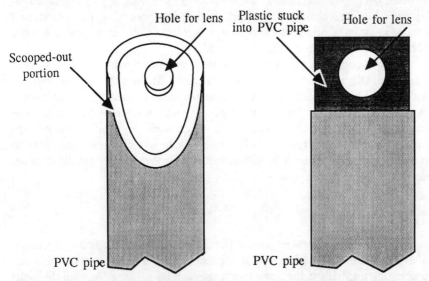

Fig. 13-4. Two ways to attach optics to PVC pipe (which you can readily stick into the sand in the sand table). No matter what mounting technique you use, be sure to keep all optics spotlessly clean.

The purpose of the lens is to expand the pencil-thin beam of the laser into an area large enough to completely illuminate the object being photographed. The area of the beam is expanded proportionately to the distance between the lens and object, and the basic 2-by-2-foot table doesn't allow much room for extreme beam expansion. If the beam is not adequately expanded, use a lens with a shorter focal length or position two negative lenses together.

The direct one-beam transmission hologram setup described later in this chapter does not require the use of mirrors or beam splitters, but multiple-beam arrangements do. The size of the mirror depends on the amount of beam spread and the size of the object, but in general, you will need one or more mirrors measuring 2 by 3 inches or larger. Some setups also require beam steering or transfer mirrors. These are used before the beam has been expanded, so they can be small. See FIG. 13-5 for ideas on how to mount mirrors and other optics.

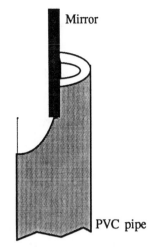

Fig. 13-5. You can mount mirrors and beam splitters to PC pipes as shown.

Holographic film

Holography requires a special ultrahigh-resolution film emulsion that is sensitive most to the wavelength of the laser you are using. Kodak Type 120 and S0-173, or Agfa-Gevaert Holotest 8E75 and 10E75, for example, are made for 632.8 nm helium-neon lasers and are most sensitive to red light—a relatively insensitive wavelength for orthochromatic film. You can handle and develop this film using a dim (7 watts or less) green safelight—a small green bug light works well as long as it is placed 5 feet or more from the sand table and film handling areas. Holographic film comes in various sizes and base thicknesses. Film measuring $2^1/_4$-inches square to 4 by 5 inches is ideal for holography.

Holographic films are available with and without antihalation (AH) backing. This backing material, which comes off during processing, prevents halos during exposure. The backing is semi-opaque, which prevents light from passing through it. Transmission holograms are made with the laser light striking the emulsion surface, so you can use a film with antihalation backing. Reflective holograms are made with light passing through the film and striking the emulsion on the other side. Unless you are eager to make exposures lasting several minutes or more, you'll want a film without the antihalation backing when experimenting with reflection holography.

When buying or ordering film, be sure to note whether the stuff you want comes with an antihalation backing. If you want to try both of the basic forms of holography (transmission and reflection) but don't want to spend the money for both kinds of films, use stock without the AH backing. Note that this film does not make the best transmission holograms, but the results are more than adequate.

Another alternative is to use glass plates for your first holograms. These consist of a photographic emulsion sprayed onto a piece of optically clear glass. The benefit of the glass is that it is easier to handle for beginners and is not as susceptible to buckling and movement during the exposure.

Where do you get holographic film? Start by opening the Yellow Pages and looking under the Photography heading. You will probably have the best luck calling those outlets that specialize in professional photography or darkroom supplies.

You might also happen to live in an area close to a Kodak or Agfa-Gevaert field office. Call and ask for help. A few companies, such as Metrologic, offer film suitable for holography; check Appendix A for the address. Lastly, write directly to the film manufacturers and ask for a list of local dealers that handle the materials you need. You might even be able to order it through the mail.

Film holder

You will need some means to hold the film in place during the exposure. One method is to sandwich the film between two pieces of glass, held together by two heavy spring clips. The glass must be spotlessly clean. The disadvantage of this method is that the contact of the glass and film can create what's known as Newton's Rings, a form of interference patterns.

If you plan on using a glass-plate film holder, remember to press the plates together firmly and keep constant pressure for 10 to 20 seconds (use wooden blocks for even pressure). This removes any air bubbles trapped between the glass and film. Snap the binder clips around the glass and position the holder in front of the object. The handles of the clips can be secured to the table by butting them against two pieces of PVC pipe. Depending on the arrangement of the optics, you might be able to locate the film holder between the PVC pipes that contain the expansion lenses.

Yet another approach is to use a commercially made film holder, the kind designed for processing plate film. The holders come in a variety of sizes and use novel approaches to loading and holding the film. Any camera store that carries professional darkroom supplies has a variety of film holders to choose from. You can also make your own, following the diagram in FIG. 13-6. Construct the holder to accommodate the size of film you are using. You might need to make several holders if you use different sizes of film.

To use any of the holders, load the film (in complete darkness or under the dim illumination of a small green safelight) into the holder, and then stick the holder in the proper location in the sand. If the holder does not allow easy mounting in the sand, attach it to small wood or plastic pieces.

Processing chemicals

You can use ordinary film developer to process most holographic film emulsions. A good choice is Kodak D-19, available in powder form at some photographic shops. If D-19 is not available, you can use most any other high-resolution film developer. Agfa, Nacco, and Kodak make a variety of high-quality powder and liquid developers that you can use.

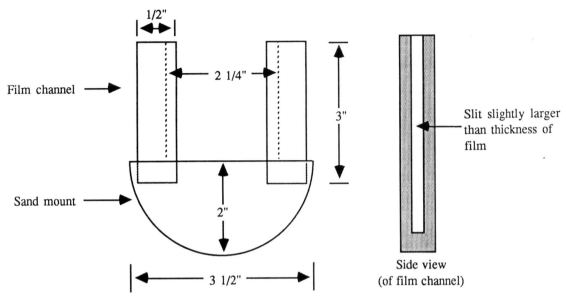

Fig. 13-6. Construction details for a homemade wooden film holder. This holder is made for 2¹/₄-by-2¹/₄ film.

Both transmission and reflection holograms require bleaching as one of the processing steps. If you can't buy premade bleach specifically designed for holography, you can mix your own following the directions that appear below. Note that transmission and reflection holograms need different bleach formulas. Chemicals for making the bleach are available at many of the larger industry photographic outlets. If you can't find what you want, dial up a chemical supply house and see if you can buy what you need in small quantities.

The remaining, optional chemicals are available at most camera stores. These include acetic acid stop bath, fixer (with or without hardener), photoflo, and hypo clear.

MAKING A TRANSMISSION HOLOGRAM

The layout in FIG. 13-7 shows how to arrange the laser, lens, photographic object, and film plate for a direct, one-beam transmission hologram. A parts list is provided in TABLE 13-2. Note that the term "transmission" has nothing to do with the arrangement of the film, optics, or object for making the hologram but rather the method of viewing the image after it is processed. You look through transmission holograms to see an image; you shine light off the emulsion side of a reflection hologram to see the picture.

You can use almost anything as the object, but for your first attempt, choose something small (about the size of a pack of cigarettes or less) that has a smooth but not highly reflective surface. A pair of dice, a coffee mug, or a chess piece make good subjects.

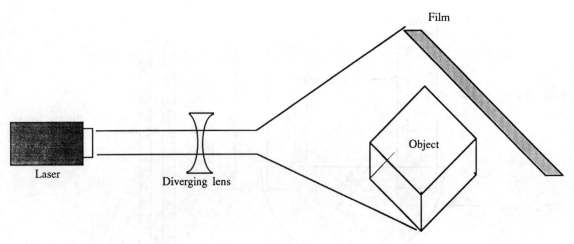

Fig. 13-7. The basic arrangement for making a single-beam transmission hologram.

1 He-Ne laser	
1 8 to 10 mm biconcave lens	*Table 13-2. Parts List for*
1 Holographic film in film holder	*Single-Beam Holography Setup*
1 Object to holograph	

Making the exposure

After you arrange the laser and optics as shown above, turn the laser on. The distance between the lens and object/film should be $1^1/2$ to 2 feet. Be sure the beam covers the object and the film, as shown in the figure. Now block the beam with a black "shutter" card.

Turn off all the lights and switch on the safelight. The safelight should emit only a tinge of illumination, barely enough for you to see your hand held out in front of you. The idea is to help you see your way in the dark, not provide daylight illumination. You can test for excessive safelight illumination by placing a piece of film on the sand table for 5 minutes. Cover half of it with a piece of black cardboard. Process the film and look for a darkening on one side. A dark portion means that the film was fogged by the safelight.

Load the film into the holder (emulsion side towards the laser; you can easily tell the emulsion side by wetting a corner of the film; the emulsion side will seem sticky), and place the holder at a 45-degree angle to the laser beam. The film should be close to the object but not touching it. Wait 10 minutes or so for the table to settle; then quickly but carefully remove the card covering the laser. After the exposure is complete, replace the card. Do *not* control the exposure by turning the laser on and off. The coherency of lasers improves after they have warmed up. For best results, allow the laser to warm up for 20 to 30 minutes before making any holographic exposure.

Exposure times depend on the power output of the laser. As a rule of thumb, allow 3 to 4 seconds for a 1 mW laser and 1 to 2 seconds for a 3 mW laser.

Processing steps

Once the hologram is exposed, place it in a lighttight container or box. Mix the processing chemicals (if you haven't already) as follows:

- D-19 developer—Use full strength as detailed in the instructions printed on the package (when using other developers, mix according to instructions to obtain the highest resolution).
- Bleach—Mix one tablespoon of potassium ferrocyanide with one tablespoon of potassium bromide in 16 ounces of water. Alternatively, you can dissolve 5 grams of potassium dichromate in 1 liter of water and then add 20 ml of diluted sulfuric acid.

Use water at room temperature (68 to 76 degrees F) and be sure the temperature is roughly the same between each chemical bath. High temperatures or wide variations in temperature can soften the film and introduce reticulation—two side effects you want to avoid. Place the developer and bleach in large plastic bowls or processing trays (available at photo stores). The chemicals have a finite shelf life of a few months when stored in stoppered plastic bottles. Use small or collapsible bottles to minimize the air content, and store the bottles in a cool place away from sunlight.

In the dark or with a green safelight, remove the film from the lighttight box and place it in the developer. Use plastic tongs to handle the film. Avoid metallic implements and your fingers. Swish the film in the developer soup and knock the tongs against the side of the tray or bowl to dislodge any air bubbles. Maintain a slow and even agitation of the film. After about 2 to 5 minutes, the image on the film will appear and development is then complete.

If you have a safelight, you can see the image appear and better judge when the development is complete. As a rule of thumb, when held up to a safelight a properly exposed and developed transmission hologram allows about 70 to 80 percent of the light to pass through it (reflective holograms about 20 to 30 percent of the light).

After development, rinse the film in running water or a tray filled with clean water. You can also use an acetic stop bath, mixed according to directions. The stop bath more completely removes trace developer and helps prolong the life of the bleach used in the next step.

After 15 seconds or so of rinsing, place the film in the bleach. Keep it there for 2 to 3 minutes. After a minute or two of bleaching, you can turn on the lights. Follow bleaching with another rinse. Use lots of water and keep the film in the wash for 5 to 10 minutes. A cyclonic rinser, used in many professional darkrooms, does a dandy job at removing all the processing chemicals.

The hologram is almost ready for viewing. Before drying it, dip the film in Kodak Photoflo solution—a capful in a pint of water is sufficient. The Photoflo reduces the surface tension of the water and helps promote even drying without spotting. It also removes minerals left from the water bath. You can also use a squeegee to remove excess liquid. Darkroom squeegees cost $5 to $10 but a new car windshield wiper blade costs $2 and is just as good.

Use clothespins or real honest-to-goodness film clips to hang up the film to dry. Place the film in a dust-free area: a specially made drying cabinet is the best place, but the shower or bathtub also works.

Wait until the film has completely dried before trying to view it (the fringes might not be clear until drying is complete). Depending on the temperature, humidity, and other conditions, it will take 30 to 120 minutes for the film to dry completely. Glass-plate film dries more quickly. Note that when wet, photographic emulsion becomes soft and easily scratched. Be sure to handle the film with care and avoid touching it (with hands or tongs) except by the edges.

To recap the processing steps:

- Develop for 2 to 5 minutes.
- Rinse for 15 seconds (water or stop bath).
- Bleach for 2 to 3 minutes.
- Rinse for 5 to 10 minutes in constantly running water.
- Dip in Photoflo for 15 seconds (optional step).
- Squeegee to remove excess liquid (don't scratch the emulsion!).
- Let dry at least 30 minutes.

Viewing the hologram

Part of the fun of holography is making the picture; the rest is seeing the result with your own eyes. If you haven't disturbed the setup used to take the picture, simply replace the film in the film holder. The emulsion should face the laser. View the image by looking through the film. The image might not be noticeable unless you move your head from side to side or up and down.

Note that if you have dismantled the sandbox setup, you must replace the laser, optics, and film holder in the same arrangement to reconstitute the image. The simple layout used in the single-beam direct transmission exposure makes it relatively easy to reconstitute the image.

PROBLEMS

Having problems? The image in the hologram isn't clear or the exposure just isn't right? Most difficulties in image quality are caused by motion. The fringes must remain absolutely still during the exposure or the hologram could be ruined. A more powerful laser can reduce the exposure time, reducing the problems of vibration. For example, a 5 to 8 mW laser might require an exposure of only 0.5 to 1 second. If at all possible, use the most powerful laser you can get your hands on, but don't give up on holography if all you own is a small 0.5 mW tube.

Like regular photography, it takes time, patience, and practice to make really good holograms. Don't expect to make a perfect exposure the first time around. Odds are that the exposure will be too short and the development too long, or vice versa. If your first attempt doesn't come out the way you want it (or doesn't come out at all), analyze what went wrong and try again.

Check the layout and be sure that the film is inserted in the holder emulsion side out (that is, towards the laser). If you use a film with an AH backing (recommended for transmission holograms), the light won't pass through to the emulsion if the film is inserted backward in the holder. Be sure that the chemicals are mixed right and that you are following the proper procedure.

OBSERVATIONS ON HOLOGRAPHY

You can learn more about holography by actually making a hologram than reading an entire book on the subject (also that's recommended, too; see Appendix B for a list of books on holography). Below are some observations that will help you better grasp the technology and artistry of holograms.

Field of vision

Here's an important point to keep in mind as you experiment further. You might have realized that the film holder represents the frame of a window onto which you can view the subject. Place your head against the film holder, close one eye, and look at the object up close. Without actually moving your eye, scan your head vertically and horizontally and note the different views you can see of the subject. They are the same views you see in a hologram. Use this technique to view the perspective and field of vision for your holograms. You can then adjust the position of the object, film, or even the optics to obtain the views you want.

Varying exposure

Unless you have a calibrated power meter and lots of experience, expect mistakes in estimating exposure times when experimenting with different types of holographic setups. The direct, one-beam setup detailed above conserves laser light energy, thus reducing exposure time. Some multibeam setups (see below) require exposure time of 15 to 25 seconds, assuming a 3 to 5 mW laser.

Light ratios

All holograms are made by directing a reference beam and an object beam onto a film plate. The reference beam comes directly from the laser, perhaps after bouncing off a beam splitter and a mirror or two. The object beam is reflected off of the object being photographed. The ratio between the reference and object beams is a major consideration. If the ratio isn't right, the hologram becomes "noisy" and difficult to see. The reflectivity of the object largely determines the ratio, but with most commonly holographed subjects, the reference-to-object light ratio for a transmission hologram is about 4:1 (four parts reference to one part object). Intermediate and advanced holography requires careful control over light ratios, which means you must take readings using a calibrated power meter (available through Metrologic) or a light meter.

You can adjust the light ratios by repositioning the optics or by using a variable-density beam splitter. A variable-density beam splitter is a wheel (sometimes a rectangular piece of glass) with an antireflective coating applied in varying density to one side. An area with little or no antireflection coating reflects little light and passes most of it. The opposite is true at an area with a high amount of AR coating. You can "dial" in the ratio of transmitted light versus reflected by turning the wheel. Note that basic holography doesn't require a variable-density beam splitter, which is good because they're expensive.

Reverse viewing

If you flip the hologram over top to bottom, you'll see the image appear in front of the film, looming toward you. You see the image inside-out as you view the image from the back side. This effect is created by light focusing in space in front of the hologram. Although the entire image can't be focused onto a plane, you can see the formation of three-dimensions by placing a piece of frosted glass at the apparent spot where the hologram appears. As you move the plate in and out, different portions of the picture will come in and out of focus.

Projected viewing

The image in a hologram can be projected in a variety of ways. Try this method: Replace the hologram so that its emulsion faces the laser. Remove the beam-expanding lens so that just the pencil-thin beam of the laser strikes the center of the hologram film. Place a white card or screen behind the film to catch the light going through the hologram. The screen or card should be located where your eyes would be if you were viewing the hologram. Watch the image that appears. You'll see the complete object, but the image will be two-dimensional.

Now move the film up and down and right and left. Notice how the picture of the object remains complete but the perspective changes. This is the same effect you get if you cut a hologram into many pieces. Each piece contains a full picture of the subject but at slightly different views.

Transfer mirrors

Some holographers prefer to place the laser in the sand table. This is perfectly acceptable as long as you mount the laser on a wood or plastic board.

Position it along one side of the table and direct the beam diagonally across the table with a transfer mirror. Aiming the beam diagonally across the surface of the sand table gives you more room and allows you to create more elaborate setups.

SPLIT-BEAM TRANSMISSION HOLOGRAM

The visual effect of a single-beam hologram is limited due to the single source of light. Objects photographed in this manner might look dark or lack detail in shadow areas. Only one side of the hologram may be illuminated, and as you move your head to see different views, the image grows dim. In portraiture photography, two, three, and sometimes four different light sources are used to illuminate different parts of the head or body. The amount of illumination from each light is carefully controlled to make the picture as pleasing as possible.

A similar approach can be used in holography by splitting the laser beam and providing two or more sources of light to illuminate the object (there is still only one reference beam). The setup in FIG. 13-8 shows how to illuminate an object by splitting the laser light and directing two expanded beams to either side of the object. See TABLE 13-3 for a list of required materials.

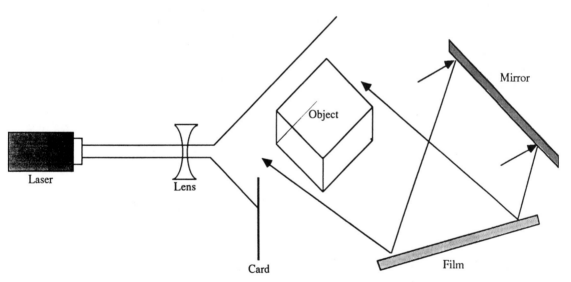

Fig. 13-8. A split-beam transmission hologram using a large mirror. The mirror provides the reference beam as well as a large portion of light for the illumination of the object.

The main consideration when splitting the light is that the reference and object beams should travel approximately the same distance from the laser. Use a fabric tape to measure distances. Accuracy of 1 or 2 inches will assure you of good results. Longer distances could exceed the coherency length of the laser and your holograms might not come out right. Note that high-power lasers generally have longer coherency lengths than low-power ones.

Table 13-3. Parts List for
Split-Beam Holography Setup

1	He-Ne laser
3	8 to 10 mm biconcave lens
1	Holographic film in film holder
4	1- to 2-inch-square front-surface mirror
2	Plate beam splitter (50:50)
2	Black blocking card
1	Object to holograph

The arrangement requires three diverging lenses, three mirrors, and two beam splitters. The first beam splitter reflects the light to a reference mirror. This light is then expanded by a biconcave lens and directed at the film. The transmitted light through the first beam splitter is divided again by a second beam splitter. These two beams are bounced off positioning mirrors, expanded by lenses, and pointed at the object.

It's very important that no light from the two object mirrors strikes the film. It is a good idea to baffle the light by placing black cards on either side of the film holder, as illustrated in the figure.

Exposure time should be similar to the one-beam method described earlier in this chapter, but you should probably make another test strip to make sure. Although roughly the same total amount of light is striking the film in both single- and multiple-beam setups, there is inherent light loss through the beam splitters as well as additional light scattering through the three lenses.

Process the film in the usual manner, and when dried, place it back in the film holder with the emulsion side facing the reference beam. Remove the object you photographed as well as the object mirrors and second beam splitter (you need only the reference beam to reconstitute the image). An image should now appear.

You can visually see how the processed hologram must be placed in the exact same position relative to the reference beam, or the image won't appear. Try turning the film in the holder. Notice that the image disappears if you rotate the film more than a few degrees in either direction.

MAKING A REFLECTION HOLOGRAM

Transmission holograms require you to shine the expanded beam of a laser through them in order to see an image. Another type of hologram that doesn't require a laser for viewing is the reflection hologram. These work by shining light (white or colored) off the diffraction grating surface of the hologram. If viewed under white light, the hologram gives off a rainbow of dazzling colors.

Mixing the chemicals

Reflection holograms are no more difficult to make than transmission holograms, although the processing chemistry is a bit different. White light reflection holograms can use Kodak D-19 developer but should be bleached using one of the following formulas:

- Mix 20 grams (or about one tablespoon) of potassium bromide and 20 grams mercuric chloride (FATAL IF SWALLOWED) in one liter of water. This stuff is dangerous and highly caustic, so never touch it with bare fingers or allow it to splash on skin, clothes, or eyes.
- Dissolve two grams of potassium dichromate with 30 grams potassium bromide in one liter of water. After these have thoroughly mixed, add two cubic centimeters of concentrated sulfuric acid (always add acid to water, not the other way around). This is also nasty stuff and burns skin if you touch it. Wear gloves and safety goggles when mixing, and use gloves or tongs when processing the film.

The best reflection holograms need a fixing step, just like regular photographic film and paper. Fixer comes premade (in powder or liquid), making it easy to use. Kodak Rapid Fixer with hardener is a good choice. Mix according to instructions.

Setting up

Figure 13-9 shows the most rudimentary arrangement for making a reflection hologram. See TABLE 13-4 for a parts list. Position the laser, lens, film, and object in a direct line. For best results, the object used in reflection-type holography should be relatively small in comparison to the film and should be placed within a few inches of the film. Larger images, or those placed far away, tend to be dark and fuzzy.

As mentioned previously, reflection holograms require a film that lacks an antihalation backing. Be sure to use this type. When you are ready to make the exposure, turn out the lights, remove the film, and place it in the holder.

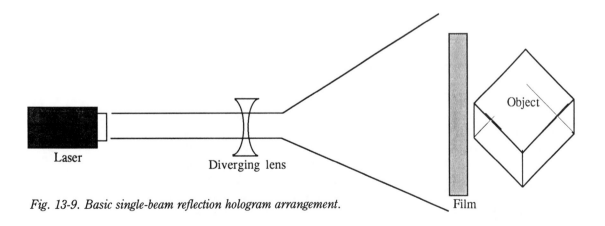

Laser

Diverging lens

Object

Fig. 13-9. Basic single-beam reflection hologram arrangement.

Film

1	He-Ne laser
1	8 to 10 mm biconcave lens
1	Holographic film in film holder
1	Object to holograph

Table 13-4. Parts List for Basic Reflection Holography Setup

If you are using a glass-plate film holder, remember to press the plates together firmly and maintain constant pressure for 10 to 20 seconds. Use wood blocks to apply even pressure on the glass and film. This removes all air bubbles trapped between the glass and film. Snap binder clips around the glass and position the holder in front of the object. If necessary, mount the film holder between two PVC or dowel pillars.

As with the transmission hologram, the beam should be expanded so that the outer $1/3$ of the diameter falls off the edge of the film; you want the inner $2/3$ of the beam, which is the brightest portion. Unlike transmission holograms, reflection holograms call for a ratio between reference and object at a more even 1:1 or 2:1. Use a photographic light meter or power meter to determine proper beam ratios.

Exposure time depends on the power output of the laser as well as the size of the film (or more precisely, the amount of beam spreading), but try using exposures of 3 to 5 seconds with a 1 mW laser, and 1 to 2 seconds for a 3 mW laser.

Processing the film

In dim or green-filtered light, dip the film in the developer tray and process for 2 to 5 minutes. As a general rule of thumb, a reflection hologram should pass about 20 to 30 percent of the light when held up to a green safelight. After developing, rinse in water or stop bath for 15 seconds.

Dip the film in the first fixer bath for 2 to 3 minutes. After fixing is complete, the room lights can be turned on (fixer renders the film insensitive to further exposure to light). Wash again in water for 15 seconds.

Place the film in the bleach mixture for 1 to 2 minutes or until the film clears (becomes transparent). Rinse once more in water for 15 seconds. Finally, place the film in a second fixer bath. Keep the film in that for 3 to 5 minutes or until the hologram turns a brown color.

Wash all the chemicals away by rinsing the film under running water for at least 5 minutes. Then, dip the hologram in Photoflo, squeegee it, then hang it up to dry. Be aware that you can't see a holographic image until the film is completely dried, so don't judge your success (or failure) at this point.

If you are impatient and can't wait for the film to dry on its own, you can hurry up the process by blow drying the film with a hair dryer. Set the dryer on no heat (air only) and gently waft it 6 to 8 inches in front of the film. The backing will dry quickly, but the emulsion takes 5 to 10 minutes to dry. Always remember to dry film in a dust-free place. Amateur photographers like to use the tub or shower in the bathroom, a place where airborne dust usually doesn't stay for long.

Here is a summary of the preceding steps:

- Develop in D-19 (or similar) developer for 2 to 5 minutes.
- Rinse in water or stop bath for 15 seconds.
- Fix in first fixer for 2 to 3 minutes (light on after fixing).
- Rinse in water for 15 seconds.
- Bleach for 1 to 2 minutes (or until the film clears).
- Rinse in water for 15 seconds.
- Fix in second fixer for 3 to 5 minutes.
- Wash thoroughly under running water for 5+ minutes.
- Dip in Photoflo for 15 seconds; squeegee.
- Dry for at least 30 minutes.

Alternate method

A slightly less complex processing method can be used to make reflection holograms. Develop the film in D-19 for 2 to 5 minutes, and then wash in running water for 5 minutes. Bleach the film using the sulfuric acid bleach described above for 2 minutes or until the film clears. Wash another 5 to 10 minutes and dry it.

VIEWING REFLECTION HOLOGRAMS

Reflection holograms don't require a laser for image reproduction. Just about any source of light will work, including sunlight or the light from an incandescent light. Avoid greatly diffused light, such as that from a fluorescent lamp, or the hologram will look fuzzy. The ideal light source is a point source, such as an unfrosted filament bulb. View the image as you tilt the hologram at angles to the light.

Note the many colors in the picture, particularly green. Although made with a red helium-neon light, the film shrinks after processing so it tends to reflect shorter wavelength of light. The amount of shrinkage varies depending on the film, but it often correlates to 50 to 100 nanometers, bringing the red 632.8 nm wavelength of the helium-neon laser to about 500 to 550 nm.

The best viewing setup for a reflection hologram is shown in FIG. 13-10. Point the light from a desk lamp straight down at the table. Tilt the hologram toward you until the image becomes clear (about 45 to 50 degrees).

IMPROVED REFLECTION HOLOGRAMS

A split-beam reflection hologram provides more even lighting and helps improve the three-dimensional quality. Figure 13-11 shows one arrangement you can use (see TABLE 13-5 for a parts list for this arrangement). Except for the angle of some of the objects, it is nearly identical to the split-beam transmission hologram setup described earlier in this chapter.

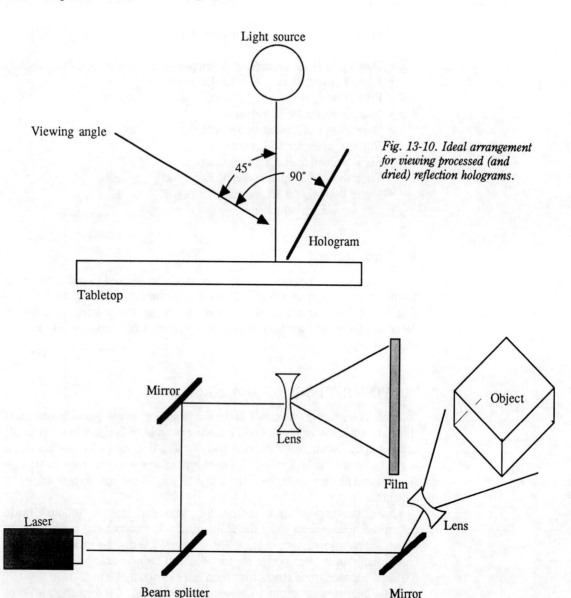

Light source

Viewing angle

45°   90°

Fig. 13-10. Ideal arrangement for viewing processed (and dried) reflection holograms.

Hologram

Tabletop

Mirror

Lens

Film

Object

Laser

Beam splitter

Lens

Mirror

Fig. 13-11. Optical arrangement for a multiple-beam reflection hologram.

1	He-Ne laser
3	8 to 10 mm biconcave lens
1	Holographic film in film holder
4	1-to 2-inch-square front-surface mirror
2	Plate beam splitter (50:50)
3	Black blocking card
1	Object to holograph

Table 13-5. Parts List for Split-Beam Reflection Holography Setup

SUMMARY OF HOLOGRAPHY TIPS

Follow these tips and tricks while you experiment with holography.

- Allow time for the table and film to stabilize before taking the exposure.
- When using glass-plate film holders, be sure the glass is perfectly clean. Press both pieces firmly together for about 30 seconds. Use blocks of wood to exert even pressure. Remove all the trapped air or the film might move during exposure.
- Be sure to position the film so that the emulsion faces the subject and/or reference beam. This isn't always necessary for reflection-type holograms, but it is a good habit.
- Observe proper lighting ratios between reference and object beams. Generally, transmission holograms have a 3:1 or 4:1 ratio between reference and object beams (but up to 10:1 is sometimes required to eliminate noise); reflection holograms have 1:1 or 2:1 ratio.
- Measure distances for reference and object beams to ensure they are approximately equal. Use a cloth or flexible tape.
- Use the proper chemicals mixed fresh (or stored properly), as per the directions. Throw out exhausted chemicals—flush them down the sink and run plenty of water to wash away the chemical residue.

14

Creating laser
light shows

If you have a laser, you are already on the road to producing your own laser light shows. A small assortment of basic accessories are all you need to make dancing, oscillating shapes on the ceiling, wall, or a screen.

This chapter details some basic approaches to affordable laser light shows. You'll learn how to produce light shows using dc motors and mirrors that make interesting and controllable "Spirograph" shapes, how to make a laser beam dance to your controls or the beat of music, and more.

THE "SPIROGRAPH" EFFECT

Imagine your laser drawing unique "atom-shaped" repeating spiral light forms, with you able to adjust their size and shape by turning a couple of knobs. The "Spirograph" light show device, named after the popular Spirograph drawing toy made by Kenner, uses three small dc motors and an easy-to-build motor speed and direction control circuit.

Depending on how you adjust the speed and direction of the motors, you alter the shape and size of the spiral light forms. And because the motors used are not constant speed, slight variations in the rotation rate cause the light forms to pulse and change all on their own. A complete parts list for the "Spirograph" light show device is provided in TABLE 14-1.

Mirror mounting

Got a penny? That and a little bit of glue is all you need to mount each mirror to a motor. The best motors to use are the 1.5-to-6-volt dc hobby motors made by Mabuchi, Johnson, and numerous other companies and sold by Radio Shack and almost every other electronics outlet in the country. Measure the diameter of the shaft; it can vary depending on the manufacturer and original

Table 14-1. Parts List for
"Spirograph" Light Show Device

3	Small 1.5 to 6 Vdc hobby motor
3	Lincoln penny
3	1-inch-diameter or square, thin, front-surface mirror
3	$3/4$-inch electrical conduit pipe hanger
3	$10/24$-by-$3/4$-inch bolt, flat washer, toothed lock washer
6	$10/24$ nut
1	8-by-24-inch pegboard ($1/4$-inch thick)
2	24-inch lengths of 2-by-2-inch framing lumber
2	4-inch lengths of 2-by-2-inch framing lumber

application for the motor. Then drill a hole in the exact center of a penny using a bit just slightly smaller than the motor shaft.

Use a drill press to hold the penny in place and prevent the bit from skipping. Drilling is easier if you turn the coin over and position the bit in the middle column of the Lincoln Memorial (if the penny is less than about 25 years old, after they changed the design of the back). Note that the newest pennies are easiest to drill because they contain aluminum inside. Don't worry; the hole can be off a few fractions of an inch, but it should not be larger than the motor shaft. If anything, strive for a press fit. File away the flash left by the bit so the surface of the penny is smooth.

Next, apply a drop of cyanoacrylate adhesive (Super Glue) to a 1-inch-square (or diameter) mirror to the center of the penny. Best results are obtained when using a fairly thin mirror and gap-filling glue. The "Hot Stuff Super 'T'" glue made by HST-2 (available at hobby stores) is a good choice. Wait an hour for the adhesive to dry and set. Repeat the procedure for the other three mirrors.

Avoid gaps between the mirrors and pennies. Although a small amount of misalignment is desirable, a large gap will cause excessive beam displacement when the motor turns. You'll see exactly why this is important once you build the Spirograph light show device.

Finally, mount the penny and mirror on the end of the motor shaft, as depicted in FIG. 14-1. Apply several drops of adhesive to the shaft and let it seep into the hole in the penny. Wait several hours for the adhesive to set completely before continuing. Alternatively, you can solder the penny to the shaft. This requires a heavy-duty soldering iron or small, controllable torch. Mount the penny on the motor shaft first and then tack on the mirror.

Mounting the motors

Ideal motor mounts can be made with 1-inch plumbing pipe hangers, sold at the neighborhood plumbing supply outlet or hardware store. The hanger is made of formed U-shaped metal with a mounting hole on one end and an adjustable open end at the other (many other styles can also be used). Secure the motor in the hanger by loosening the bolt on the end, slipping the motor in, then finger tightening the bolt.

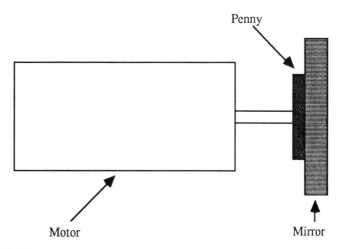

Fig. 14-1. Mirror and penny mounting detail for the dc motors used in the "Spirograph" light show device.

Secure the hanger on an 8-by-24-inch piece of $1/4$-inch hardwood pegboard, as shown in FIG. 14-2. Add wood blocks to the underside of the pegboard to make an optical breadboard, as shown in the figure. Arrange the hangers as shown in FIG. 14-3, and lightly secure the hangers to the pegboard using $10/24$ by $1/2$-inch bolts and matching hardware. Use flat and split washers as indicated in the figure to prevent movement when the motors are turning (and vibrating).

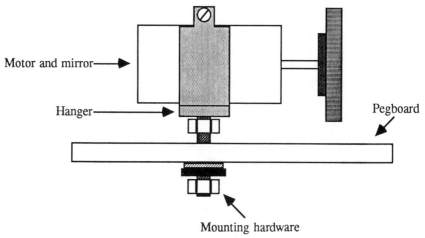

Fig. 14-2. How to mount the motors to a pegboard base.

Building the motor control circuit

The motor control circuit allows you to individually control each motor. You have full command over the speed and direction of each motor by flicking a switch and turning a dial.

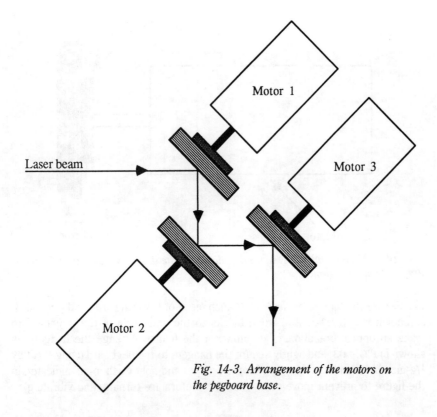

Fig. 14-3. Arrangement of the motors on the pegboard base.

The schematic for the motor control circuit is shown in FIG. 14-4 (parts list in TABLE 14-2). The illustration shows the circuit for only one motor; duplicate it for the remaining two motors. The prototype used a $3^{1}/_{2}$-by-4-inch perforated board and wire-wrapping techniques. Your layout should provide room for the electronics, switches, potentiometers, and transistors on heatsinks (the latter are very important). Lay out the parts before cutting the board to size. Power is provided by a 6 Vdc battery pack consisting of four alkaline D cells.

The double-pole, double-throw switches allow you to control the direction of the motors or turn them off. The potentiometers let you vary the speed of the motors from full to about $1/2$ to $2/3$ normal. Different speeds are obtained by varying the "on" time, or duty cycle, of the motors. The more the duty cycle approaches 100 percent, the faster the motor turns. The design of this circuit does not allow the motors to turn at drastically reduced speeds, which in any case is not really desirable to achieve the spiral light form effects.

Alternative speed control circuits are shown in FIGS. 14-5 and 14-6. Figure 14-5 details a similar control circuit using 2N3055 heavy-duty transistors. This circuit also works on the duty-cycle principle (more accurately referred to as pulse width modulation). These transistors should be placed in a suitable aluminum heatsink with proper case-to-heatsink electrical insulation. The 2N3055s

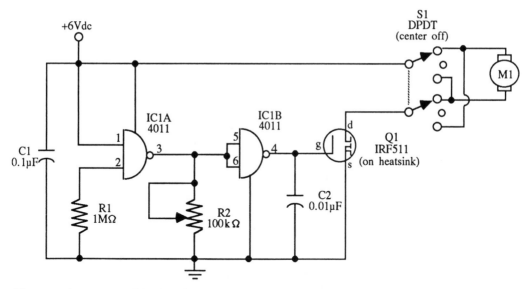

Fig. 14-4. An easy-to-build pulse-width-modulation speed control circuit and direction switch for a small dc motor. Transistor Q1 must be mounted on a heatsink.

Table 14-2. Parts List for Motor Speed Control

R1	1 MΩ
R2	100 kΩ
C1	0.1 μF disc
C2	0.01 μF disc
IC1	4011 CMOS NAND gate IC
Q1	IRF511 MOSFET power transistor (or equivalent)
S1	DPDT switch, center off
Misc.	Heatsink for Q1

All resistors are 5 to 20 percent tolerance, 1/4 watt. All capacitors are 10 to 20 percent tolerance, rated at 35 volts or more.

and heatsinks require more space than the IRF511 power MOSFET transistors used in the schematic outlined earlier, so make the board larger.

Figure 14-6 shows a basic approach using 2-to-5-watt power potentiometers. Be sure to use pots rated for at least 2 watts or you run the risk of burning them out. Bear in mind that the potentiometer approach consumes more power than the pulse-width-modulated systems. No matter how fast the motors are turning, a contact amount of current is always drawn from the batteries. Current not used by the motors is dissipated by the potentiometer as heat. Parts lists for the alternative speed control circuits are provided in TABLE 14-3.

Mount the circuit board on one end of the optical breadboard using 10/24 by 1/2-inch bolts and 10/24 nuts. Make sure there is sufficient space between the bottom of the board and the wire wrap posts and top of the optical breadboard.

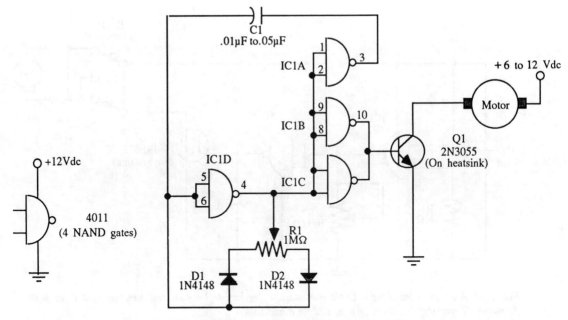

Fig. 14-5. An alternate method of providing pulse width modulation using a 2N3055 power transistor.

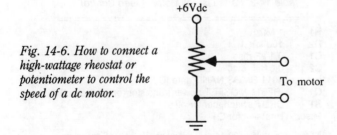

Fig. 14-6. How to connect a high-wattage rheostat or potentiometer to control the speed of a dc motor.

Table 14-3. Parts List for
Alternate Motor Speed Controls

Alternate #1

M1	1.5 to 6 Vdc hobby motor
R1	1 MΩ potentiometer
C1	0.01 or 0.05 µF capacitor
IC1	4011 CMOS NAND gate IC
D1,D2	1N4148 diode
Q1	2N3055 power transistor
Misc.	Heatsink for Q1

Alternate #2

M1	1.5 to 6 Vdc hobby motor
R1	10 kΩ potentiometer, 3 to 5 watts

All resistors are 5 to 10 percent tolerance, 1/4 watt.
All capacitors are 10 to 20 percent tolerance, rated at
35 volts or more.

Mounting the laser

Cut a 3-inch length of 2-inch plastic PVC pipe lengthwise in half. Using 10/24 by 1/2-inch bolts and hardware, mount the two halves on the optical breadboard. Insert a worm-gear pipe clamp 2 1/2 to 3 inches in diameter under the PVC half before tightening the nuts and bolts. Use silicone adhesive to attach rubber feet above the four bolt heads. The rubber feet serve as a cushion between the laser tube and bolts as well as to increase the height of the beam over the optical bench. PVC pipe laser holder is designed for a cylindrical laser head. If you are using a bare laser tube, install it in a suitable enclosure.

Aligning the system

Slip the laser into the holder and tighten the two clamps. The distance between the front of the laser and the motors is not critical, but be sure that there is no chance that the mirror on the first motor will touch the end of the laser. Turn the laser on but do not apply power to the motor control circuit. Rotate the hangers on each motor so that the beam is deflected from mirror 1 to 2 to 3.

Fine tune the alignment by rotating each motor 90 degrees. The misalignment inherent in the mirror mounting should displace the beam on the mirrors. Avoid fall-off where the beam skips off the mirror. Beam fall-off causes a void in the spiral when the motors turn.

If you cannot align the motors so that the beam never falls off the mirrors, check the gap between the mirrors and pennies. Place the motor with the largest gap at the end of the chain—as motor number 3. If beam fall-off is still a problem, try mounting a mirror and penny on a new motor.

Place all three switches to their center position and apply power to the motor control circuit. Flick switch #1 up or down and rotate the potentiometer. The motor should turn. If the motor whines but refuses to turn, flick the switch off, turn the pot all the way on, and reapply power. The motor should turn.

Test the speed control circuit by turning the pot. The motor should slow down by an appreciable amount (you'll be able to hear the decrease in speed). If nothing happens, double check your work. A motor that won't change speed could have improper wiring or a bad MOSFET transistor. A blown transistor can cause the motor to spin at about a constant 80 percent of full speed. Next reverse the motor by moving the switch to the opposite position. The motor should momentarily come to a halt, turn in its tracks, and go the other way.

Turn the first motor off and repeat the testing procedure for the other two. After all motors check out, turn them back on and point the #3 mirror so that the beam falls on a wall or screen. Watch the spiral light form as all three motors turn. Do you notice any beam fall-off? If so, *stop all the motors* and readjust them. Note that the motors vibrate a great deal at full speed, and that can cause them to go out of alignment. When you get the motors aligned just right, tighten the hangers to prevent them from coming loose.

Test the different types of light forms you can create by turning off the #1 motor and using just #2 and #3. Depending on how the direction is set on the

Fig. 14-7. The "atom" laser lightform made with the Spirograph device.

motors, you should see an "orbiting atom" shape on the screen, as depicted in FIG. 14-7. If the form looks more like constantly changing ellipses, reverse the direction of one of the motors. Adjust the speed control on both motors and watch the different effects you can achieve. Now try the same thing with motor #1 and #3 on. Try all the combinations and note the results.

What happens if the light form doesn't show up or appears very small, even when the screen is some distance from the light show device? This can occur if the mirror is precisely aligned with the rotation of the motor. Although this is rare using the construction technique as outlined, it can happen. You can see how much each motor contributes to the creation of the light form by turning on each one in turn. You should see a fairly well-formed circle on the screen. The mirror is too precisely aligned if a dot appears instead of the circle. Replace the mirror and motor with another one and try again.

Note that the size of the circle does not depend on where the beam strikes the mirror. The circle is the same size whether the beam hits the exact center of the mirror or its edge.

Notes on using and improving the "Spirograph" device

Here are some notes on how to get the most from the spiral light form device. Keep the mirrors clean and free of dust or the light forms will appear streaked and blurred. *Never* adjust the position of the motors when they are turning. The mirrors are positioned close together, and moving the motors could cause the glass to touch. The mirrors will shatter and fragments of glass will fly in all directions if they touch when the motors are turning. It is a good idea to use protective goggles when adjusting and using the spiral light show device.

Cheap dc motors such as those used in this project make a lot of noise. You might want to use higher quality motors if you plan on using the Spirograph maker in a light show. Get ones with bearings on the shaft. You can also place the device in a soundproof box. Provide a clear window for the beam to come out.

The Spirograph device is designed for manual control. With the right interface circuit, you can easily connect it to a computer for automated operation. See Chapter 25, Computer control of robots, for ideas on how to control motors via a computer.

You can obtain even more light forms by adding a fourth motor. Try it to see what happens. Also, don't be shy about turning some of the motors off. Some of the most interesting effects are achieved with just two motors.

EXPERIMENTING WITH GALVANOMETERS

Professional light shows don't use mirrors and pennies mounted on cheap motors but rather a unique electromechanical device called the *galvanometer*. A galvanometer—or "galvo" for short—provides fast and controllable back-and-forth oscillation. Mount a mirror on the side of the shaft and the reflected light forms a streak on the wall. Position two galvanometers at a 90-degree angle, apply the right kind of signal, and you can project circles, ovals, spirals, stars, and other multidimensional geometric shapes.

What is a galvanometer?

Most electronics buffs are familiar with the basic galvanometer movement of an analog meter. The design of the movement is shown in FIG. 14-8. A coil of wire is placed in the circular gap of a magnet. Applying current to the coil causes it to turn within the electromagnetic field of the magnet. The amount of turning is directly proportional to the amount of current that is applied. The needle of the meter is attached to the coil of wire.

Some meter movements are designed so that the needle rests in the center of the scale. Applying a positive or negative voltage swings the needle one direction or the other. Say that at full deflection, the meter reads + and −5 volts. When charged with +5 volts, the needle swings all the way to the right. When charged with −5 volts, the needle sways all the way to the left. Current under five volts (positive or negative) causes the needle to travel only part of the way right or left.

Meter movements are designed for precision and are not capable of moving much mass. But by using a stronger magnet and a larger coil of wire, a galvanometer can be made to motivate a larger mass. Heavy-duty scanning galvanometers are often made to actuate the needles in chart recorders, and they

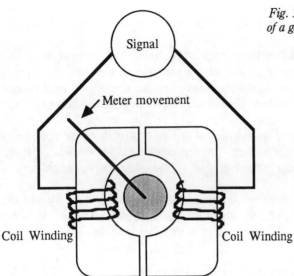

Fig. 14-8. The basic operation of a galvanometer.

have more than enough "oomph" to rack a small first-surface mirror back and forth. Some galvo manufacturers, most notably General Scanning, make units specifically designed for high-speed laser light deflection. These are best suited to laser light show applications, but their cost is enormous. Surplus galvanometers, which are available from several sources including Meredith Instruments and General Science & Engineering cost from $30 to $60 apiece; new high-speed models cost upwards of $350.

You can use either commercially made galvanometers for the projects that follow or make your own using small hobby dc motors. Be aware that commercially made galvos work much better than homebrew types, but if you are on a budget and simply want to experiment with making interesting light show effects, the dc motor version should prove to be more than adequate.

Using commercially made galvos

Figure 14-9 shows a typical commercially made galvanometer. The General Scanning model GVM-735 galvanometers illustrated in the picture are actually Cadillacs among scanners, so the units are not representative of typical quality. There are other makers of fine galvanometers including C.E.C., Minneapolis-Honeywell, and Midwestern.

Driving galvanometers

Galvanometers can be driven in a variety of ways including power op amps, audio amplifiers, and transistors. A basic, no-frills drive circuit appears in FIG. 14-10 (refer to TABLE 14-4 for a parts list). The input can be an audio signal from the LINE OUT jack of a hi-fi or an unamplified input from a frequency generator (more on these later). You *can* apply an amplified signal to the input of the drive circuit, but the op amp will clip the output if the input is excessively high.

Fig. 14-9. A commercially made precision galvanometer.

Fig. 14-10. Driver circuit for operating a galvanometer from a line-level audio source. Build two circuits for controlling two galvanometers.

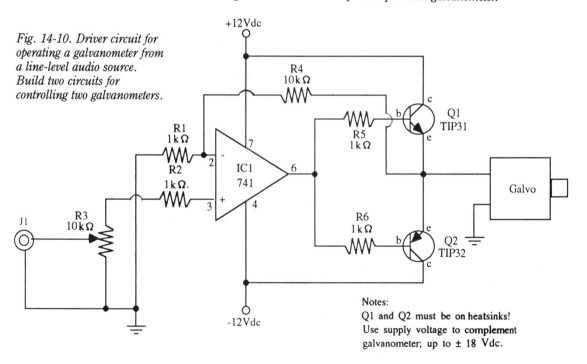

Notes:
Q1 and Q2 must be on heatsinks!
Use supply voltage to complement galvanometer; up to ± 18 Vdc.

Table 14-4. Parts List
for Galvanometer Drive

R1,R2,R5,R6	1 kΩ
R3	10 kΩ potentiometer
R4	10 kΩ
IC1	LM741 op amp IC
Q1	TIP 31 npn transistor
Q2	TIP 32 pnp transistor
J1	1/8-inch jack
G1	Galvanometer
Misc.	Heatsinks for Q1 and Q2

All resistors are 5 to 10 percent tolerance, 1/4 watt.

The two drive transistors, a complementary pair consisting of TIP31 and TIP32 power types mounted on a heatsink, interface the output of the op amp to the coil of the galvanometer. The circuit works with a variety of voltages from ±5 volts to ±18 volts. Most scanners operate well with supply voltages of between ±5 and ±12 volts. Check the specifications of your galvanometers and make sure you don't exceed the rated voltage.

If anything, operate the galvos at a reduced voltage. They will still operate satisfactorily but the rotor might not deflect the full amount. This is not a problem for most applications, including laser light shows, where full deflection is not always desired. One by-product of full deflection is a "ringing" that occurs when the rotor hits the stop at the ends of both directions of travel. The ringing appears in the laser light form as glitches or double streaks.

To make two-dimensional shapes, you need two galvanometers positioned 90 degrees apart, as illustrated in FIG. 14-11. Mount mirrors on the shafts using aluminum or brass tubing. Add a set screw (see FIG. 14-12) so that you can tighten the mirror mounts on the rotor shaft of the galvanometer.

You can use any number of mounting techniques to secure the galvos to an optical breadboard or table, but the mounts you use must be sturdy and stable. Vibrations from the galvos can be transferred to the mounts, which can shake and disturb the light forms. Build a separate drive circuit for both galvanometers and enclose it in a project box (or you can include the driver circuit in a larger do-everything light show console). I built the prototype drive circuit on a universal breadboard PCB and had plenty of room to spare. The enclosure measured 4^{3}/8 by 7^{3}/4 by 2^{3}/8 inches. Subminiature 1/8-inch phone jacks were provided for the audio inputs and scanner outputs and potentiometers were mounted for easy control of the input level.

To test the operation of the galvanometers and drive circuits, plug in the right and left channels of the hi-fi and turn the gain controls (R1 for both drives) all the way up. The galvos should shake back and forth in response to the music. Shine a laser beam at the mirrors so that the light bounces off one, is deflected by the other, and projects on the wall. The light forms you see should undulate in time to the music.

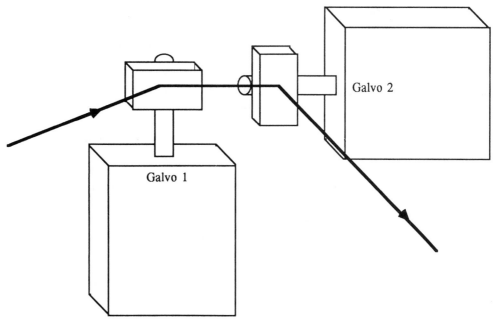

Fig. 14-11. How to arrange two galvanometers to achieve full X- and Y-axis deflection. Place the mirrors of the galvanometers close together to counter the effects of beam deflection.

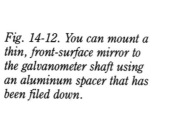

Fig. 14-12. You can mount a thin, front-surface mirror to the galvanometer shaft using an aluminum spacer that has been filed down.

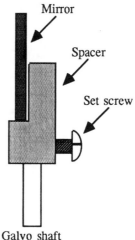

Providing an audio source

The test performed explained previously only lets you test the operation of your galvanometer setup. With the arrangement detailed above, the light form is always squeezed into a fairly tight line that crawls up and down the wall at a 45-degree angle. Full flexibility of a pair of galvanometers physically set apart 90 degrees requires an audio source that has two components—both of which are set 90 degrees apart in phase.

Audio signals are sine waves, and sine waves are measured not only by frequency and amplitude but phase. The phase is measured in degrees and spans from 0 to 360 degrees. Figure 14-13 shows two sine waves set apart 90 degrees. Notice that the second wave is a quarter step (90 degrees) behind the first one.

If you could somehow delay the sound coming from one channel of your stereo, you could broaden the 45-degree line into a full two-dimensional shape. The closer the delay is to 90 degrees out of phase, the more symmetrical the light form is. Imagine a pure source of sine waves—a sine wave oscillator. The oscillator is sending out waves at a frequency of 100 Hz. It has two output channels called sine and cosine. Both channels are linked so they run at precisely the same frequency, but the cosine channel is delayed precisely 90 degrees. The light form projected on the screen is now a perfect circle.

The circuit in FIG. 14-14 provides such a two-channel oscillator. Two controls allow you to change the frequency and "symmetry" of the sine waves. The symmetry (or phase) control counters the destabilizing effects caused by rotating the frequency knob. This circuit, with parts list provided in TABLE 14-5, is designed so that the resistors and capacitors are the same value. Changing the value of one resistor throws off the balance of the circuit, and the symmetry control helps rebalance it. Note that the frequency and symmetry controls provide a great deal of flexibility in the shape of the projected beam. Fiddling with these two potentiometers allows you to create all sorts of different and unusual light forms.

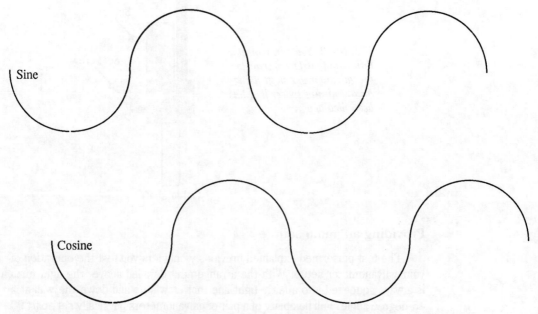

Fig. 14-13. Two sine waves where the bottom sine wave is delayed 90 degrees from the top wave.

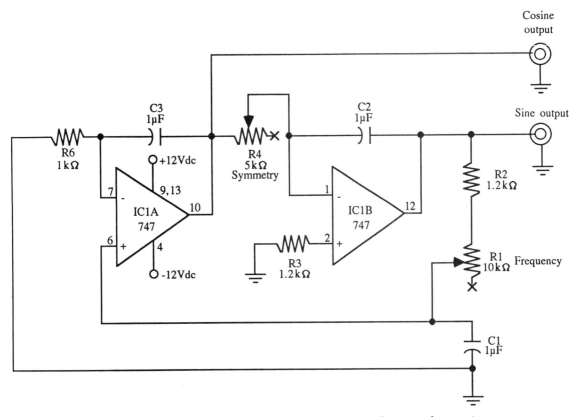

Fig. 14-14. One way to implement a sine/cosine audio generator for operating two galvanometers.

Table 14-5. Parts List for Sine/Cosine Generator

R1	10 kΩ potentiometer
R2,R3	1.2 kΩ
R4	5 kΩ potentiometer
R5	1 kΩ
C1 – C3	1 μF polarized electrolytic
IC1	LM747 dual op amp IC
J1,J2	1/8-inch miniature phone jack

All resistors are 5 to 10 percent tolerance, 1/4 watt.
All capacitors are 10 to 20 percent tolerance, rated at 35 volts or more.

The circuit is designed around the commonly available 747 dual op amp (two LM741 op amps in one package). You can use almost any other dual op amp, such as the LM1458 or LF353, but test the circuit first on a breadboard. For best results, use a dual op amp.

You can obtain even more outlandish light forms when combining the sine and cosine signals from two separate oscillator circuits. On a perf board, combine two oscillators and mix them using two LM741 op amps. An overall circuit

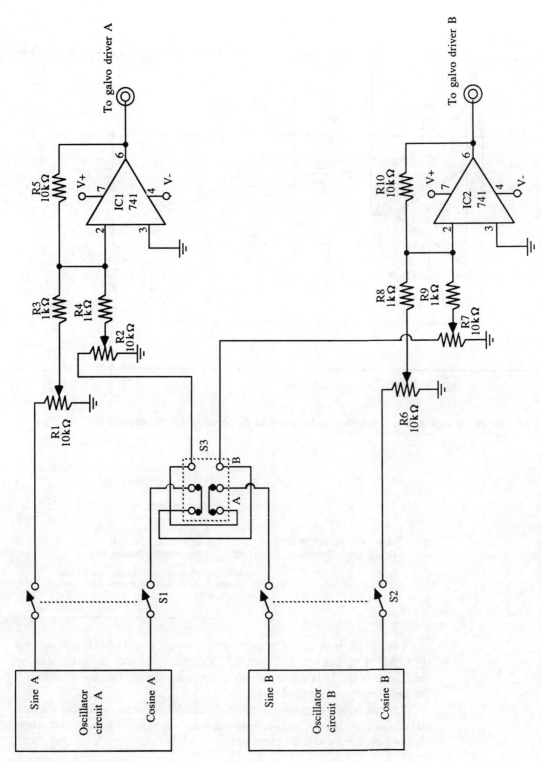

Fig. 14-15. A schematic diagram for designing a two-channel sine/cosine audio generator with dual op amp mixers (note: use separate op amps for the two mixer channels).

design is shown in FIG. 14-15. One set of switches allows you to turn either oscillator on and off, and the input controls let you individually control the amplitude from each sine/cosine channel (for a total of four inputs). A parts list for the general circuit is in TABLE 14-6.

Table 14-6. Parts List for
Complete Light Show Circuit

2	Sine/cosine generators (see FIG. 14-14)
R1,R2,R6,R7	10 kΩ potentiometer
R3,R4,R8,R9	1 kΩ
R5,R10	10 kΩ
IC1,IC2	LM741 op amp IC
J1,J2	1/8-inch miniature phone jack
S1 – S3	DPDT switch

All resistors are 5 to 10 percent tolerance, 1/4 watt.

The other set of switches lets you flip between mixing the sine inputs together or criss-crossing them so that the sine channel of one oscillator mixes with the cosine of the other oscillator and vice versa.

When the switch is in the "pure" position (sine with sine and cosine with cosine), you obtain rounded-shape designs, such as spirals, circles, and concentric circles. When the switch is in the "criss-cross" position (sine with cosine for both channels), you obtain pointed shapes, which include diagonals, stars, and squares.

Like the drive circuit, place the oscillator with all its various potentiometers in a project box or tuck it inside a console. Provide two 1/8-inch jacks for the outputs for the two galvos.

Using the oscillator

Connect the outputs of the oscillator to the inputs of the drive circuit. Apply power to both circuits and rotate the mixer input controls (R3, R4, R7, and R8) to their fully on positions. Flick on switch #1 so that only the signals from one oscillator are routed to the mixer amps. Slowly turn the control knobs until the galvanometers respond.

If the galvanometers don't seem to respond, temporarily disconnect the jumpers leading between the oscillator and drive circuits and plug an amplifier into one of the oscillator output channels. You should hear a buzzing or whining noise as you rotate the frequency and symmetry controls. If you don't hear a noise, double-check your wiring and be sure the mixer controls are turned up. When turned down, no signals can pass through the mixing amps.

Aim a laser at the mirrors and watch the shapes on a nearby wall or screen. Get the feeling of the controls by turning each one and noting the results. With the symmetry control turned down and the frequency control almost all the way down, you should see a fairly round circle on the screen. If the circle looks like

Fig. 14-16. The basic circle produced by turning on one sine/cosine generator and adjusting the frequency and symmetry controls to produce a set of pure sine waves.

an egg, adjust the mixer controls to decrease the X or Y dimension, as shown in FIG. 14-16.

If the egg is canted on a diagonal, the phase of the cosine channel is not precisely 90 degrees. Try adjusting the symmetry control and fine tuning it with the frequency control. There should be one or more points where you achieve proper phase between the sine and cosine channels.

Now turn off the first oscillator and repeat the testing procedures for the second oscillator. It might behave slightly different than the first. This is to be expected due to the tolerance of the components. If you need more precise control over the two oscillators, use 1-percent tolerance resistors and high-Q capacitors.

For the really interesting light forms, turn on both oscillators and adjust the controls to produce various symmetrical and asymmetrical shapes. At many settings, the light forms undulate and constantly change, and at other settings, the shape remains stationary and appears almost three-dimensional. Figures 14-17 and 14-18 show sample light forms created with the circuit and galvanometers described above.

Alternate between the "pure" and "criss-cross" settings by flicking switch #3. Note the different effects you create when the switch is in either position.

Powering the oscillator and drive circuits

So far we've discussed using galvo oscillator and drive circuits but have paid no attention to the power supply requirements. Although you can build your

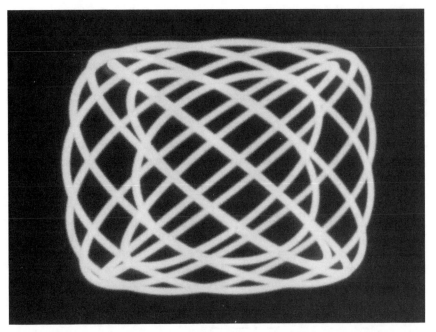

Fig. 14-17. A clearly definable "Lissajous" figure, made with the galvanometer light show device.

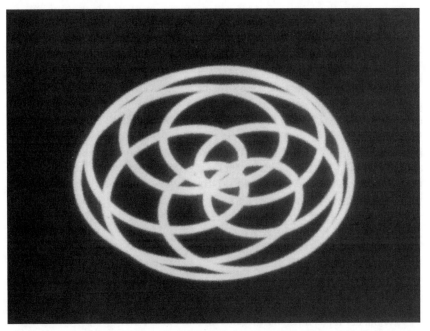

Fig. 14-18. One of the many spiral light forms created by turning on both sine/cosine generators.

own dual-polarity power supply to run the galvo system, I strongly recommend that you use use a well-made commercial supply, one that has very good filtering. Sixty-cycle hum, caused by insufficient filtering and poor regulation, can creep into the op amps and make the galvos shudder continuously.

Output voltage and current depend on the galvanometers you use. I successfully used two General Scanning GVM 734 galvos with a power supply that delivered ±5 volts at about 250 mA for each polarity. You might need a more powerful power supply if you use different galvanometers (some require ±12 at one amp or more). The power supply I used for the prototype system was surplus from a Coleco computer system at $12.95. Look around and you should be able to find an equally good deal.

Useful modifications and suggestions

Not shown in the driver circuit above is an additional switch that allows you to change the polarity to one of the galvanometers. This provides added flexibility over the shape of the light forms. Wire the switch as shown in FIG. 14-19.

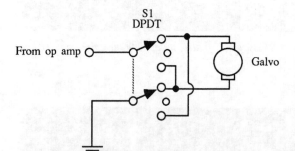

Fig. 14-19. Wire a DPDT switch to one of the galvanometers to reverse its direction in relation to the other galvanometer.

The drive circuit detailed above might not provide adequate power for all galvanometers. If the galvos you use just don't seem to be operating up to par, use the drive circuit shown in FIG. 14-20, developed by light show producer Jeff Korman (see parts list in TABLE 14-7). This new driver is similar to the old one but provides better performance at low frequencies. Korman also uses a special-purpose sine/cosine oscillator chip made by Burr-Brown. The chip, called the 4423, is available directly from Burr-Brown or a local electronics distributor that stocks Burr-Brown components. Its cost is high ($25 to $35), but it provides extremely precise control over frequency without the worry of knocking the signals out of 90-degree phase.

Other useful tips:

- Whenever possible, try to reduce the size of the light forms; it makes them appear brighter.
- Connect an audio signal into one channel of the drive circuit and connect the sine/cosine oscillator into the other channel. The light form will be modulated in 2-D to the beat of the music.

- You can record a galvanometer light show performance by piping the output of the oscillators into two tracks of a four-track tape deck. Use the remaining two tracks to record the music. When played back, the galvanometers exactly repeat your original recording session. This is how most professional light show producers do it.

- Try adding one or two additional oscillators. Combine them into the mixing network by adding a 10 kΩ pot (for volume control) and a 10 kΩ input resistor. Wire in parallel as shown in the FIG. 14-15 schematic.

- Triangle or sawtooth (ramp) waves create unusual pointed-star shapes, boxes, and spirals. You can build triangle and sawtooth generators using op amps, but an easy approach is to use the Intersil 8038 or Exar XR-2206 function generator ICs.

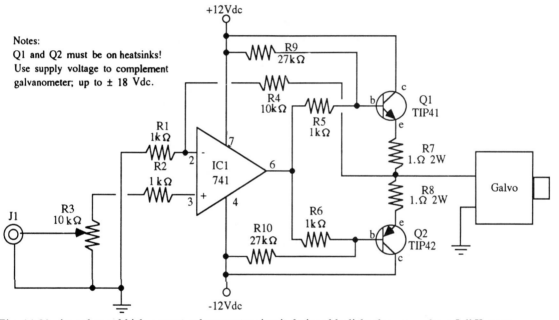

Fig. 14-20. *An enhanced high-current galvanometer circuit designed by light show consultant Jeff Korman.*

Table 14-7. Parts List for Enhanced Galvanometer Driver

R1,R2,R5,R6	1 kΩ
R3	10 kΩ potentiometer
R4	10 kΩ
R7,R8	0.1 Ω, 2 watt
R9,R10	27 kΩ
IC1	LM741 op amp IC
Q1	TIP 41 npn transistor
Q2	TIP 42 pnp transistor
J1	1/8-inch jack
G1	Galvanometer
Misc.	Heatsinks for Q1 and Q2

All resistors are 5 to 10 percent tolerance, 1/4 watt, unless otherwise indicated.

Making your own galvos

A set of commercially made galvanometers can set you back $100 to $200, even when they are purchased on the surplus market. If you are interested in experimenting with laser graphics and geometric designs but are not interested in spending a lot of money, you can make your own using small dc motors.

Follow the diagram shown in FIG. 14-21 to make your own galvanometers. You can use almost any 1.5-to-6-volt dc motor, but it should be of fairly good quality. Test the motor by turning the shaft with your fingers. The rotation should be smooth, not jumpy. Measure the diameter of the outside casing. It should be about 1 inch. Refer to TABLE 14-8 for a list of required parts.

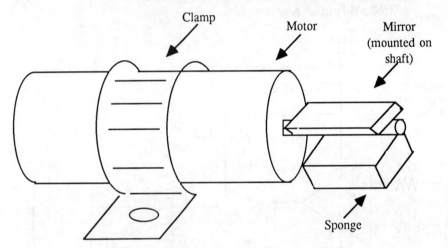

Fig. 14-21. Basic arrangement for making homemade galvanometers using small dc motors. The sponge prevents the mirror and shaft from turning more than 20 to 25 degrees in either direction and helps damp vibrations.

Table 14-8. Parts List for
Dc Motor Galvanometer

1	Small 1.5 to 6Vdc hobby motor
1	$3/4$- to 1-inch pipe clamp
1	2-by-3-inch $1/8$-inch-thick acrylic plastic
1	$8/32$ by $1/2$-inch bolt, nut, washer
1	$1/2$-by-$3/4$-inch, thin, front-surface mirror
1	Lincoln penny
1 ea.	Small piece of sponge, antistatic foam

Use a high-wattage soldering iron or small brazing torch to solder a penny onto the side of the motor shaft. For best results, clean the penny and coat it and the motor shaft with solder flux. Make sure the penny is thoroughly heated before applying solder, or the solder might not stick. You'll need to devise some sort of clamp to hold the penny and motor while soldering because you need both hands free to hold the solder and gun.

Let the work cool completely, and then mount a small $1/2$-by-$3/4$-inch front surface mirror to the front of the penny. You can use most any glue: Duco adhesive or gap-filling cyanoacrylate glue are good choices. Note that the mirror should not be exceptionally thick. You get the best results when the mirror is thin because the motor has less mass to move so it can vibrate faster.

Snap the motor into a $3/4$-inch electrical conduit clamp and move it into position (the clamp has a 1-inch opening and holds most small hobby motors). If the clamp is too small, widen the opening by gently prying it apart with a pair of pliers. Mount the motor and clamp to a 2-by-3-inch acrylic plastic base ($1/8$-inch in thickness is fine). Drill holes as shown in FIG. 14-22. Use a $6/32$-by-$1/2$-inch bolt and a $6/32$ nut to secure the pipe clamp to the base.

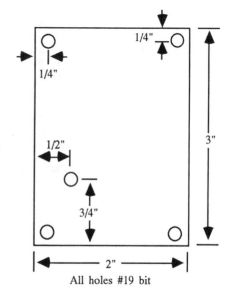

Fig. 14-22. Cutting and drilling template for the base for the homemade motor galvanometers.

Use a treated, synthetic sponge and cut into two 1-by-$1/2$ inch pieces. The sponge should be soft but will dry out when left overnight. After the sponge has dried out, compress it and secure it under the penny using all-purpose adhesive. Now slide a 1-by-$1/2$-inch piece of antistatic foam in the gap between the sponge and penny. The fit should be close but not overly tight. If you need more clearance, compress the sponge by squeezing it more. The finished dc motor galvanometer is shown in FIG. 14-23.

As an alternative, cut a piece of $1/2$-inch foam and stick it under the penny. Try different foams to test their "suppleness." The foam should be soft enough to let the penny and mirror vibrate but not so soft that it acts as a tight spring and bounces the penny back after only a small movement.

Using the motor/galvanometers

Attach leads to the motor terminals and connect the home-made galvanometers to the drive circuit and oscillator detailed earlier in this chapter. Repeat

Fig. 14-23. A completed motor galvanometer secured to the base with a 3/4-inch electrical conduit clamp. Make two motor galvanometers for your laser light show system.

the testing procedures outlined for commercially made galvos. Position the motors so that they are 90 degrees off axis and shine a laser off both mirrors. You should see shapes and patterns as you adjust the controls on the oscillator.

The motor/galvanometers can be mounted in a variety of ways. One approach is to use metal strips bent 90 degrees at the bottom. Drill matching holes in the strip and attach the base of the motor/galvos to the strips using 6/32 hardware. You can also secure the base of each motor/galvanometer using 1/2-inch galvanized hardware brackets (available at the hardware store).

The light forms might not be perfectly symmetrical. Depending on the motors you used and the type and thickness of foam backing you installed, one motor could vibrate at a wider arc than the other. Try adjusting the foam and sponge on both motors to make them vibrate the same amount.

15

Light wave communications

Imagine living 100 years ago when the concept of the "wireless" was new and relatively unknown. Today, most of our daily lives revolve around communications using radio waves: we use advanced forms of the wireless to watch television, listen to the radio, contact a "good buddy" via CB radio, or even talk around the world using microwave and satellite links.

Radio waves (which include radio, TV, and microwave) are part of the electromagnetic spectrum, and so is light. Far above the microwave frequencies and past the large block of infrared radiation are the greens, blues, reds, and other colors we associate with visible light, a tiny patch in the wide expanse of the electromagnetic spectrum.

Perhaps you've asked yourself these questions: If radio waves can be used to transmit sound and pictures, why not beams of light? If they can send a pop music broadcast from the local AM radio station, why can't you pipe your favorite tunes along the pencil-thin beam of a helium-neon laser? Or if the indiscriminate emissions of radio waves are detectable using snooper scopes, can't you use a narrow and direct laser light link to send sensitive computer data from one location to another?

You can do all of these things and more. Because light is at such a high frequency in the electromagnetic spectrum, it is an even *better* medium for communications than radio waves. Lasers especially are perfect instruments for communications links because they emit a powerful, slender beam that is not affected by interference and is nearly impossible to intercept.

Here in this chapter you'll learn the basics of light wave communications using ordinary light-emitting diodes as well as helium-neon and semiconductor lasers. You'll discover the different ways light can be modulated and cajoled.

LIGHT AS A MODULATION MEDIUM

Higher frequencies in the radio spectrum provide greater bandwidth. The bandwidth is the space between the upper and lower frequencies that define an information channel. Bandwidth is small for low-frequency applications such as AM radio broadcasts, which span a range 540 kHz to 1600 kHz. That's a little more than one megahertz of bandwidth, so if there are 20 stations on the dial, there's only 50 kHz per DJ.

Television broadcasts, including both VHF and UHF channels, span a range from 54 MHz to 890 MHz with each channel taking up 6 MHz. Note that the 6 MHz bandwidth of the TV channel provides 100 times more room for information than the entire AM radio band. That way, television can pack more data into the transmission.

Microwave links, which operate in the gigahertz (billions of cycles per second) region, are used by communications and telephone companies to beam thousands of phone calls in one transmission. Many calls are compacted into the single microwave channel because the bandwidth required for one phone conversation is small compared to the overall bandwidth provided by the microwave link.

Visible light and near-infrared radiation has a frequency between about 430 to 750 terahertz (THz)—or 430 to 750 trillion cycles per second. Thanks to the immense bandwidth of the spectrum at these high frequencies, one light beam can simultaneously carry all the phone calls made in the United States or almost 100 million TV channels. Of course, what to put on those channels is another thing!

Alas, all of this is theoretical. Transmitters and receivers don't yet exist that can pack data into the entire light spectrum; the current state of the art cannot place intelligent information at frequencies higher than about 25 to 35 gigahertz (25 to 35 billion cycles per second). It could take a while for technology to advance to a point where the full potential of light beam communications can be realized.

Even with these limitations, light transmission offers additional advantages over conventional techniques. Light is not as susceptible to interference from other transmissions, and when squeezed into the arrow-thin beam of a laser, it is highly directional. It is difficult to intercept a light beam transmission without the intended receiver knowing about it. And unlike radio gear, experimenting with even high-power light links does not require approval from the Federal Communications Commission. Businesses, universities, and individuals can test light wave communications systems without the worry of upsetting every television set, radio, and CB in the neighborhood (however, FDA regulations must be followed).

On the down side, light is greatly affected by weather conditions, and unlike low frequencies, such as AM radio, it does not readily bounce off of objects. Radar (low-band microwave) pierces through almost any weather condition and bounces off of just about everything.

EXPERIMENTING WITH A VISIBLE LED TRANSMITTER

It's easy to see how light wave communication links work by experimenting with a system designed around the common and affordable visible light-emitting diode. The LED provides a visual indication that the system is working and allows you to see the effects of collimating and focusing optics.

The LED communications link, like any other, consists of a transmitter and receiver. An LED is used as the transmitting component and a phototransistor is used as the receiving component. To facilitate testing, a radio or cassette player is used as the transmission source. You listen to the reception at the receiver using headphones. Later projects in this chapter show how to substitute the radio or cassette player for a microphone.

Just about any LED will work in the circuit shown in FIG. 15-1, but if you want to operate the link over long distances (more than 5 or 10 feet), you should use a high-output LED, as discussed in the next section. And after you test the visible LED, you can exchange it with or more high-output infrared LEDs to extend the working distance. Of course, you enjoy the greatest range using an infrared of visible laser. (We'll get to that later in this chapter.)

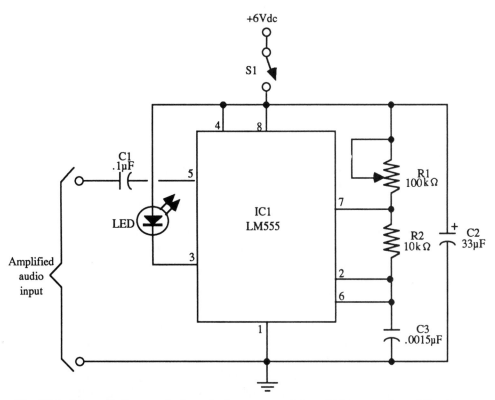

Fig. 15-1. Schematic diagram for the pulse frequency modulated LED transmitter. Adjust frequency by rotating R1. With components shown, frequency range is between 8 and 48kHz.

Building the transmitter

The transmitter, with parts indicated in TABLE 15-1, is designed around a 555 timer IC. The 555 generates a modulation frequency upon which the information you want to send is placed. The output frequency of the 555 changes as the audio signal presented to the input changes. This modulation technique is commonly referred to as *pulse frequency modulation* (PFM) and is shown diagrammatically in FIG. 15-2. The signal can be received using a simple amplifier, as shown later in this section, but a receiver designed to "tune in" to the PFM signal is desired for best response. Advanced receivers are discussed later.

R1	100 kΩ potentiometer
R2	10 kΩ
C1	0.1 µF disc
C2	33 µF polarized electrolytic
C3	0.0015 µF mica or high-Q disc
IC1	LM555 timer IC
LED1	Light-emitting diode (see text)
S1	SPST switch

All resistors are 5 to 10 percent tolerance, 1/4 watt.
All capacitors are 10 to 20 percent tolerance, rated at 35 volts or more, unless otherwise indicated.

Table 15-1. Parts List for LED Transmitter

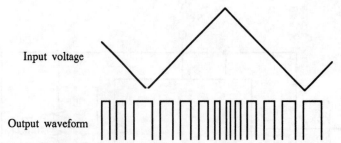

Input voltage

Output waveform

Fig. 15-2. Comparison of input voltage and width of the output waveform.

Construct the transmitter in a small project box. Power comes from a single 9-volt transistor battery. The switch lets you turn the circuit on and off and the potentiometer allows you to vary the relative power delivered to the LED. In actuality, adjusting the pot changes the modulation frequency, which in turn changes the pulse width, which in turn changes the current delivered to the LED.

In any case, the entire range is beyond human hearing and above the audio signals that you will be transmitting. You can readily increase the modulation frequency to the upper limit of the components used in the transmitter and receiver, but lowering them into the 20 to 20,000 Hz region of the audio spectrum produces an annoying buzz. Any sourcebook on using the LM555 timer IC can show you how to calculate output frequency for astable operation.

Mounting details are provided in the following two figures; parts are shown in TABLE 15-2. Solder an LED to the terminals of a 1/8-inch phone plug jack, and mount the jack in the base of a 3/4-inch PVC end plug, as shown in FIG. 15-3. If the plug is rounded on the end, file it flat with a grinder or file. Lightly countersink the hole so that the shaft of the phone jack is flush to the outside of the plug. Countersinking also helps the shaft of the jack to pole all the way through the thick-walled PVC fitting.

Table 15-2. Parts List for Plug and Box Transmitter

1	3/4-inch schedule 40 PVC end plug
1	1/8-inch miniature phone jack
1	1/8-inch miniature phone plug
1 ea.	Project box, knob for potentiometer, 6 Vdc battery holder (four AA).

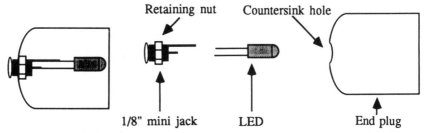

Fig. 15-3. How to mount the LED in a PVC end plug. The same approach is used for the receiver phototransistor.

The transmitter and LED connect via a 1/8-inch plug that is mounted so it extrudes through the project box as detailed in FIG. 15-4. Use a 5/16-inch 18 nut to hold the plug in place. The 1/8-inch mini plug used in the prototype is threaded for 5/16-inch 18 threads, but not all plugs are the same. Check yours first.

Attach the transmitter into the LED by plugging it in. Install a 9-volt battery and turn the transmitter on. The LED should glow. You won't be able to test the transmitter circuit until you build the receiver, which is covered in the next section.

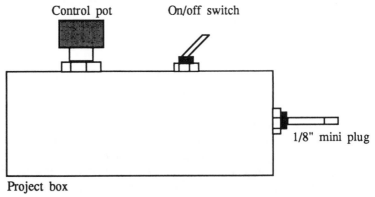

Fig. 15-4. The project box, shown with a 1/8-inch mini plug for connecting to the LED.

Building the receiver

The receiver, shown in FIG. 15-5, is designed around the common LM741 op amp and an LM386 audio amplifier. See TABLE 15-3 for a parts list. Power is supplied via two 9-volt batteries (to provide the 741 with a dual-ended supply). A switch turns the circuit on and off (interrupting both positive and negative battery connections) and a potentiometer acts like a volume/gain control.

You can listen to the amplified sounds through headphones or a speaker, or you can connect the output of the receiver to a larger amplifier. A good, handy outboard amplifier to use is the pocket amp available at Radio Shack. The pocket amp accepts an external input and has its own built-in speaker.

Construct the receiver in a plastic project box. The one used for the prototype measured $2^3/4$ by $4^1/8$ by $1^9/16$ inches and was more than large enough to accommodate the circuit, batteries, switch, potentiometer, and output jack.

The receiving phototransistor is built into a PVC end plug in the same manner as the transmitter LED described earlier. Mount the phototransistor as shown, being sure to note the orientation of the transistor leads and jack terminals. Although the circuit will work if you connect the phototransistor backwards, sensitivity will be greatly reduced.

Connect the receiver to the phototransistor, install two batteries, plug in a set of headphones, and turn the power switch on (don't put the headphones on just yet). Adjust the potentiometer midway through its travel and point the phototransistor at an incandescent lamp. You should hear a buzzing sound through the headphones (the buzzing is the lamp fluctuating under the 60-cycle current).

If you don't hear the buzz, adjust the volume control until the sound comes in. Should you still not hear any sound, double-check your wiring and the batteries. Even with the phototransistor not plugged in, you should hear background hiss. No hiss might mean that the circuit is not getting power or the headphone jack is not properly wired.

The receiver can be used with the LED lightwave link as well as all the other communications projects in this chapter (as well as most of those in the remainder of this book). Its wide application makes it an ideal all-purpose universal laser beam receiver. When I refer to the ''universal receiver,'' this is the one I'm talking about.

Using the light wave link

Once the receiver checks out, you can test the transmitter. Switch off the lights or move to a darkened part of the room. Turn on the transmitter source (radio, tape player) and aim the transmitter LED at the receiver phototransistor. Adjust the controls on the receiver and transmitter until you hear sound. You might hear considerable background hiss and noise caused by other nearby light sources. If you use the communications link outdoors in sunlight, the infrared radiation from the sun might swamp (overload) the phototransistor, drastically reducing or cutting off the sound completely. The transmitter and receiver works best in subdued light.

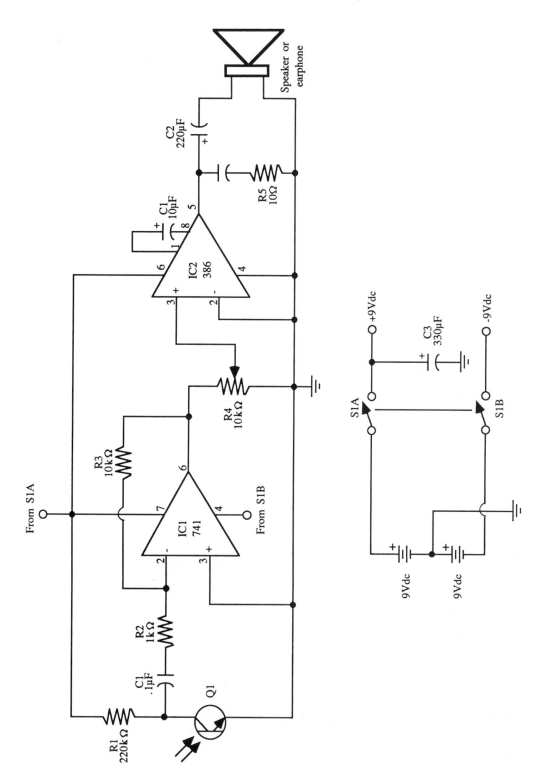

Fig. 15-5. The universal laser light communications receiver. The output of the LM286 audio amplifier can be connected to a small 8-ohm speaker or earphone. Two 9-volt batteries provide power. Decrease R1 to lower sensitivity; increase R3 to increase the gain of the op amp (avoid very high gain or the op amp might oscillate). Circuit courtesy Forrest Mims III.

*Table 15-3. Parts List
for Universal Receiver*

R1	220 kΩ
R2	1 kΩ
R3	10 kΩ
R4	10 kΩ potentiometer
R5	10 Ω
C1	0.1 μF disc
C2	220 μF polarized electrolytic
C3	10 μF polarized electrolytic
C4	100 μF polarized electrolytic
IC1	LM741 operational amplifier IC
IC2	LM386 audio amplifier IC
Q1	Infrared phototransistor
S1	DPDT switch

All resistors are 5 to 10 percent tolerance, 1/4 watt. All
capacitors are 10 to 20 percent tolerance, rated at 35
volts or more.

Test the sensitivity and range of the communications link by moving the receiver away from the transmitter. You will find that depending on the output of the LED, the range will be limited to about 5 feet before reception drops out.

Extending the range of the link

Most all phototransistors are most sensitive to infrared light. The peak spectral sensitivity depends on the makeup of the transistor, but is generally between about 780 and 950 nm, in the near infrared portion of the spectrum. A red LED has a peak spectral output of about 650 nm, considerably under the sensitivity of the phototransistor. A solar cell offers a wider spectral response and can provide greater range. The best type of solar cell to use is the kind encased in plastic, like the phototransistor (many have a built-in lens). Connect the cell in the circuit as shown in FIG. 15-6.

The solar cell is sensitive to a wide range of colors. The light spectrum above or below the red radiation from the LED isn't needed for reception, so block it with a red filter. Test the effectiveness of the filter by temporarily taping it to the front of the solar cell.

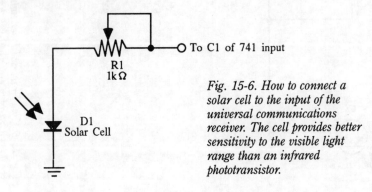

Fig. 15-6. How to connect a solar cell to the input of the universal communications receiver. The cell provides better sensitivity to the visible light range than an infrared phototransistor.

You can also use the filter with the phototransistor to help limit the incoming radiation to the red wavelengths. Even though the phototransistor is designed to be most sensitive to near-infrared radiation, it can still detect light at other wavelengths, especially red. One or two layers of red acetate placed over the phototransistor can increase the range in moderate light conditions by several feet.

The best way to increase the working distance of the communications link is to add lenses to the LED and/or the phototransistor. The PVC end plug makes it fairly easy to add lenses to both transmitter and receiver components. Mount a simple biconvex or plano-convex lens in the end plug as shown in FIG. 15-7. The PVC rings hold the lens in place and let you easily adjust the distance between the lens and phototransistor. If the lens has a focal length of more than 10 to 15 mm, attach a coupling to the end plug (see FIG. 15-8) and stuff the lens in the coupling. Again, use PVC rings to hold the lens in place.

Be sure that you position the lens at the proper focal point with respect to the *junction* of the LED or phototransistor.

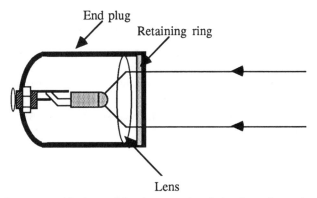

Fig. 15-7. A lens placed in front of the phototransistor helps focus the received light and improves reception.

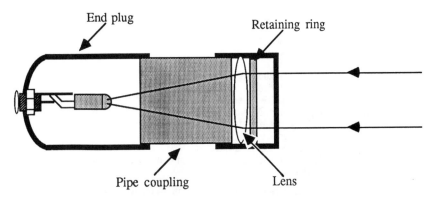

Fig. 15-8. A pipe coupling allows you to focus the lens for optimum reception. You can use a similar arrangement for adding polarizers and a second focusing lens.

You can see the effect of the lens on the transmitted light by pointing the LED against a lightly colored wall. At close range and with the lens properly adjusted, you should actually see the junction of the LED projected on the wall (assuming you are not using an LED with a diffused case). You might also see a faint halo around the junction; this is normal and is caused by light emitted from the sides of the LED.

With the lens(es) attached, try the light wave link again and test its effective range. With extended range comes increased directionality, so you must carefully aim the transmitter element at the receiver. A simple focusing lens on both receiver and transmitter should extend the working distance to 100 feet or more. Test the real effectiveness of the system at night outside. The darkness can also help you better aim the transmitter. At 100 feet, the light from the LED will be dim, but you should be able to spot it if you know where to look for it.

Note that the plastic case of the LED and phototransistor acts as a kind of lens and that it can alter the effective focal length of the system. Experiment with the position of the lens until the system is working at peak performance.

LONG-RANGE LIGHT WAVE COMMUNICATIONS LINK

With the right high-output LED, transmitter, and receiver circuit, you can talk over long distances on a shaft of light. The following plans show how to build a receiver and transmitter that lets you communicate more than $1/2$ mile. Though the circuit is a bit complex, fortunately as of this writing it is available in kit form from General Science and Engineering (see Appendix A). See the parts lists in TABLES 15-4 and 15-5.

Table 15-4. Parts List for Long-Range Transmitter

R1,R4	47 kΩ
R2,R3	470 kΩ
R5,R6	33 Ω
R7,R9	1 kΩ potentiometer
R8	10 kΩ potentiometer
C1	0.002 μF disc
C2	0.1 μF disc
C3	180 pF disc
C4	10 μF polarized electrolytic
C5	100 μF polarized electrolytic
C6	1.2 μF polarized electrolytic
C7	30 μF polarized electrolytic
IC1	5532 low-noise amplifier IC
Q1	A7937 transistor
LED1	High-output LED (see text)
J1	Miniature phone jack (for electret condenser microphone)
S1	SPST switch

All resistors are 5 to 10 percent tolerance, $1/4$ watt. All capacitors are 10 to 20 percent tolerance, rated at 35 volts or more.

Table 15-5. Parts List for Long-Range Receiver

R1	3.5 MΩ to 10 kΩ resistor (see text)
R2	3.4 kΩ
R3	1 kΩ
R4,R10	35 Ω
R5	100 kΩ
R6,R8	5 kΩ potentiometer
R7	1 MΩ
R9	107 kΩ
R11	10 kΩ
R12	27 Ω
C1,C7,C15,C16	10 μF polarized electrolytic
C2,C11,C12	0.01 μF disc
C3	0.47 μF disc
C4	10 μF polarized electrolytic
C5,C8	220 pF disc
C6	1.2 μF polarized electrolytic
C9,C10	100 μF polarized electrolytic
C13,C14	6.8 μF polarized electrolytic
C17	0.3 μF disc
IC1,IC2	5534 low-noise amplifier
Q1	PF5102 FET transistor
Q2	2N4410 npn transistor
Q3	2N4248 pnp transistor
D1	Phototransistor
S1	SPST switch
S2	DPDT switch
J1	Miniature jack (for output)
Misc.	Two 9-volt transistor battery clips, two 9-volt transistor batteries, case, focusing lenses

All resistors are 5 to 10 percent tolerance, $1/4$ watt. All capacitors are 10 to 20 percent tolerance, rated at 35 volts or more.

Building the transmitter

Figure 15-9 shows a schematic for the transmitting link of a long-range light wave communications device. The constant-current transmitter provides an input for an ordinary dynamic microphone. The transmitting light-emitting diode is a very high-output visible LED—in the 2 to 5 candle power range. The brighter the LED, the longer the range. You can obtain very high-output visible LEDs from a number of sources including Radio Shack and General Science & Engineering. The secret behind the transmitter is the NE5532 IC dual low-noise operational amplifier and the 7937 3-amp npn transistor. These components should not be substituted.

You can build the transmitter using almost any construction technique, but printed circuit boards and universal solder boards work the best.

Building the receiver

The schematic for the long-range receiver circuit is shown in FIG. 15-10. Photodiode D1 of the receiver intercepts the light from the LED of the transmitter section. The signal is then preamplified by a PF5102 field-effect transistor and then amplified by a pair of single low-noise op amps.

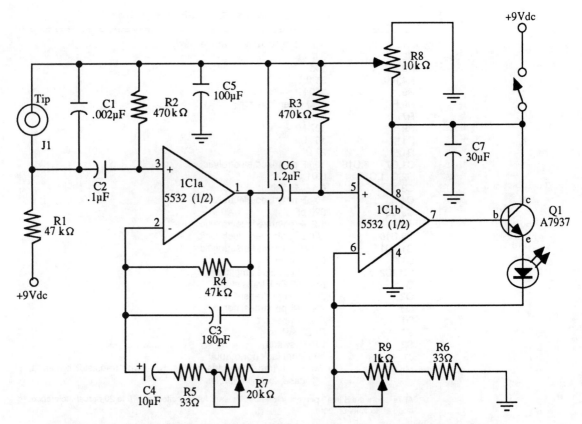

Fig. 15-9. Schematic diagram for the long-distance LED transmitter. Circuit courtesy Roger Sonntag.

Power supply filtering and decoupling is important to reduce noise. Capacitors C10 through C14, along with chokes L1 and L2, form a complete filtering and decoupling system. These components straddle the two 9-volt transistor batteries. Note that additional power is obtained by two N-size alkaline batteries.

The value of resistor R1 can be between 3.4 megohms to 150 kilohms. The higher the resistance, the greater the gain (and hence sensitivity and range), but bandwidth will suffer. A lower value decreases the gain and sensitivity but provides a wider bandwidth. Don't use a potentiometer here because it could introduce noise.

As with the transmitter, you can build the receiver using almost any construction technique, but printed circuit boards and universal solder boards are highly recommended. Keep component leads as short as possible.

Enclosures for the transmitter and receiver

For carefree operation, place the transmitter and receiver boards in suitable plastic project boxes. Provide access to the batteries because these go

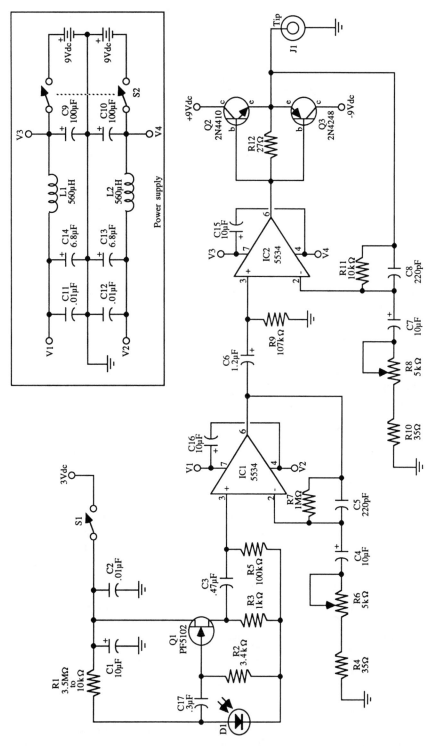

Fig. 15-10. Schematic diagram for the long-distance LED receiver. Note the power supply decoupling circuitry. Circuit courtesy Roger Sonntag.

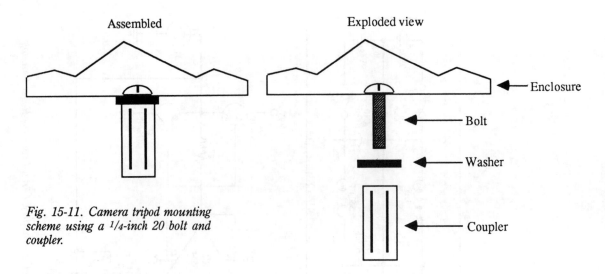

Assembled   Exploded view

Enclosure

Bolt

Washer

Coupler

Fig. 15-11. Camera tripod mounting scheme using a 1/4-inch 20 bolt and coupler.

dead after a few dozen hours of tinkering. Drill holes on the transmitter project box for the on/off switch (S1) and a larger hole for the intensity control (R9). You can also use a combination pot and switch to combine the functions of S1 and R9. Drill holes on the receiver box for power switches S1 and S2 as well as gain controls R6 and R8.

Both receiver and transmitter enclosures should include mounting hardware for small camera tripods. Drill a 1/4-inch hole on the bottom of the enclosure, and pass a 1-by-1/4-inch 20 bolt through it. Tighten with a 1/4-inch 20 threaded nut and a 1/4-inch 20 coupler, as shown in FIG. 15-11. The mounting screw on the camera tripod attaches to the open end of the coupler. You might also want to mount a $4 \times$ rifle scope to the top of the receiver enclosure. With some careful alignment, you can use the scope as a telescopic viewfinder for aiming the receiver.

Enclosures also allow you to add lenses to the high-output LED of the transmitter and photodiode of the receiver. You can use almost any lens for the LED and photodiode. Try a 25 to 50 mm focal length biconvex or plano-convex mirror 25 mm in diameter in front of both the LED and photodiode, as shown in FIG. 15-12.

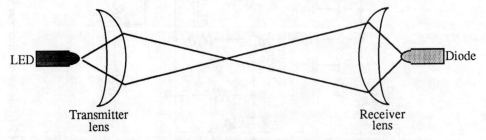

LED   Diode

Transmitter lens   Receiver lens

Fig. 15-12. Arrangement of lenses for the transmitter LED and receiver diode. Place the transmitter lens at its focal length to the LED and place the receiver lens at its focal length to the diode.

Be sure to place the lens the proper distance from the LED and photodiode (distance equal to the focal length of the lens). Ensure that the components are in-line with the optical axis of the lens. A tube-mounting system, like that in FIG. 15-13, lets you adjust the placement of the lenses for best results.

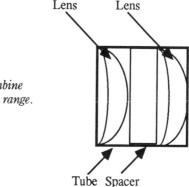

Lens   Lens

Fig. 15-13. How to combine two lenses for increased range.

Tube Spacer

Using the long-range communications link

You need two people to fully test the long-range communications link. Testing is easier at dusk or dark because it's easier to see the light emitted from the transmitter. Place the transmitter and receiver on tripods and "eyeball" them until they are aligned. If the distance is less than about 100 feet, turn the transmitter on and watch the faint glow of light as it falls on the photodiode of the receiver.

Plug a microphone into the transmitter and dial the gain controls (R7 and R8) to their midpoints. Dial intensity control R9 to its midpoint as well. On the receiver, dial both gain controls, R6 and R8, to minimum. Plug a set of headphones into the receiver, but wear them above your ears instead of directly over your ears.

Have someone speak through the microphone while you listen on the headphones. Slowly increase gain controls R6 and R8 on the receiver to about their midpoints or until you start to hear sound. The two controls affect the gain of the op amps; R6 controls the first stage and amplifies the signal from 100 to 1000 times and should be used as an overall sensitivity control. Potentiometer R8 amplifies the signal an additional 10 to 100 times and can be used as an overall volume control. The two controls permit precise control of the gain of the circuit and help prevent oscillations.

You can increase the intensity of the transmitted beam by adjusting potentiometer R9 on the transmitter. You can adjust the gain of the transmitter by tweaking R7 and R8. Once the system is aligned and adjusted, increase the distance in 100- to 200-foot increments. When necessary, make adjustments in gain controls to pick up a clear voice.

After building and testing the long-range light wave communications link, here are some projects you can undertake to further your experimentation.

- Use the light wave link as a wireless public address (PA). Set up a microphone and transmitter on a podium or stand and aim it at a receiver and amplifier system. Make sure nothing blocks the beam from transmitter to receiver. Unlike a regular radio link, the light wave communications link doesn't need FCC certification.
- Build two links for a two-way communications system. As long as you take care to reduce unwanted reflections, you don't have to worry about conflicting radio frequencies.
- Use just the receiver to listen to modulated light. You can listen to the 120-cycle hum from a desklamp or the several-thousand-cycle buzz from multiplexed fluorescent/LED displays on microwave ovens, VCRs, and digital clocks. The designer of the light wave communications link, Roger Sonntag of General Science & Engineering, even suggests listening to the wings beating on small flying insects. As sunlight shines through their wings, the receiver picks up the slight variations in light intensity.

The light wave receiver exhibits extraordinary high gain and low noise. Within reason, you can substitute the photodiode for a number of other sensing elements. Try wrapping 20-gauge wire around an antenna element as shown in FIG. 15-14 to listen to the induced electromagnetic field of 20 Hz to 20 kHz (hearing range). Connect a microphone in place of the photodiode and you have a super-sensitive listening device.

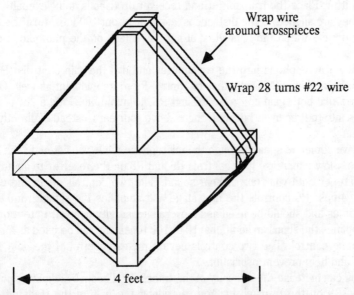

Wrap wire
around crosspieces

Wrap 28 turns #22 wire

|← 4 feet →|

Fig. 15-14. You can receive very low frequency (VLF) signals—from between about 10 and 40 kHz—using the long-distance receiver and this simple wire antenna. Typical of VLF signals are "pops" from distant thunder storms and "whistlers," which are of an unknown origin.

ELECTRONIC MODULATION OF HELIUM-NEON LASERS

Although it might seem otherwise, it's fairly easy to modulate the beam of a helium-neon laser and using only a handful of parts at that. Two approaches are provided here: both have an effective bandwidth of around 0 Hz to 3 kHz, making them suitable for most voice and some music transmission schemes.

Transformer

See the parts list in TABLE 15-6. A transformer placed in line with the high-voltage power supply and cathode of the tube can be used to vary the current supplied to the tube. This causes the intensity of the beam to vary. This is amplitude modulation, the same technique used in AM radio broadcasts.

Table 15-6. Parts List	1 70-volt PA transformer
for Transformer Modulator	1 1/8-inch miniature jack
	2 5-way binding posts (25 amp)
	1 Project box

Although you can use a number of transformers as the modulating element, Dennis Meredith of Meredith Instruments suggests you use a PA power output transformer. It's ideal for the job because of its high turns ratio (the ratio of wire loops in the primary and secondary). You wire the transformer in reverse to the typical application: the speaker terminals from a hi-fi or amplifier connect to the "output" of the transformer and the laser connects to the "input."

PA transformers are available from almost any electronics parts store, including Radio Shack, who offers a good one for under $5. PA transformers are rated by their voltage, usually either 35 or 70 volts. Get the higher voltage rating. There are several terminals on the transformer. Connect the speaker terminals to the common and 8 Ω terminals; connect the laser cathode, as shown in FIG. 15-15, to the common and one of the wattage terminals. Experiment with the wattage terminal that yields the most modulation. The prototype seemed to work best using the 5-watt terminal.

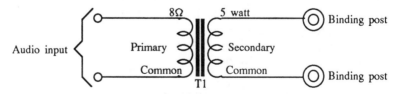

Fig. 15-15. Wiring diagram for the He-Ne laser transformer modulator.

The cathode passes some current, and touching its leads can cause a shock. Isolate the transformer and wires in a small project box. Five-way binding posts (fancy banana jacks) are used for the cathode connections; the audio input is an 1/8-inch miniature phone jack.

Transistor

Who wants to lug around a bulky and heavy transformer when you can provide modulation to the He-Ne tube using a simple silicon transistor? This next mini project provides a seed that you can use to design and build an all-electronic analog or digital laser communications link. I have not fully tested the upward frequency limits of the transistor modulator, but I successfully passed a 4 kHz tone through the prototype circuit using a 2 mW He-Ne tube.

You can employ just about any transistor, but I found for experimental purposes, a common 2N2222 signal transformer is adequate. Connect the transistor between the high-voltage power supply and cathode of the laser tube, as shown in FIG. 15-16.

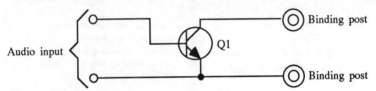

Fig. 15-16. Wiring diagram for the He-Ne laser transistor modulator. Experiment with different transistors and test the results. Both the transistor and transformer schemes require a well-amplified audio signal.

While the 2N2222 transistor is adequate for short-term testing, a high-voltage transistor mounting on a suitable heatsink is better. A number of high-voltage transistors are readily available such as those used as replacements for the horizontal output section of a TV. Radio Shack sells such a transistor, the 2SC1308, as do most other electronics stores.

The transistor is all you need for the basic setup, but you might want to add a 390 Ω resistor to the base of the transistor. To make the ''circuit'' more permanent, mount it on a small piece of perf board or wire it into one of your He-Ne laser enclosures.

Apply a well-amplified signal to the base of the transistor and aim the laser at the universal laser light receiver. You should hear sound. If the sound is weak, double-check your wiring and try turning up the volume. You might need one or two watts of power to produce a measurable amount of modulation.

You can build a completely portable He-Ne laser modulation system using a Walkman cassette player, IC amplifier, helium-neon tube, and 12-volt power supply. Suitable amplifier circuits appear later in this chapter. Put everything in a box and sling it around your shoulder.

ELECTRONIC MODULATION OF CW DIODE LASER

A constant-wave diode laser can be modulated using the circuit provided earlier in this chapter for the LED transmitter. Although it's always better to limit current to the laser using feedback from the monitor photodiode, this system provides a safety net because the laser is driven with pulses at the modulation frequency of about 40 kHz.

Laser diodes exhibit a great deal of divergence, so collimating optics are necessary if you want to use one in a light wave communications project. Many surplus cw diode lasers come with collimating optics or have suitable optics available for them (most are pulled from existing equipment such as compact disc players or bar code scanners). Alternatively, you can build your own collimator using a simple biconvex lens.

Mount the battery pack, modulating circuit, and laser in a project box. The box used in the prototype measured 6¹/4 by 3³/4 by 2 inches. Construction details for the project are shown in FIG. 15-17; the parts list is included in TABLE 15-7. Note the ¹/4-inch 20 bolt and threaded coupler. The coupler provides a socket that mates with most photographic tripods.

Be aware that aiming a laser diode is tough at best. Although the laser emits a deep red glow, the visible illumination is not enough to see in anything but absolute darkness. Looking directly into the laser for any length of time is decidedly a bad idea: the collimating lens acts to focus the light in a narrow beam. Don't let the red glow of the laser fool you. Your eye loses its sensitivity as it approaches the near-infrared band, but its susceptibility to damage from radiation is not lessened.

You can try aiming the laser using trial and error, but the results might be frustrating. Another approach is to use an infrared imaging card such as the one sold by Kodak. The card is coated with a substance that's sensitive to infrared

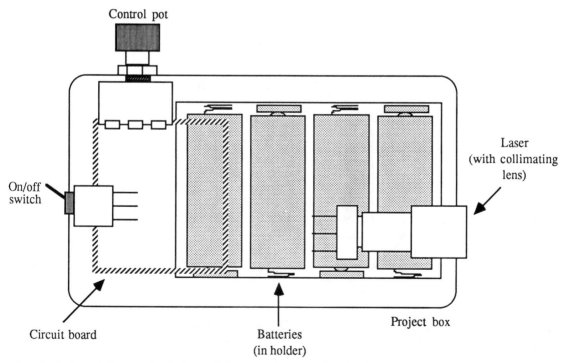

Fig. 15-17. Layout diagram for the laser diode transmitter. Use C or D size batteries for long-life performance.

Table 15-7. Parts List for CW
Modulated Laser Diode Transmitter

1	Laser diode with collimating optics (see Chapter 11)
1	Transmitter board (see FIG. 15-1)
1	Control potentiometer and knob
1	SPST switch
1	Battery holder—C size
1	Project box (approximately 3³/₄ by 6¹/₄ by 2 inches)

light. Directing an infrared source at the card causes it to glow. Before you use the card, you must first "charge" it under the white light of the sun or a desk lamp.

Still another approach is to use an infrared inverter tube. These are expensive "see-in-the-dark" devices used by police, soldiers, and voyeurs. The tube—usually mounted in a pair of binoculars, goggles, or glasses—blocks most visible light and amplifies infrared light. They are normally used with a separate infrared light source, but in this instance, the diode laser provides the needed IR radiation. Chapter 19 provides several approaches to building your own see-in-the-dark night scopes.

The phototransistor used with the receiver can by swamped by the power emitted by the diode laser. Use a pair of polarizers, as shown in FIG. 15-18, to control the amount of infrared radiation that reaches the phototransistor. A filter placed in front of the phototransistor can also increase sensitivity and reduce

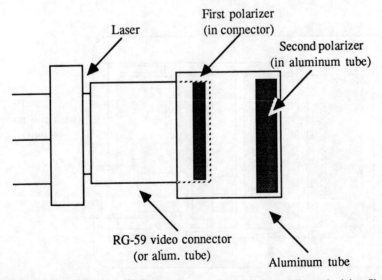

Fig. 15-18. The output beam of the laser can be controlled by adding polarizing films to an RG-59 video connector and brass tube, as shown in this idea from Forrest Mims. The same method can be used on the receiver phototransistor.

background noise. Infrared filters are available at most photographic stores and surplus is another good source. Many IR filters appear dark red or purple or even completely black. You might not be able to see through the filter, but it is practically transparent to near-infrared radiation.

AUDIO AMPLIFIER CIRCUITS

The universal receiver described at the beginning of this chapter has a built-in LM386 integrated amp. The sound output is minimal, but the chip is easy to get, cheap, and can be wired up quickly. It's perfect for experimenting with sound projects.

If you need more sound output or must amplify the audio input for the transformer or transistor modulator, try the circuit in FIG. 15-19 (see TABLE 15-8 for a parts list). You can use it instead of the LM386 in the receiver or in addition to it. The circuit is designed around an LM383 8-watt amplifier IC. The IC comes mounted in a TO-220-style transistor package, and you should use it with a suitable heatsink.

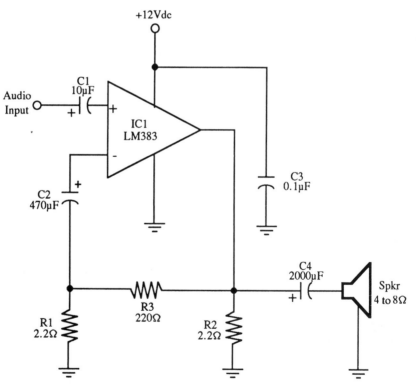

Fig. 15-19. An 8-watt audio amplifier designed around the LM383 integrated amplifier IC. The IC must be installed on a suitable heatsink.

Table 15-8. Parts List for
8-watt Audio Amplifier

R1,R2	2.2 Ω
C1	10 μF polarized electrolytic
C2	470 μF polarized electrolytic
C3	0.1 μF disc
C4	2000 μF polarized electrolytic
IC1	LM383 audio amplifier IC
SPKR	8 Ω speaker

All resistors are 5 to 10 percent tolerance, 1/4
watt. All capacitors are 10 to 20 percent tolerance, rated at 35 volts or more.

Figure 15-20 (parts list in TABLE 15-9) shows a higher output 16-watt version
using two LM383s. Note that the LM383 IC is functionally identical to the
TDA2002 power audio amplifier.

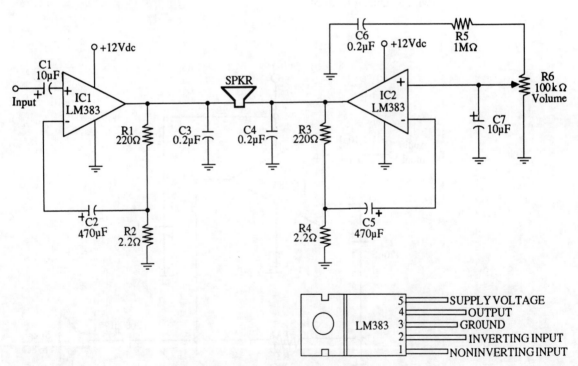

*Fig. 15-20. A 16-watt audio amplifier designed around two LM383 integrated amps. The ICs must be installed on
a suitable heatsink.*

Table 15-9. Parts List for
16-watt Audio Amplifier

R1,R3	220 Ω
R2,R4	2.2 Ω
R5	1 MΩ
R6	100 kΩ potentiometer
C1,C7	10 μF polarized electrolytic
C2,C5	470 μF polarized electrolytic
C3,C4,C6	0.2 μF disc
IC1,IC2	LM383 audio amplifier IC
SPKR	8 Ω speaker

All resistors are 5 to 10 percent tolerance, $1/4$ watt. All capacitors are 10 to 20 percent tolerance, rated to 35 volts or more.

16

Working with fiberoptics

Fiberoptics are everywhere. Many telephone calls—both local and long distance—are now carried at least part way by light shuttling through a strand of plastic or glass. Fiberoptics are now used on some high-end audio systems as a means to prevent digital signals from interfering with analog signals. And fiberoptics sculptures, in vogue in the late 1960s but coming back in style today, look like high-tech flowers that seem to burst out with brightly colored lights.

An optical fiber is to light what PVC pipe is to water. Though the fiber is a solid, it channels light from one end to the other. Even if the fiber is bent, the light will follow the path, altering its course at the bend, and traveling on. Because light acts as the information carrier, a strand of optical fiber no bigger than a human hair can carry the same information as about 900 copper wires. This is one reason why fiberoptics is used increasingly in telephone communications.

Laser light exhibits unique behavior when transmitted through optical fiber. In this chapter, you'll learn how to work with optical fibers, how to interface fibers to common LEDs and lasers, and many interesting applications of fiberoptic links.

HOW FIBEROPTICS WORK

The idea for optical fibers is over 100 years old. British physicist John Tyndall once demonstrated how a bright beam of light was internally reflected through a stream of water flowing out of a tank. Serious research into light transmission through solid material started in 1934, when Bell Labs was issued a patent for the light pipe.

In the 1950s, the American Optical Corporation developed glass fibers that transmitted light over short distances (a few yards). The technology of fiberoptics really took off in about 1970 when scientists at Corning Glass Works developed long-distance optical fibers.

All optical fibers are composed of two basic components, as illustrated in FIG. 16-1: the *core* and the *cladding*. The core is a dense glass or plastic material where the light actually passes through as it travels the length of the fiber. The cladding is a less dense sheath, also of plastic or glass, that serves as a refracting medium. An optical fiber might or might not have an outer jacket such as a plastic or rubber insulator used as protection.

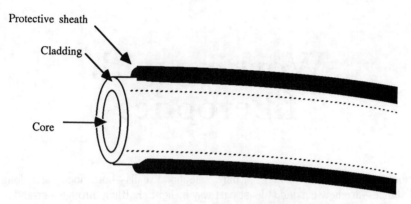

Fig. 16-1. Design of the typical optical fiber.

Optical fibers transmit light by *total internal reflection* (TIR). Imagine a ray of light entering the end of an optical fiber strand. If the fiber is perfectly straight, the light will pass through the medium just as it passes through a plate of glass. But if the fiber is bent slightly, the light will eventually strike the outside edge of the fiber.

If the angle of incidence is great (greater than the critical angle), the light will be reflected internally and will continue its path through the fiber. But if the bend is large and the angle of incidence is small (less than the critical angle), the light will pass through the fiber and be lost. The basic operation of fiberoptics is shown in FIG. 16-2.

Note the *cone of acceptance*; the cone represents the degree to which the incoming light can be off-axis and still make it into the fiber. The *angle of acceptance* (usually 30 degrees) of an optical fiber determines how far the light source can be from the optical axis and still manage to make it into the fiber. Though the angle of acceptance might be great, fiberoptics perform best when the light source (and detector) are aligned to the optical axis.

Optical fibers are made by pulling a minute strand of glass or plastic through a small orifice. The process is repeated until the strand is just a few hundred (or less) micrometers in diameter. Although single strands are sometimes used in special applications, most optical fibers consist of many strands bundled and fused together. There might be hundreds or even thousands of strands in one fused optical fiber bundle. Separate fused bundles can further be clustered to produce fibers that measure 1/16 inch or more in diameter.

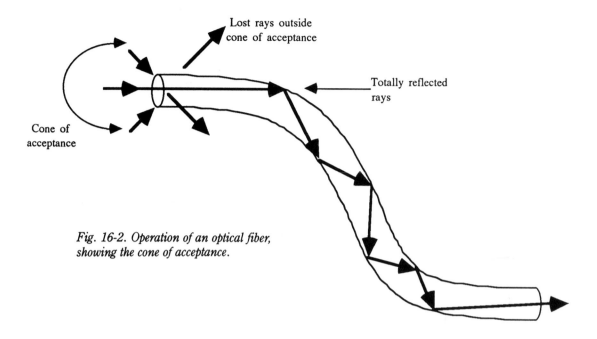

Fig. 16-2. Operation of an optical fiber, showing the cone of acceptance.

TYPES OF OPTICAL FIBERS

The classic optical fiber is made of glass, also called *silica*. Glass fibers tend to be expensive and are more brittle than stranded copper wire. But they are excellent conductors of light, especially light in the infrared region between 850 and 1300 nm.

Less expensive optical fibers are made of plastic. Though light loss through plastic fibers is greater than with glass fibers, they are more durable. Plastic fibers are best used in communications experiments with near-infrared light sources—the 780 to 950 nm range. This nicely corresponds to the output wavelength and sensitivity of commonplace infrared emitters and detectors.

Optical fiber bundles might be *coherent* or *incoherent*. The terms don't directly relate to laser light or its properties but to the arrangement of the individual strands in the bundle. If the strands are arranged so that the fibers can transmit an image from one end to the other, it is said to be coherent. The vast majority of optical fibers are incoherent, which means any image or particular pattern of light is lost when it reaches the other end of the fiber.

The cladding used in optical fibers might be one of two types—*step-index* and *graded-index*. Step-index fibers provide a discrete boundary between more dense and less dense regions between core and cladding. They are the easiest to manufacture, but their design causes a loss of coherency when laser light passes through the fiber; that is, coherent light goes in, and largely incoherent light comes out. The loss of coherency, which is due to light rays traveling slightly different paths through the fiber, reduces the efficiency of the laser beam. Still, it offers some practical benefits, as you'll see later in this chapter.

There is no discrete refractive boundary in graded-index fibers. The grading acts to refract light evenly, at any angle of incidence. This preserves coherency and improves the efficiency of the fiber. As you might have guessed, graded-index optical fibers are the most expensive of the bunch. Unless you have a specific project in mind (like a 10-mile fiber link), you shouldn't need graded-index fibers, even when experimenting with lasers.

WORKING WITH FIBEROPTICS

Where do you buy optical fibers? While you could go directly to the source, such as Dow Corning, American Optical, or Dolan-Jenner, a cheaper and easier way is through electronic and surplus mail order. Radio Shack sells a 5-meter length of jacketed glass fiber; Edmund Scientific offers a number of different types and diameters of optical fibers. You can buy most any length you need, from a sampler containing a few feet each of several types to a spool containing thousands of feet of one continuous fiber.

A number of surplus outfits, such as Jerryco and C&H Sales, offer optical fibers from time to time (stocks change, so write for the latest catalog before ordering). You might not have much choice over the type of fiber you buy, but the cost should be more than reasonable.

Although it's possible to use an optical fiber by itself, serious experimentation requires the use of fiber couplings and connectors. These couplings are mechanical splices used to connect two fibers together or to link a fiber to a light emitter or detector. Be aware that good fiberoptic connectors are expensive. Look for the inexpensive plastic types meant for non-military applications. You can also make your own homebuilt couplings. Details follow later in this chapter.

Optical fibers can be cut with wire cutters, snippers, or even a knife. But exercise care to avoid injury from shards of glass that could fly out when the fiber is cut (plastic fibers don't shatter when cut). Wear heavy cotton gloves and eye protection when working with optical fibers. Avoid working with fibers around food serving or preparation areas because the bits of glass could inadvertently settle on food, plates, or eating utensils and cause bodily harm.

One good way to cut glass fiber is to gently nick it with a sharp knife or razor and then snap it in two. Position the thumb and index finger of both hands as close to the nick as possible and break the fiber with a swift downward motion (snapping upwards increases the chance of glass shards flying towards your face or body).

Whether snapped apart or cut, the end of the fiber should be prepared before splicing it to another fiber or connecting it to a light emitter or detector. The ends of the cut fiber can be polished using extra fine grit aluminum oxide wet/dry sandpaper (330 grit or higher). Wet the sandpaper and gently rub the end of the fiber on it. You can obtain good results by laying the sandpaper flat on a table and holding the fiber in your hands. Rub in a circular motion and take care to keep the fiber perpendicular to the surface of the sandpaper. If the fiber is small for you to handle, mount it in a pin vise.

Inspect the end of the fiber with a high-powered magnifying glass (a record player stylus magnifier works well). Shine a light through the opposite end of the fiber. The magnified end of the fiber should be bright and round. Recut the fiber if the ends look crescent shaped or have nicks in them.

FIBEROPTIC CONNECTORS

Commercially made fiberoptic connectors are pricey, even the plastic AMP Optimate Dry Non-Polish (DNP) variety. A number of mail-order firms, such as Digikey-Key and Jameco, offer splices, for joining two fibers, and connectors, for attaching the fiber to emitters and detectors. Depending on the manufacturer and model, the connectors are made to work with either the round- or flat-style phototransistors and emitters.

If you're working with lasers, you can make your own connectors and splices for homebrew laser experiments. Figure 16-3 shows several approaches.

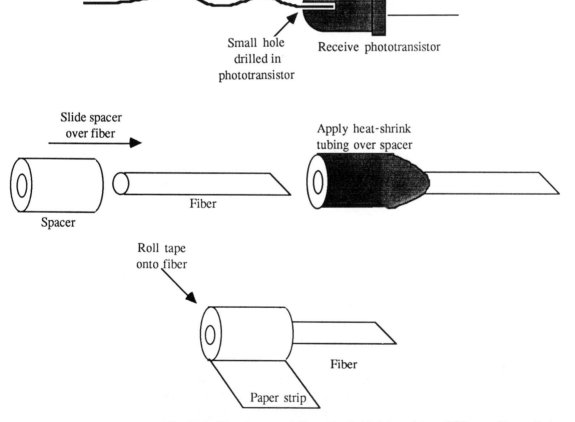

Fig. 16-3. Ways to connect fiberoptics to phototransistors, LEDs, and laser diodes.

An easy way to splice fibers is to use small heat-shrink tubing. Cut a piece of the tubing to about $1/2$ inch. After properly cutting (and polishing) the ends of the fiber, insert them into the tubing, and apply light heat to shrink it to fit. Best results are obtained when the tubing is thick-walled.

Optical fibers can be directly connected to photodiodes by drilling a hole in the casing, inserting the fiber, and bonding the assembly with epoxy. Be sure that you don't drill into the semiconductor chip itself or you'll ruin it. Keep the drill motor at a fairly slow speed to avoid melting the plastic casing. Work slowly.

A strand of optical fiber can be held in place using a pin vise (remove the outer jacket, if any, and tighten the chuck around the fiber) or by using solderless insulated spade tongues. These are designed for terminating copper wire, but can be successfully used to anchor almost any size of optical fiber to a bulkhead. The laser, be it He-Ne or semiconductor, can then be aimed directly into the cone of acceptance of the fiber.

Spade tongues, as shown in FIG. 16-4, are available in a variety of sizes to accommodate different wire gauges. Use #6 (22 to 18 gauge) for small optical fibers and #8 (16 to 14 gauge) for larger fibers. Secure the fiber in the spade tongue by crimping with a crimp tool. Do not exert too much pressure or you will deform the fiber. If the fiber is loose after crimping, dab on a little Super Glue epoxy to keep everything in place.

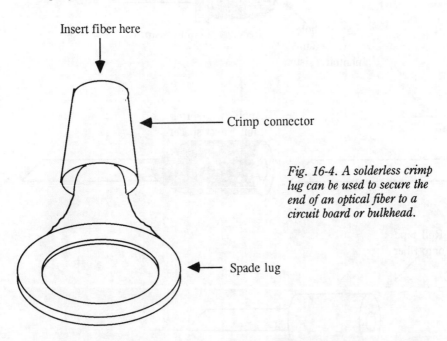

Insert fiber here

Crimp connector

Fig. 16-4. A solderless crimp lug can be used to secure the end of an optical fiber to a circuit board or bulkhead.

Spade lug

The FLCS package, shown in FIG. 16-5, is a low-cost fiberoptic connector available from a variety of sources, including Radio Shack, Circuit Specialists, and many Motorola semiconductor representatives. It can be easily adapted for use with laser diodes by removing the back portion. This exposes the optical fiber.

Fig. 16-5. The popular plastic "FLCS" optical fiber connector.

After removing the emitter diode, file or grind the back end of the connector, as shown in FIG. 16-6. You can also drill out the back of the connector with a 5/32-inch bit. Mount the connector and laser on a circuit board or perf board, being careful to align the laser so that its beam directly enters the end of the fiber.

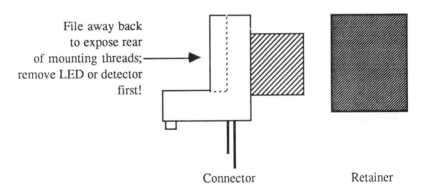

File away back to expose rear of mounting threads; remove LED or detector first!

Connector   Retainer

Fig. 16-6. How to modify an "FLCS" connector to open the back portion. With the back open, you can shine a laser beam inside the connector and down the optical fiber.

BUILD AN OPTICAL DATA LINK

A number of educational fiberoptic kits are available (see Appendix A for sources) and at reasonable cost. For example, the Edu-Link (available through

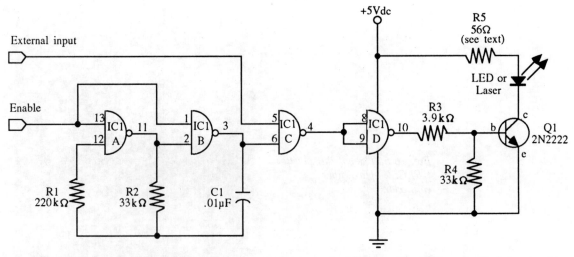

Fig. 16-7. Schematic for experimenter's data transmitter using optical fibers and an LED or double-heterostructure laser diode. Transmission frequency of the free-running oscillator is approximately 3 kHz.

Advanced Fiberoptics Corp., Circuit Specialists, Edmund, and others) contains pre-etched and drilled PCBs for a small transmitter and receiver, along with a short length of jacketed plastic fiber. The emitter LED and photodetector are housed in plastic connectors, and the circuits provide input and output pins for sending and receiving digital data.

You can use the Edu-Link or the circuits shown in FIGS. 16-7 and 16-8 to build your own fiberoptic transmitter and receiver. Parts lists for the two circuits are provided in TABLES 16-1 and 16-2. Note that the Edu-Link, as well as

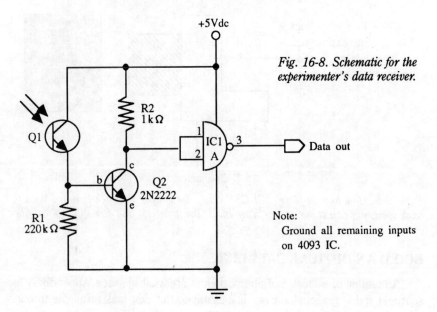

Fig. 16-8. Schematic for the experimenter's data receiver.

Note:
Ground all remaining inputs on 4093 IC.

Table 16-1. Parts List for Optical Fiber Transmitter	R1	220 kΩ
	R2	33 kΩ
	R3	3.9 kΩ
	R4	33 kΩ
	R5	56 Ω (nominal)
	C1	0.01 μF disc
	IC1	4093 CMOS NAND gate IC
	Q1	2N2222 transistor
	Misc.	Laser diode

All resistors are 5 to 10 percent tolerance, 1/4 watt. All capacitors are 10 to 20 percent tolerance, rated at 35 volts or more.

Table 16-2. Parts List for Optical Fiber Receiver	R1	220 kΩ
	R2	1 kΩ
	IC1	4093 CMOS NAND gate IC
	Q1	Infrared phototransistor
	Q2	2N2222 transistor

All resistors are 5 to 10 percent tolerance, 1/4 watt.

other fiberoptic data communications kits, use a resistor to limit current to the emitter LED. The value of the resistor is typically 100 to 220 ohms. At 100 ohms and a 5 Vdc supply, current to the emitter LED is 35 milliamps (assuming a 1.5-volt drop through the LED).

If you're working with lasers, that isn't enough to operate a dual-hetero-structure laser diode, so the resistor must be exchanged for a lower value. A 56-ohm resistor provides about 62 mA of current to the laser diode; a 47-ohm resistor delivers about 74 mA of current. A feedback mechanism for controlling the output of the laser diode is not strictly required because the device is used in pulsed mode.

If you are using the Edu-Link system and have not yet assembled the boards, construct the transmitter and receiver circuits. If you're using a a diode laser instead of the standard emitter, don't mount the emitter LED on the board. Rather, connect a Sharp LT020, LT022, or equivalent low-power dh laser diode to the transmitter circuit. Dismantle the emitter and modify the connector for use with the laser diode, as detailed in the section above.

The transmitter circuit of FIG. 16-7 includes a built-in oscillator. Trigger it by placing the ENABLE and DATA IN lines high. Next, connect a logic probe to the DATA OUT pin of the receiver. Apply power to the two circuits and watch for the pulses at the DATA OUT pin. If pulses are not present, double-check your work and make sure that the fiberoptic connection between the transmitter and receiver is aligned and secure and that the output laser diode is properly aligned to the fiber. Remove power and disconnect the logic probe.

In normal operation, place the ENABLE pin low and connect any serial data to the DATA IN pin. Be sure the incoming signal is TTL compatible or does not exceed the supply voltage of the transmitter. You can use the circuit to transmit and receive ASCII computer data, remote control codes, or other binary information.

The CMOS chips in the transmitter and receiver can be operated with supplies up to 18 volts, but care must be taken to adjust the value of R5, the resistor that limits current to the LED or laser diode. An increase in supply current must be matched with an increase in resistance, or the laser diode could be damaged.

Use Ohm's Law to compute the required value (R = E/I). Assume a 1.2- to 1.5-volt drop through the diode.

EXAMPLE:
   Supply voltage = 12 volts
   Voltage drop through laser diode = 1.5 volts
   Working voltage = 10.5 volts (12 – 1.5)
   Desired current (60 to 80 mA) = 70 mA
   Resistor to use = 150 ohms (10.5 / 0.070 = 150)

MORE EXPERIMENTS WITH LIGHT AND FIBEROPTICS

Besides data transmission, fiberoptics can be used to:

- Transmit analog data (modulate an LED, diode laser or He-Ne laser, and pass it through the fiber)
- Detect vibration and motion
- Route light to remote locations
- Separate a single beam of light into several shafts of light

This is only a partial listing; there are literally dozens of useful and practical applications for fiberoptics. Some hands-on projects follow.

Vibration and movement detection

A fiberoptic strand doesn't make the best medium for transmitting laser-light analog data. Why? The fiber itself can contribute to noise. As mentioned earlier, when a beam of coherent laser light is passed through a conventional step-index optical fiber, the rays travel different paths and the light that exits is largely incoherent. You can see the effects of this interference by shining a helium-neon laser through an optical fiber. Point the exit beam at a white piece of paper and you'll see a great deal of speckle. The speckle is the constructive and destructive interference, created inside the fiber, as the laser light rays travel from one end to the other.

This interference—which is most prominent in low-cost plastic optical fibers—is normally an undesirable side effect. However, it can be put to good use as a vibration and motion detection system. Connecting a phototransistor to the exit end of the fiber lets you monitor the light output. Movement of the fiber causes a change in the way the light is reflected inside, and this changes the coherency (or incoherency, depending on how you look at it) of the beam. A simple audio amplifier, connected to the phototransistor, lets you hear the movement.

Figure 16-9 shows a setup you can use to test the effects of fiberoptic vibration and motion. A parts list for the system is provided in TABLE 16-3. More advanced projects using this technique can be found in Chapter 17, Experiments in laser seismology. The noise is sometimes a hiss and sometimes a "thrum." Depending on the length of the fiber and type of motion, you might also hear low- and high-pitched squeals. These squeals can change pitch as the fiber or phototransistor is moved slowly.

The squeals are caused by the *Doppler effect* and *optical heterodyning*, a process whereby two rays of light at slightly different frequencies meet. The two basic (or fundamental) frequencies mix together, creating two additional frequencies. One is the sum of the two fundamental frequencies and the other is the difference.

Optical heterodyning is most conspicuous when only two coherent rays of light meet. In an optical fiber, dozens and even hundreds of rays of internally reflected laser light might meet at the phototransistor, and the result might sound more like cacophonous noise than a distinct tone.

You might also notice a varying tone when sampling the beam directly from the laser. Even though lasers are highly monochromatic, they can still emit several frequencies of light, each spaced only fractions of a nanometer apart. As these frequencies meet on the surface of the photodetector, they cause heterodyning or *beat frequencies*. When the difference frequency is 20 kHz or less, you can hear them. You can precisely measure the difference frequencies using an oscilloscope. The tones heard when sampling the beam directly from a laser are most prominent with short tubes and when they are first turned on.

Whether you are listening to the change in noise level or varying tone when an optical fiber is moved, you can apply the technique in a number of different ways:

- Place a length of fiber around the perimeter of your home. Motion caused by cars, people, and animals will be transferred to the fiber and detected as a change in sound level or pitch.
- A similar vibration sensor can be built into an antitheft device for an automobile. Wind a coil of optical fiber around a piece of wood or metal and attach the assembly to the frame of the car. Any motion (collision, towing, or jacking) will be detected. A change in sound level trips the alarm.
- A short strand of optical fiber suspended in air or a liquid can detect motion of the surrounding medium. For example, you can build an alarm system that alerts you if water in a pipe or trough stops flowing.

A

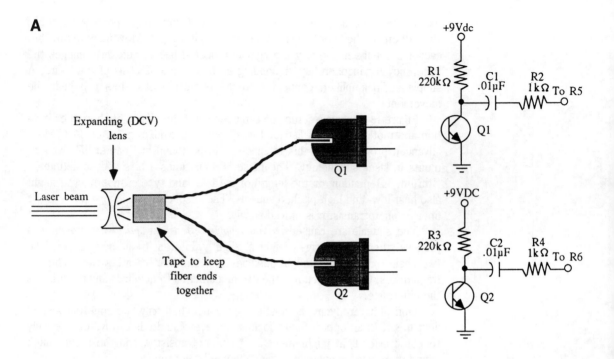

B

Notes:
Vary R5 and R6 to adjust
sound levels from Q1 and Q2.

Fig. 16-9. Hookup diagram for connecting a pair of phototransistors to an op amp and audio amplifier. (A) Physical connection showing beam expansion into both fibers and phototransistor hookup; (B) op amp and audio amplifier.

Table 16-3. Parts List for Vibration Detection System

R1,R3	220 kΩ
R2,R4	1 kΩ
R5,R6,R8	10 kΩ potentiometer
R7	1 MΩ potentiometer
C1,C2	0.01 μF disc
C3	220 μF polarized electrolytic
IC1	LM741 operational amplifier IC
IC2	LM386 audio amplifier IC
Q1,Q2	Infrared phototransistors
J1	Audio output jack (¹/4 or ¹/8 inch)
Misc.	Double concave (DCV) expansion lens (approx. 6 to 10 mm diameter)

All resistors are 5 to 10 percent tolerance, ¹/4 watt. All capacitors are 10 to 20 percent tolerance, rated at 35 volts or more.

Separating beam with optical bundles

By grouping together one end of two or more fused bundles, you can separate the beam of an LED or laser into many individual sub-beams. The light enters the common end (where all the bundles are tied together) and exits the opposite end of each individual fiber. Some optical fibers come premade with four or more grouped strands (used most often in automotive dashboard application) or you can make your own.

The pencil-thin beam of the typical He-Ne laser is too narrow to enter all the fibers at once, so the beam must be expanded. Place a biconcave or plano-concave lens in front of the entrance to the bundles. Adjust the distance between the lens to the bundle until the beam is spread enough to enter all the fibers (see FIG. 16-10).

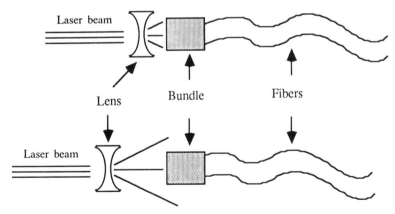

Fig. 16-10. Expand the beam by moving the lens with respect to the fiber bundle.

Split bundles can be used to experiment with optical heterodyning as well as to split the beam of one laser into several components. Each beam can be used in a separate optical fiber system. For example, you must use a four-fiber split bundle to provide illumination for four fiberoptic intrusion detection systems. Each subsystem is placed in a quadrant around the protected area and has its own phototransistor, making it easier to locate the area of disturbance.

17

Experiments in laser seismology

Earthquakes are among the most frightening natural phenomenon, feared most for their stealthy suddenness. Even though geologists and seismologists have been measuring earthquakes for decades with sophisticated instruments on land, in the air, and even in space, predicting tremors is an inexact science. The best seismologists can do is warn that a "big earthquake is due soon"—expect it anytime between now and the next century.

Fortunately, massive earthquakes on land are rare. Many of the largest earthquakes occur out at sea, and while they can cause enormous tidal waves (such as the Japanese tsunami), earthquakes at sea seldom topple buildings or swallow up people. Earthquakes on the West Coast in southern California are almost a dime a dozen, but most of them are so faint that they cannot be felt.

Earthquakes can happen anywhere, and even tremors that occur hundreds of miles away can be detected with the proper instruments. Most seismographs use complex and massive electromagnetic sensors to detect earthquakes, both near and far, but you can readily build your own compact seismograph using a laser and a coil of fiberoptics. When constructed properly, a laser/fiberoptic seismograph can be just as sensitive as an electromagnetic seismograph costing several thousand dollars.

This chapter covers construction details of two types of laser/fiberoptic seismographs as well as useful information on how to attach the seismograph to a personal computer.

THE RICHTER SCALE

A number of scales are used to quantify the magnitude of earthquakes. The best known and most used is the Richter scale, named after Charles F. Richter, a pioneer in seismology research. The Richter scale is a logarithmic measuring system that has a scale from 1 to 10 (and theoretically from 10 to 100), where each increase of 1 represents a tenfold increase in earthquake magnitude. However, the actual energy released by the earth during the quake can be anywhere between about 30 and 60 times for each increase of one digit. Each numeral is further broken down into units of 10, so earthquakes are often cited as 4.2 or 5.6 on the Richter scale.

To give you an idea of how the Richter scale works (and why it can cause confusion), consider the difference between a 3.0 and 5.0 earthquake. Both are small and therefore difficult to detect without instruments (although some people say they can feel the swaying motion of a 4.0 earthquake). But since magnitude is increased logarithmically by a factor of 10 per numeral and the actual energy released can be up to 60 times as great, the difference between a 3.0 and 5.0 earthquake is quite large. That is, the 5.0 earthquake is 10 × 10 (or 100) times more powerful than the 3.0 earthquake.

Similarly, a 6.0 earthquake can cause extensive structural damage and close down buildings for repair, and an earthquake measuring 7.0, if it continues for any length of time, can result in massive destruction of buildings, bridges, and roads. An earthquake measuring 9.0 or 10.0 would level any town.

Most earthquakes occur at a *fault*, a crack or fissure in the earth's crust. The majority of earthquakes occur at the boundaries of crustal plates. These plates slip and slide over the earth's inner surface. Sudden motion in these plates is released as an earthquake. Major faults, such as the San Andreas in California, create many thousand ''mini faults'' or fractures that spread out like cracks in dried mud. These fractures are also responsible for earthquakes. In fact, the two largest earthquakes in southern California, occurring in 1971 and 1987, were caused by relatively small faults laying outside the San Andreas line.

Though faults are long—some measure thousands of miles—the earthquake occurs at a specific location along it. This location is the epicenter. The magnitude of earthquakes are measured at the epicenter (or more accurately, at a standard seismographic station distance of 100 km or 62 miles), where the amount of released energy is the greatest. The shock waves from the earthquake fan out and lose their energy the further they go. Obviously, there can't be a seismograph every 50 or even 100 miles along a fault to measure the exact amplitude of an earthquake. When the epicenter is some distance from a seismograph, its magnitude is inferred based on its strength at several nearby seismograph stations, past earthquake readings, and the geological makeup of the land in between.

This accounts for the uncertainty of the exact magnitude of an earthquake immediately after it has occurred and why different seismologists can arrive at different readings. It takes some careful calculations to determine an accurate Richter scale reading for an earthquake, and the precise measurement could be debated for months or even years after the tremor.

HOW ELECTROMAGNETIC SEISMOGRAPHS WORK

The most common seismograph in use today is the electromagnetic variety, which uses a sensing element not unlike a dynamic microphone. Basically, the case of the seismograph is a large and heavy magnet. Inside the case is a core, consisting of a spool of fine wire. During an earthquake, the spool bobs up and down, inducing an electromagnetic signal through the wires. A similar effect occurs in a dynamic microphone. Sound vibrates a membrane, which causes a small voice coil (spool of wire) to vibrate. The voice coil is surrounded by a magnet, so the vibrational motion induces a constantly changing alternating current in the wire. As the sound varies, so does the polarity and strength of the alternating current.

To prevent accidental readings of surface vibration, the seismograph is buried several feet into the ground and sometimes attaches to the bedrock. In cases, it is encased in concrete or secured to a cement piling sunk deep into the ground. Wires lead from the seismograph to a reading station, which might be directly above or several miles away. Telephone lines, radio links, or some other means connect distant seismographs to a central office location.

The signal from the seismograph is amplified and applied to a galvanometer on a chart recorder (the galvanometer is similar to the movement on a volt-ohmmeter). The galvanometer responds to electrical changes induced by the moving core of the seismograph. The bigger the movement, the larger the response. Attached to the galvanometer is a long needle that applies ink to a piece of paper wound around a slowly rotating drum.

An advance over the chart recorder is the computer interface. The pulses from the seismograph are sent to an analog-to-digital converter (ADC), which connects directly to a computer. The ADC transforms the analog signals generated by the seismograph into digital data for use by the computer. Software running on the computer records each tremor and can perform mathematical analyses.

LASER/FIBEROPTIC SEISMOGRAPH BASICS

The laser/fiberoptic seismograph (hereinafter referred to as the laser/optic seismograph) doesn't use the electromagnetic principle to detect movement in the earth. Though there are several ways you can implement a laser/optic seismograph, we'll concentrate on just one that offers a great deal of flexibility and sensitivity. The system detects the change in coherency through a length of fiberoptics.

As discussed in Chapter 16, Working with fiberoptics, when a laser beam is transmitted through a step-index optical fiber, some of the waves arrive at the other end before others. This reduces the coherency of the beam in proportion to the design of the fiber, its length, and the amount of curvature or bending of the fiber. Given enough of the right optical fiber, a laser beam could emerge at the opposite end that is totally incoherent.

It is not our intent to completely remove the coherency of a laser beam but just to alter it slightly through a length of 10 or 20 feet of fiber. Movement or vibration of the fiber causes a displacement of the coherency, and that displacement can be detected with a phototransistor. You can even hear this change in coherency by connecting the phototransistor to an audio amplifier. The ''hiss'' of the light coming through the fiber changes pitch and makes odd thuds, pings, and thrums as the fiber vibrates. The sound settles as the fiber stops moving or vibrating. In a way, the optical fiber makes a unique form of interferometer that settles quickly after the external vibrations have been removed.

Reducing local vibrations

The laser/optic seismograph is susceptible to the effects of local vibrations, movement caused by people walking or playing nearby, passing cars, trucks, and trains, or even the vibration triggered by the sound of a jet passing overhead.

To be most effective, the seismograph should be placed in an area where it won't be affected by local vibrations. Those living on a ranch or the outskirts of town will have better luck at finding such a location than city dwellers or those conducting earthquake experiments in a school or other populated area.

Even if you can't move away from people and things that cause vibration, you can reduce its effects by firmly planting the seismograph in solid ground. Avoid placing it indoors, especially on a wooden floor. Most buildings are flexible, and not only do they readily transmit vibrations from one location to the next, they act as a spring and/or cushion to the movement of an earthquake, improperly influencing the readings.

The cement flooring or foundation of the building is only marginally better. Small vibrations easily travel through cement, so if you attach your seismograph to the floor in your room, you are likely to pick up the movement of people walking around in the livingroom and kitchen.

The best spot for a seismograph is attached to a big rock out in the back yard, away from the house. Lacking a rock, you can fasten the seismograph to a cement piling and then bury some or all of the piling into the ground. You can also spread out four to eight cement blocks (about 75 cents each at a builder's supply store) and partially bury them in the ground. Fill the center of the blocks with sand and mount the seismograph on top. Other possible spots include (test first):

- The base of a telephone pole
- A heavy fence post
- A brick retaining wall or fence
- The cement slab or a separate garage, work shop, or tool shed

CONSTRUCTING THE SEISMOGRAPH

Cut a piece of optical fiber (jacketed or unjacketed) to 15 feet long. Polish the ends as described in Chapter 16. Using a small bit (one that matches the diameter of the fiber), drill a hole in the top of a phototransistor. Be sure to drill directly over the chip inside the detector, but do not pierce through to the chip. Epoxy the fiber in place. Alternatively, you can terminate the output end of the fiber using a low-cost FLCS-type connector. You can also use a modified FLCS connector for the emitter end of the fiber (as detailed in Chapter 16) or one of the other mounting techniques described.

Put the fiber aside and construct the base following the diagram in FIG. 17-1. A parts list is included in TABLE 17-1. You can use metal, plastic, or wood for the base, but it should be as dimensionally sturdy as possible. The prototype used 3/16-inch thick acrylic plastic. Cut the base to size and drill the post and mounting holes as shown. Insert four 1/4-inch 20 by 3-inch bolts in the post holes. Starting at one post, thread the fiber around the posts in a counterclockwise direction (see FIG. 17-2). Leave 1 to 2 feet on either end to secure the laser and photodiode. If the fiber slips off the posts, you can secure it using dabs of epoxy.

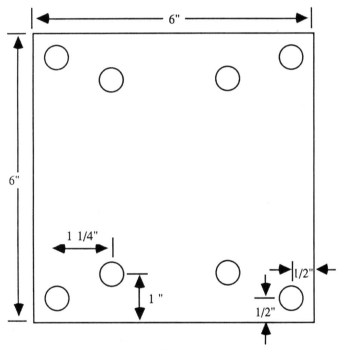

Fig. 17-1. Drilling and cutting guide for the laser/optic seismograph laser.

Table 17-1. Parts List for Fiberoptic Seismograph

1	6-inch-square acrylic plastic base (3/16 inch thick)
4	3-by-1/4-inch 20 carriage bolt, nuts, washers
1	Laser diode on heatsink
1	Sensor board (see FIG. 17-4)
Misc.	15 feet (approx.) jacketed or unjacketed fiberoptics, connector for receiver photodiode, connector or attachment for laser to fiber

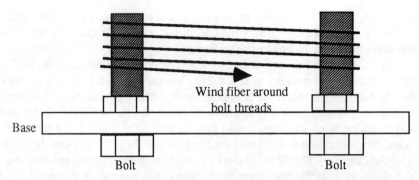

Fig. 17-2. How to wind the optical fiber around the bolts.

Mount the photodetector and laser diode (on a heatsink, as shown in FIG. 17-3) in the center of the platform. Alternatively, you can use a He-Ne laser as the coherent light source. Mount the laser tube securely on a separate platform and position the end of the optical fiber so that it catches the beam.

You can use the universal laser light detector presented in Chapter 15, Lightwave communications, to receive and amplify the laser light intercepted by the phototransistor. You can use either a pulsed or cw drive power supply for the laser diode. Schematics for these drives appear in Chapter 11, Working with laser diodes. In either case, be sure that the laser diode does not receive too

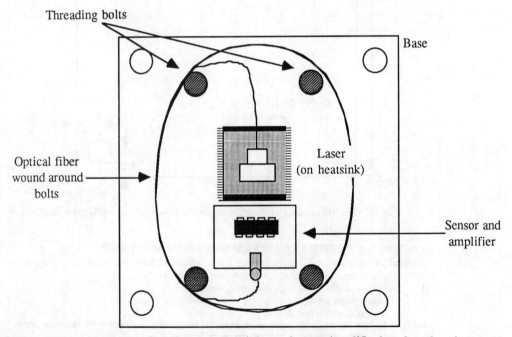

Fig. 17-3. Arrangement of fiber, laser (on heatsink), and sensor/amplifier board on the seismograph base.

much current. In pulsed mode, the laser doesn't operate at peak efficiency, but it is not required in this application.

Note that when using He-Ne tubes, you don't need excessive power—a 0.5 or 1 mW He-Ne is more than enough. This simplifies the power supply requirements and allows you to operate the seismograph for a day or two on each charge on a pair of lead-acid or gelled electrolyte batteries.

Test the seismograph by turning on the laser and connecting the output of the amplifier to a speaker or pair of headphones. A small vibration of the base should cause a noticeable thrum or hiss in the audio output. You might need to adjust the control knob on the amplifier to turn the sound up or down.

The mounting holes allow you to attach the seismograph to most any stable base. Whatever base you use, it should be directly connected to the earth. Use cement or masonry screws to attach the seismograph to concrete or a concrete pylon. When you get tired of listening to earthquake vibrations, connect the output of the amplifier to a volt-ohmmeter or centered current or voltmeter. Remove the ac coupling capacitor on the output of the amplifier.

COMPUTER INTERFACE OF SEISMOGRAPH SENSOR

A small handful of readily available electronic parts are all that is necessary to convert the voltage developed by the solar cell or phototransistor into a form usable by a computer. The circuit shown in FIG. 17-4 uses the TLC548 serial ADC, connected to a Commodore 64. The Commodore 64 provides the timing pulses, so only a minimum number of parts are required. See TABLE 17-2 for a parts list for the circuit.

Construct the TLC548 circuit on a perforated board using soldering or wire-wrapping techniques. You can measure the voltage directly from the phototransistor, or you can add the LM331 circuit if you are experimenting with the coherency change seismograph.

Software

The software is relatively simple. The BASIC program listing appears below.

```
10   POKE 56579, 255
20   POKE 56577, 0
30   POKE 56589, 127
40   FOR N = 0 TO 7
50   POKE 56577, 0
60   POKE 56577, 1
70   NEXT N
80   IF (PEEK (56589) AND 8) = 0 THEN 80
90   N = PEEK (56588)
100  PRINT N;
110  POKE 65677, 2
120  GOTO 40
```

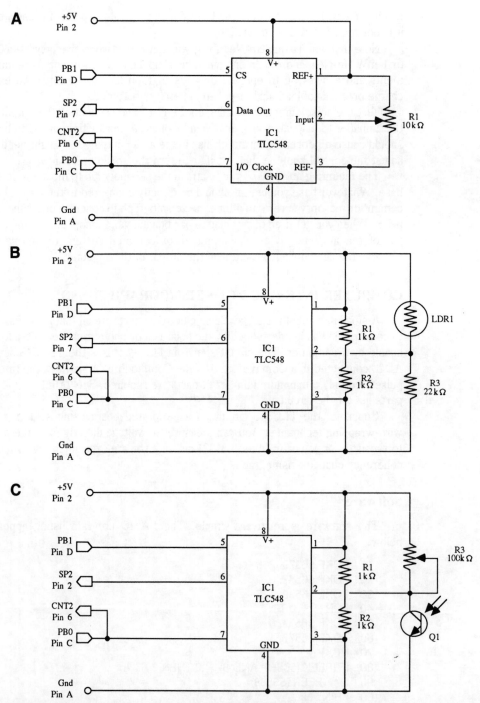

Fig. 17-4. Hookup diagrams for connecting the TLC548 serial analog-to-digital converter to a Commodore 64 computer. (A) Test circuit (vary R1 and watch change in values). (B) Interfacing the circuit with a photoresistor. (C) Interfacing the circuit with a phototransistor. Adjust R3 to vary the sensitivity.

Table 17-2. Parts List for Commodore 64 Seismograph ADC

Basic
IC1   TLC548 serial ADC
R1   10 kΩ potentiometer

Light-Dependent Resistor
IC1   TLC548 serial ADC
R1,R2   1 kΩ resistor
R3   22 kΩ resistor

Phototransistor
IC1   TLC548 serial ADC
R1,R2   1 kΩ resistor
R3   100 kΩ potentiometer
Q1   Infrared phototransistor
Misc.   12/24 pin connector for attaching to Commodore 64 User Port

You'll probably want to collect a number of samples and either print them out for future reference or graph them in a chart. Such programs are beyond the scope of this book, but if you are interested in pursuing the subject, you can check back issues of magazines that cater to owners of the Commodore 64.

18

Working with
xenon flash tubes

"Say cheese!" We're all familiar with the ritual of posing for a picture: you stand awkwardly with a silly grin on your face for an interminable period until the brilliant discharge of the camera's flash indicates the picture has been taken and the anguish of posing is over.

During that brief microsecond when the room fills with the bright white light of the flash and your image is indelibly embossed on film, have you ever wondered how the electronic strobe apparatus worked? Many pocket cameras now come with built-in miniature electronic flashes that are scarcely larger than a roll of film. How do they cram so much light energy into such a little package?

This chapter details the inner workings of electronic strobes using xenon-filled flash tubes. You'll learn the basics of operation and how to build your own variable-speed repeating strobe light.

THE XENON LAMP

A xenon lamp ordinarily consists of a tube capped on both ends with metal electrodes and filled with xenon gas. The shape of the tube can take many forms: straight (as with a fluorescent light tube), spiral, U-shaped, or almost any other configuration desired by a manufacturer.

There are two types of xenon flash lamps: *continuous* and *pulsed*. Continuous xenon tubes work on the same principle as the carbon arc, but instead of the arc electrodes being exposed to air, they are sealed in a glass envelope containing xenon gas (see FIG. 18-1). Pulsed xenon tubes emit a brief but exceptionally bright flash of light. As shown in FIG. 18-2, pulsed tubes lack the tungsten discharge terminals as found in continuous xenon tubes, and they are generally simpler in design. However, pulsed tubes must be far more rugged than their continuous counterparts. The reason is that the sudden shock wave caused by the near-instantaneous discharge of light can readily rupture a thin

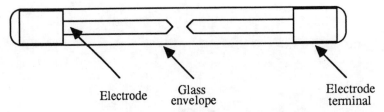

Fig. 18-1. *Construction of a continuous-output xenon tube.*

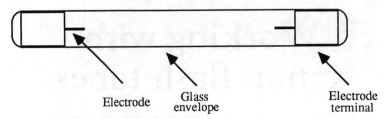

Fig. 18-2. *Construction of a pulsed-output xenon flash tube.*

glass envelope. We'll concentrate mainly on the pulsed type of xenon lamp (also called the *flash tube* or *strobe*) because it offers the most versatility to the weekend gadgeteer.

One of the properties of the xenon flash tube is the color of its light. The wavelengths of light emitted by the flash tube are very constant over the visible region. When a xenon tube fires, you see the light not as individual colors of the spectrum, but as a brilliant white flash. All the colors of the rainbow contribute to the white lightning.

POWERING THE LAMP

The xenon flash tube requires rather high voltages to operate. Because we only need the high voltages for a brief period of time, however, the power supply requirements aren't as stringent as they are, say, for a helium-neon laser tube.

The typical xenon tube power supply consists of three parts: high-voltage generator, discharge capacitor, and trigger, shown in the block diagram in FIG. 18-3. All work together to provide a brief pulse of energy that is released by the xenon tube in the form of light.

High-voltage generator

Flash tubes require several hundred volts to even begin to generate light. The high voltages are required to ionize the xenon gas trapped inside the glass envelope of the tube. In many xenon flash tube circuits, high voltage is generated from a dc battery cell (1.5 to 12 volts on average) and stepped up with an

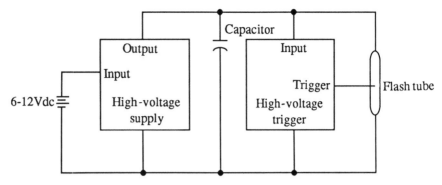

Fig. 18-3. Block diagram of a xenon flash power supply.

oscillator and transformer. Such a circuit trades low voltage and relatively high current for high voltage at very low current (on the order of a few microamps or milliamps).

Discharge capacitor

The discharge capacitor in the average strobe light circuit typically has a value of a 10 to 200 microfarads and a breakdown voltage of 300 to 400 volts. The capacitor serves to simply collect the energy provided by the high-voltage power supply. The discharge capacitor is required because the high-voltage stepup transformer can't instantaneously deliver enough energy directly to the xenon tube. Depending on the value of the capacitor and the efficiency of the power supply, enough energy can be stored in the cap in a 2 to 5 second period to fire the tube.

The output of the discharge capacitor is connected directly across the xenon tube. But as explained below, the gas in the tube is not yet conductive, so the charge in the capacitor is not spent.

Trigger

Simply supplying a few hundred volts to a xenon tube isn't enough to make it flash. The breakdown voltage of xenon gas is in the order of 10,000 to 20,000 volts, and until the gas reaches its breakdown voltage, no electricity can pass through it.

The 10 to 20 kV is considerably higher than the voltage created by the high-voltage power supply and discharge capacitor. So to produce a flash, most xenon circuits incorporate an additional triggering device that delivers a pulse of 10 to 15 kV directly to the tube. Once triggered by this high voltage, the xenon gas becomes ionized, so it conducts the few hundred volts of electricity present at its terminals. That, in turn, releases the energy stored in the discharge capacitor, and a brilliant flash of white light ensues.

Trigger circuits consist of two main components: a trigger capacitor (usually less than 0.47 microfarads) and a trigger transformer. The trigger cap initially stores high voltage provided by the power supply and conducts it through the high turns ratio trigger transformer when a flash of light is needed.

A basic circuit

Flash circuits are readily available as salvage from old photographic strobe lights and discarded cameras. Even if you can't find an old strobe light or camera to tear apart for its flash circuit and tube, you can buy a cheap electronic flash for about $10 at many photo stores, so there's hardly a reason to spend good time and money in building your own single-shot flash lamp. The schematic in FIG. 18-4, adapted from Forrest Mims' column on photoflash circuits in the August 1988 issue of *Modern Electronics*, shows how you can build your own flash lamp with readily available parts. TABLE 18-1 provides a parts list for the circuit.

A high potential of about 150 volts is developed by the two transistors set to oscillate, and pulsed dc is delivered to the 6.3-volt (secondary) terminals of a stepdown power transformer. The high-voltage side of the circuit consists of a diode to rectify the voltage to dc, two capacitors to smooth and store the voltage, a flash tube, a trigger transformer, and a trigger cap.

In addition, a bleeder resistor (R3) is added in parallel with the two discharge caps and the xenon tube to slowly release the energy stored in the capacitors when the circuit is turned off. The bleeder resistor is necessary to prevent accidental shock from the discharge capacitors when servicing or experimenting with the circuit.

To use the flash, close switch S1 to turn the circuit on. Wait several seconds for the discharge capacitors to come to full power, and then press momentary switch S2. Closing S2 completes the circuit between the trigger capacitor and trigger transformer, thus sending a pulse of high voltage to the tube.

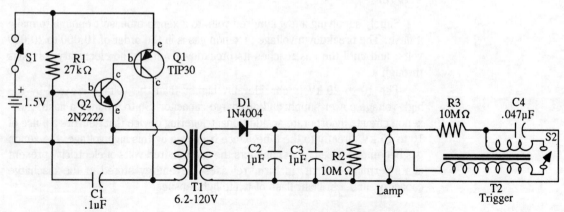

Fig. 18-4. Schematic for a single-shot xenon flash. Circuit courtesy Forrest Mims III.

Table 18-1. Parts List for Single-Shot Flash

R1	27 kΩ
R2,R3	10 MΩ
C1	0.1 μF disc
C2,C3	1.0 μF high voltage flash capacitor (450 V or higher)
C4	0.047 μF
Q1	TIP30 transistor
D1	1N4004 diode
T1	6.3-to-120-volt step-up/step-down transformer
T2	HV flash trigger transformer
S1	SPST switch
S2	SPST momentary switch (normally open)
Misc.	Flash lamp, battery clip, 1.5-volt flashlight battery

All resistors are 5 to 10 percent tolerance, 1/4 watt. All capacitors are 10 to 20 percent tolerance, rated at 35 volts or more, unless otherwise noted.

Note that the anode of the trigger transformer is connected *around* the flash tube. Though the anode of the trigger transformer is not electrically connected to the tube, it is sufficient to ionize the gas within it. This type of triggering, called *external* triggering, is just one method of activating xenon flash tubes. For more details on additional triggering methods, consult Mims' ''Electronic Notebook'' column in the July and August 1988 issues of *Modern Electronics*.

BUILDING THE VARIABLE-RATE STROBE

Unlike the single-flash xenon circuits, a variable-speed strobe light can be quite expensive, especially if it delivers more than just a weak glow of light. The circuit shown in FIG. 18-5 (with parts list in TABLE 18-2) can be used with almost any xenon flash tube and produces brilliant flashes of light, enough to use the project as an outdoor signal warning, repeating photographic strobe, or even as a landing beacon at a small airport.

Before getting into building and using the circuit, you should be aware of some good news and some bad news. The bad news is the project requires the use of some rather hard-to-find parts. These parts include the inverter transformers (two or four needed), trigger transformer, and inductor.

The good news is this project is available in kit form from a number of mail-order companies including Edmund Scientific, All Electronics, and its designer, General Science & Engineering (see Appendix A for addresses). Cost is typically under $10, so you are really better off building it from the kit rather than finding the parts yourself. The kit lacks sufficient instructions, though, so use the text that follows to help guide you during assembly and testing.

If you are building the strobe from scratch, you can use miniature inverter transformers purchased from surplus outfits or salvaged from old flash units. The transformers typically have five or six terminals, arranged as shown in FIG. 18-6. Try connecting the transformer as shown to see if your circuit works properly. You can also use a volt-ohmmeter to test the resistance of the windings.

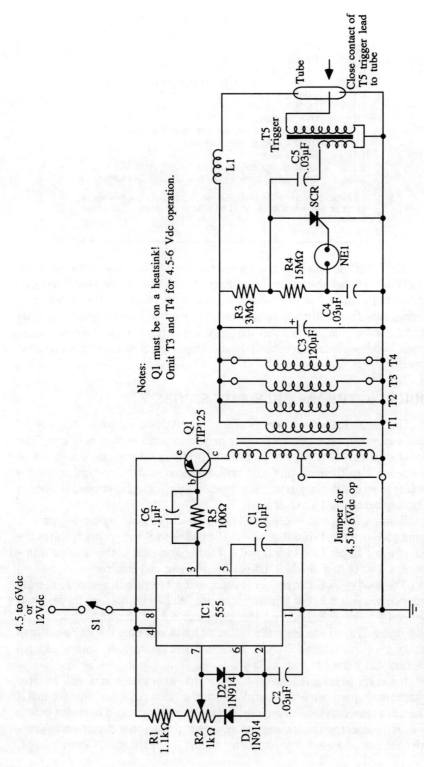

Fig. 18-5. Schematic for a variable-rate xenon strobe. Circuit courtesy Roger Sonntag.

Table 18-2. Parts List for Variable-Rate Strobe

R1	1.1 kΩ
R2	1 kΩ potentiometer
R3	3 MΩ
R4	15 MΩ
R5	100 Ω
C1	0.01 μF disc
C2,C4,C5	0.03 μF disc, 1 kV or higher
C3	120 μF HV polarized electrolytic
C6	0.1 μF disc
D1,D2	1N914 diode
Q1	TIP125 transistor
T1 – T4	Inverter transformers for flash (see text)
T5	Flash trigger transformer
SCR	Silicon controlled rectifier, 400 V at 3 A
NE1	Neon lamp
L1	Inductor (see text)
Misc.	Flash lamp

All resistors are 5 to 10 percent tolerance, 1/4 watt. All capacitors are 10 to 20 percent tolerance, rated at 35 volts or more, unless otherwise noted.

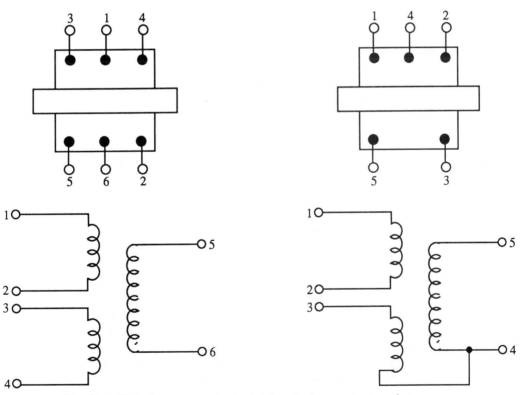

Fig. 18-6. Typical arrangements of miniature dc-dc converter transformers.

The lower resistance indicates the primary; the somewhat higher resistance indicates the secondary.

At the heart of the circuit is a 555 timer IC operating in astable mode. It has a free-running frequency of about 0.1 Hz to 7 Hz, as set by potentiometer R2. The output of the 555 is applied to power transistor Q1, which switches an on/off voltage to the primaries of the inverter transformers. The circuit operates with 4.5 to 6 volts, or 12 volts. When using 4.5 to 6 volts, only add two inverter transformers; when using 12 volts, add all four transformers.

The primaries of the inverter transformers are connected in series, as shown in the schematic. Note that if you use only two transformers, you need to add a jumper to complete the connection of the primaries to ground.

The secondaries of the transformers are connected in parallel along with discharge capacitor C3 (diode D3 is used to rectify the high voltage). The trigger circuit consists of a neon lamp, SCR, capacitor, and trigger transformer. The neon lamp is used to trigger the gate of the SCR, which when conducting completes the circuit between the trigger capacitor and trigger transformer. The result is that the tube fires.

Coil L1 helps improve the performance of the tube. You can omit L1, but the output of the tube won't be nearly as bright. Coil L1 also provides a nifty way to experiment with electromagnetic projectile launching, as discussed later in this chapter.

If you're building the project from scratch, you need to construct L1 yourself. Build the coil by hand as follows. Use 14- to 16-gauge enameled wire.

1. Slide a #8 fender washer onto a 1-inch #8 stove bolt.
2. Starting near the fender washer, wind seven turns onto the bolt. Make the winds as close as possible; don't follow the threads.
3. Put another fender washer on the other side of winding, and spin a #8 nut on to keep the washer in place.
4. Add a second layer of wire by winding another seven turns over the first seven. Keep the windings neat by cinching up the fender washers as required.
5. Continue layering the wire around the bolt until the winding is 5/8 inch in diameter.
6. Remove the nut and first fender washer. Wrap a layer of electrical tape around the winding. Then remove the winding from the bolt by unscrewing it.

From the center of L1, pull out about 1 inch of wire. From the outside of L1, pull out about 1/4 inch. Scrape at least 1/8 inch of insulation off both ends. The short, outside length attaches to the junction of D3, R3, and C3. The long, center length is soldered to the end of the flash tube.

Although trigger transformers come in many sizes and forms, almost any of the miniature variety will work in this circuit. It can be either the three- or four-wire type. The transformer should be rated to at least 6 kV. If the transformer is the three-wire type, two leads are used for the primary and one is for the high-voltage secondary. Connect the primary between ground and C5, and position the secondary on the center of the xenon tube. If the transformer is the four-wire type, two leads are used for the primary and two for the secondary. Connect one lead each from the primary and secondary to ground. Connect the remaining primary to C5, and position the remaining secondary on the center of the xenon tube.

Deciding which is the primary and which is secondary of a trigger transformer can be tough if you didn't get any specifications with it. Generally speaking, the high-voltage output of the trigger transformer is usually indicated with a red dot. Also, the primary uses heavier wire than the secondary.

When positioning the high-voltage output terminal, be sure it touches the glass of the xenon tube. The closer the terminal wire is to the tube, the brighter the flash (indeed, if the wire is any more than $1/16$ inch or so from the tube, no flash will occur). Be sure that the high-voltage terminal does not touch the metal ends of the flash tube.

You can build the variable-rate strobe using almost any construction method except "rat's nest" point to point. Install all components on some type of perforated board, such as a universal solder board. Because of the high voltages and currents used, you can't wire-wrap the circuit.

To use the strobe, place the flash tube behind a clear, strong, protective piece of plastic. This is necessary in case the tube breaks or shatters when first turned on. Keep your hands away from the tube when operating it so broken glass can't explode outward and injure you. It's a good idea to wear protective goggles when experimenting with this project.

Connect it to a battery source following the proper polarity. Because the inverter transformers operate on a linear scale and the input voltage is multiplied to obtain the high-voltage output, be sure not to exceed the recommended operating voltage. Otherwise, you could exceed the breakdown voltage of C3 and damage it and other components.

With power applied, avert your eyes away from the tube and turn the power switch to ON. You should see the room light up in a brilliant flash. If nothing happens after a few seconds, the circuit is not operating properly. Disconnect it from the battery and use an insulated jumper to *short the ends of C3* together (*never* touch C3 or its surrounding components until you have discharged it). Inspect your wiring and fix any mistakes. Check to see if the output terminal from the trigger transformer is touching the glass envelope of the xenon flash tube.

If all is working properly, adjust R2 and watch the repetition rate increase and decrease. Be sure to look away from the flash because the bright light can be very dangerous to your eyes. If you begin to see bright spots or start to feel nauseous, turn the flash off and add a dark cover over the tube. Some flash rates can affect persons subject to epileptic seizures. If you use the project in the presence of others, keep the flash rate as slow as possible.

Going further

Now that the variable rate flash is built, you should put it in a protective housing and cover the flash tube as a guard against injury in case the tube ruptures. Any plastic project box will do. Cut out a hole to expose the flash tube, and glue on a piece of clear plastic as a protective cover.

Because xenon flash tubes emit all the visible colors of the rainbow, you can add transparent gels in front of the tube to produce various colors. For example, you might want to use a bright red gel as a warning beacon for when your car breaks down. Because the flash is so bright, you can use it during day or night.

AN UNUSUAL APPLICATION

If you have been up on modern warfare, you've probably heard of a weapon called a rail gun. This unique armament uses electromagnetic force to hurl projectiles through the air at many times the speed of sound. This electromagnetic force is created by a high-voltage power supply, a triggering circuit, and a coil of wire.

As indicated in the schematic for the variable-rate strobe kit, L1 is used to inductively couple the high voltage generated by discharge capacitor C3 to the xenon tube. Because a rather large amount of high voltage passes through L1, it acts as a high-powered solenoid. You can use L1 to "launch" small ring-shaped metallic projectiles several feet in the air. One caveat: Though the force of the projectiles is not very strong, you should exercise care so that metal items are launched away from you. Never point the rail gun at someone else, or injury could result. Experiment with the rail gun outdoors so that the projectiles don't knock against walls, ceilings, or furniture.

To experiment with the homemade rail gun, position L1 so that it rests flat on the circuit board. Cut a length of $5/8$-inch I.D. diameter thin-walled steel tube to about $1/2$ inch. Remove the burrs with a small round file. Dry fit the tube over L1. If the coil is too big, take a few windings off it and retape it. Or, use a slightly larger brass tube.

Rotate rate potentiometer R1 to its minimum setting (slow flash rate). With the flash circuit off, place the piece of tube over L1. Aim the assembly away from you and momentarily turn the circuit on. At the moment the flash goes off, the tube will blast off from L1.

Try thicker- or thinner-walled metal tubing. A good source for tubes of all lengths and diameters is the local hobby store. Try brass, copper, and aluminum tubes to see what effects the coil has on them.

19

See-in-the-dark viewer

The electromagnetic spectrum cuts a wide swath that includes radio waves, microwaves, x-rays, cosmic rays, and infrared, ultraviolet, and visible light. The human eye can perceive only a very small portion of this range (visible light), so we are blind to over 98 percent of the spectrum.

Though our eyes alone can't perceive the other wavelengths, we can tune in or view our world using the invisible portions of the electromagnetic spectrum with the proper machinery. An x-ray machine peers through the skin and muscle of a leg to examine the broken bone inside; a radio latches onto a specific wavelength of broadcast signal and relays news, sports, music, and other programming. And an infrared viewing device sees in the dark by picking up infrared radiation that our eyes can't see.

Infrared or "see-in-the-dark" viewers have been around since World War II and have seen extensive use in military and industrial applications. Cost has been the limiting factor in the use of infrared (IR) viewing; the typical military IR scope retails for about $1,000 or even more if it comes with fancy rangefinder or zoom features.

With about $100, you can construct your own infrared viewer and see in total "darkness." Most of the materials are available on the surplus market, and construction is straightforward. This chapter details how to build your own IR viewer, how to provide extra infrared illumination, and even how to convert a regular video camcorder into a night scope.

Because of the variety of parts available (lens, power supplies, and so forth), the instructions in this chapter are necessarily generic. You'll need to make adjustments as you go along to conform to the parts you are using. But keep in mind that part of the fun of gadgeteering is creating things on your own.

ABOUT INFRARED RADIATION

Infrared radiation is the same as visible light—they are both components of the electromagnetic spectrum—but it has a longer wavelength. The word "infrared" means *below red*, and as shown in FIG. 19-1, you'll find the infrared band located under the red wavelengths of visible light.

Visible light encompasses a ribbon of the spectrum from about 400 to 700 nanometers (abbreviated nm, which is billionths of a meter). Many people can perceive light a little beyond 700 nm, but that light is dim and generally visible only when directly viewing the source. The start of the infrared band is loosely defined as the point where the human eye can no longer discern light, but that's different for everyone. For most people, the threshold of visible light is in the region of 720 to 750 nm.

Electronic imaging tubes of various designs, including television vidicons, have long been sensitive to light beyond 750 nanometers. In fact, filters are regularly placed in front of the imaging tube to block infrared light so that the televised scene appears as it would if you saw it with your own eyes.

Infrared scopes commonly use a specially manufactured imaging tube (see FIG. 19-2) that has a spectral response of about 500 (green) to 1,200 nanometers (infrared). Coupled with an infrared light source (usually attached to the scope), the imaging tube is able to view objects that are ordinarily too dark to be seen by the human eye.

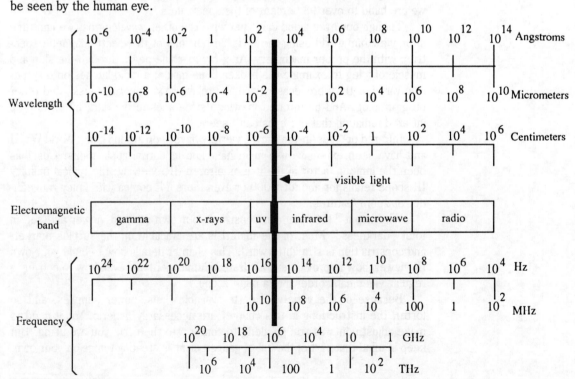

Fig. 19-1. The electromagnetic spectrum, showing the bands from radio to gamma rays, along with their associated wavelengths and frequencies.

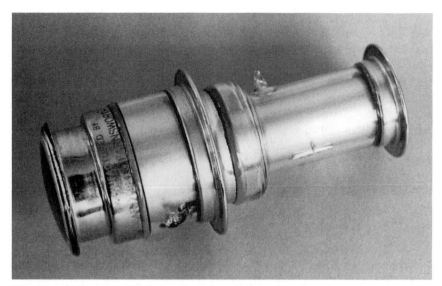

Fig. 19-2. The model 6032 infrared converter tube, originally manufactured by RCA/Fransworth and now occasionally available on the surplus market.

The construction of the image converter tube, as shown in FIG. 19-3, is unique. It combines a target faceplate, like a TV camera tube, and a viewing plate, like a TV set. Both functions are combined in one glass module for reduced weight and size. The converter tube shown in FIGS. 19-2 and 19-3 is the RCA/Farnsworth Type 6032, which is no longer manufactured but available in surplus.

Though the tubes are surplus (most are new but some are used), they aren't cheap. You should expect to pay $50 to $90 for a good tube, although you could luck out (as I did) and find the tubes at a local surplus store for under $25

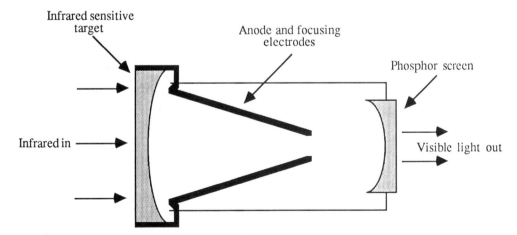

Fig. 19-3. The inner components of an infrared converter.

(I purchased two used tubes for $10 each). Mail-order sources include Meredith Instruments and Fair Radio Sales. Refer to Appendix A for more names and addresses.

The 6032 IR tube operates under very high voltages. The accelerator grid requires a potential of about − 12,000 volts, while the focusing grid requires about − 10,500 volts. One power supply is used to develop − 12,000 volts, and that main potential is then reduced to power the focus grid. The 6032 converter tube uses electrostatic focusing, which means that you electronically focus the tube by varying the voltage to the focus grid. This adds some complexity to the building of your see-in-the-dark scope.

The 8598 is another infrared converter tube available through some surplus outlets. It costs more than 6032 but it doesn't require separate accelerator and focusing voltages. You need only apply one high voltage to the accelerator grid, and the proper focusing voltage is developed internally. The 8598 converter is also smaller than 6032, so you can make more compact IR scopes with it.

CONSTRUCTION OF THE IR SCOPE

The IR see-in-the-dark viewer is composed of four major parts:

- The converter tube (typically an RCA/Farnsworth Type 6032)
- A high-voltage power supply
- Lens for focusing the tube
- Lens for focusing the eyepiece

These components are combined in an all-plastic enclosure. The power supply module described below operates from low-voltage dc, so you can use your IR viewer ''in the field'' without worry of finding an electrical outlet.

Building the power supply

The discrete power supply is based on the flyback transformer design first introduced in Chapter 3, Plasma Sphere Experiments. Refer to that chapter for parts and construction details. TABLE 19-1 provides a list of parts for the complete power supply.

The flyback transformer alone doesn't develop enough voltage to operate the IR converter tube. You can readily step up the voltage by building a voltage tripler, as shown in FIG. 19-4. The tripler consists of six high-voltage capacitors

Table 19-1. Parts List for Infrared Scope Power Supply

R1 – R10,R14	22 MΩ
R11,R12	6.8 MΩ
R13	12 MΩ
C1 – C6	150 pF disc, 3 kV
D1 – D6	High-voltage diodes (3 kV or higher)
1	Flyback transformer high-voltage power supply (from Chapter 3)

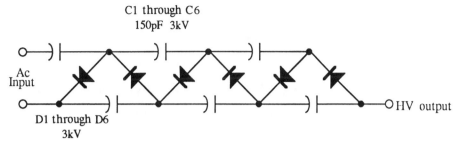

C1 through C6
150pF 3kV

Ac
Input

D1 through D6
3kV

HV output

Fig. 19-4. A high-voltage tripler.

and six high-voltage diodes. The orientation of the diodes presents a negative potential at the output. That is, the output is about −12,000 volts (give or take—the IR tube can take up to 20,000 volts); the voltage is negative with respect to ground.

Figure 19-5 shows a recommended layout for the tripler. A printed circuit board is recommended because that helps reduce the chance of high-voltage arcing, which can produce excessive amounts of ozone and impair the operation of the converter tube. Alternatively, you can build the tripler on perf board. Be sure lead lengths are as short as possible and avoid placing the components too close to one another.

The focusing grid of an IR tube requires approximately −10,500 volts. Rather than build a separate power supply or voltage multiplier, the easiest method is to construct a divider network consisting of high-value resistors. A suitable schematic is shown in FIG. 19-6. The resistor ladder offers some flexibility in selecting the best focus voltage for the particular tube you are using. By substituting R12 and moving the focus wire between the junctions of R11/R12 or R12/R13, you can choose the voltage that offers the greatest image sharpness. Note that the R12 uses Molex sockets so that you can change the value of the resistor without the hassle of desoldering.

Optics

The converter tube requires two lens assemblies, as described in TABLE 19-2. The objective lens fits over the front of the converter and focuses the

Capacitors

Input

Diodes

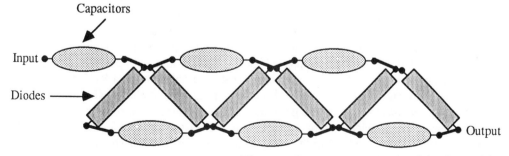

Output

Fig. 19-5. Suggested layout for the high-voltage tripler.

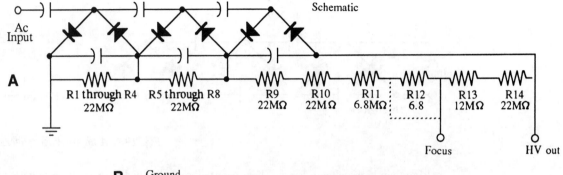

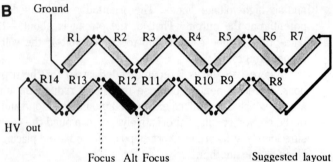

Fig. 19-6. Resistor ladder network for the see-in-the-dark scope power supply (shown with the high-voltage tripler, with required connections). (A) shows the schematic, and (B) shows the suggested parts layout for the ladder. Note that R12 should be removable and that the focus tap should be easily attached to either side of R12.

Table 19-2. Parts List for Infrared Scope Optics

1	For imaging lens: 40 to 100 mm variable- or fixed-focus lens (camera, projector, enlarger, etc.)
1	For eyepiece: two 25 mm positive lenses, plastic or metal barrel, retaining rings; or eyepiece from telescope

scene onto the target of the tube. You can make your own lens assembly with individual glass lenses, but an easier method is to salvage the lens from an old 35 mm slide projector. The lens must have a focus barrel or else you'll need to fashion a way to move the lens toward or away from the converter tube.

Another good source for objectives is surplus or cosmetically damaged video camcorder lenses, as shown in FIG. 19-7. These cost in the neighborhood of $20 to $75, depending on the complexity of the lens. A lens with servos intact for auto-focus, auto-iris and power zoom costs top dollar, while a lens with manual zoom and focus might set you back no more than about $35 to $40.

The eyepiece lens focuses the image created by the converter tube so your eye can view it. The eyepiece can be built following FIG. 19-8 using two 25 mm diameter positive lenses. Construct the eyepiece in a metal or plastic barrel, as shown.

Fig. 19-7. A camcorder lens, purchased surplus. This lens is designed for use with a camcorder sporting a 2/3-inch solid-state imager, making it ideal for use with a 6032 image converter (although there is some "tunneling" caused by the lens vignetting the edges of the target plate).

Fig. 19-8. Construction and lens placement for the eyepiece. You can also use the eyepiece from a microscope or telescope.

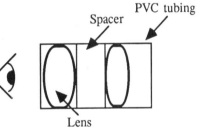

Scope assembly

With the power supply complete and the lenses at hand, you can now build your see-in-the-dark scope. You'll need an enclosure and IR tube to complete the necessary components (see TABLE 19-3 for a parts list). The exact dimensions of your scope depend on the component parts you have including the flyback transformer, voltage multiplier, and lens. You should be able to fit it all in a project box measuring 6 by 8 by 3 inches. If necessary, build the box yourself using acrylic plastic.

Table 19-3. Parts List for Infrared Tube and Enclosure

1	6032 infrared imaging tube
1	Project box, 6-by-8-by-3-inches (nominal)
Misc.	Construction hardware, hookup wire, clips for attaching leads to imaging tube, etc.

Assemble the parts in the box as suggested in FIG. 19-9. Allow room for clearance between the image tube and the lenses. Secure all components using nylon or plastic hardware to help reduce the chance of accidental shock.

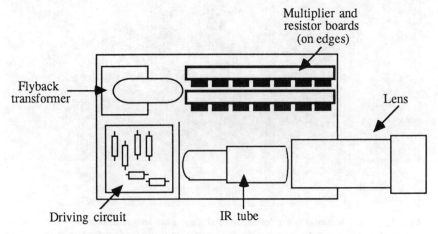

Fig. 19-9. Suggested parts layout for the see-in-the-dark scope.

The most crucial components are the image tube, objective lens, and eyepiece. By necessity, these can only be placed a certain distance apart, so plan the design of your scope using the suggested dimensions as shown. The distance between objective lens and image tube depends on the lens you use. Even if the objective has its own focus barrel, you'll need to be able to slide the entire lens in the scope housing. Lenses for 35 mm slide projectors are often grooved with a spiral around the base. Use this spiral in your scope as a focusing aid.

Because the image converter uses metal/glass seals that can be damaged if exposed to high temperatures, it's not a good idea to solder hookup wires directly to the tube. Instead, fashion clips to fit around the barrel terminals of the tube as shown in FIG. 19-10. The clips should be made of thin, springy metal.

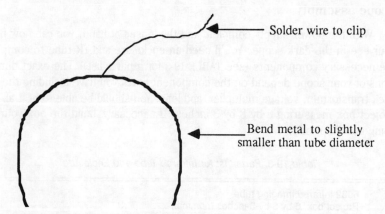

Fig. 19-10. Bend a piece of springy metal wire to a clip shape, and solder the end of a wire to establish an electrical connection.

Attach the high-voltage and ground leads to these clips, and then clip the clips over the tube, as shown in FIG. 19-11. Attach the focus wire only to the tube clip. Leave the other end free for the time being.

Testing and operation

With the parts suitably mounted to the enclosure, you can test your scope and its infrared viewing capabilities.

First, start by inserting the 6.8 MΩ resistor for R12 in the sockets provided. Cut the lead lengths to no more than 1/4 inch and fit the leads as far as they will go in the socket. Attach the loose end of the focus wire to a small alligator clip. Position the clip on one side of R12 and close the box for protection.

Apply power to the scope and turn the power switch on. Listen for cracking or hissing (signs of arcing). If you hear anything, turn the scope off and carefully lift off the cover. Turn the scope back on and look for arcing. *Keep your fingers*

Fig. 19-11. A close-up view of an image converter tube with a wire clip installed.

away from all components while the unit is on. Otherwise, you could receive a potentially FATAL shock. Apply corona dope or high-voltage putty to those areas that are arcing. If you find arcing at the alligator clip that attaches the focus wire to R12, apply a small amount of high-voltage putty around the exposed parts. Replace the cover for the next step.

If you are testing the scope in daylight, cut a piece of thin cardboard or heavy paper to fit neatly around the front of the objective lens. Inside this piece, cut a smaller circle to allow some light to pass, as shown in FIG. 19-12. Tape the cardboard or paper to the front of the lens.

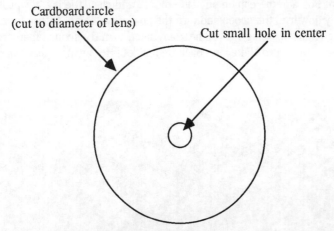

Cardboard circle
(cut to diameter of lens)

Cut small hole in center

Fig. 19-12. Cut a piece of cardboard or stiff paper into a circle the size of the front of the objective lens. Cut a smaller hole in the center of the disc to let in some light. Tape the disc to the front of the lens and use it to adjust the scope in subdued daylight conditions.

With the scope turned on, peer through the eyepiece and adjust it so that grain of the phosphor coating of the image converter is as sharp as possible. Don't worry about focusing an image at this point; just concentrate on focusing the eyepiece to the back of the tube.

If the front lens has its own focus barrel, focus it to infinity. Position the lens as close to the front of the image tube as possible. Peer through the objective and slowly move the objective lens away from the front of image tube. At some point, the image should become sharper (but not necessarily absolutely sharp). Then go out of focus again as you continue to move the lens. Stop the lens when the image is at its sharpest.

If you feel the image as seen through the eyepiece is clear, you might not need to adjust the electrostatic focusing of the tube. But should the focus still be soft (as it usually is), you'll need to fine tune the focus voltage to bring the image into clarity. Turn the scope off, remove the cover, and attach the focus wire (the one with the alligator clip) to the other side of R12. Replace the cover, turn the scope back on, and try refocusing the image with the lens.

Should the image look the same or worse, you'll need to replace R12 with a higher or lower value. Two values to try are 4.7 or 8.2 megohms. Making sure that the scope is turned off (and battery removed for complete safety), remove the 6.8-megohm value of R12 and replace it with a 4.7-megohm resistor. Attach the alligator clip firmly to one side of the resistor (reapply the high-voltage putty if necessary), and try again. With each new resistor you use, be sure to try the focus with the alligator clip on either side of R12.

After some trial and error, the focus should be reasonably sharp and you'll be able to see outlines of nearby objects (but not those closer than about 4 to 5 feet, depending on the minimum focus range of the objective lens you are using).

Note that the Type 6032 converter tube lacks sufficient resolution to display objects with the same kind of clarity people are used to with television or film. The resolution of the tube is only 18 line pairs per millimeter, and this greatly limits the apparent sharpness of images. Color print film has a resolving power of around 100 line pairs per millimeter. Also, the quality of lenses you use for the objective and eyepiece play a major role in the sharpness of the picture. You might want to try other lenses to see which ones work the best.

Once the scope has been focused, solder R12 in its sockets. Remove the alligator clip from the end of the focus wire, and solder the wire directly to R12. Tighten the lens mounts to prevent them from moving. You should not cement the lenses in place, however, as you might want to readjust the focus of the objective and eyepiece later on. Remove the cardboard cover over the lens; it's not needed for normal operation.

You are now ready to use your IR scope to view images in full darkness. But you need an infrared light source to do it. See the section Providing Infrared Light later in this chapter for more details.

Alternative power supplies

Modular high-voltage power supplies designed for television sets can often be put to use in IR scopes. Such a power supply is shown in FIG. 19-13. The supply operates from 20 to 24 Vdc and produces as much as +15 kV. The power supply is linear: output voltage decreases proportionally depending on the input voltage. These power supplies generate 6 to 8 kV with an input voltage of about 10 to 12 Vdc.

For best results, test the output of the modular high-voltage supply with a high-voltage probe and digital meter; avoid guessing. The output of some modular power supplies is negative with respect to ground, and others are positive. Generally, those supplies that generate a negative voltage with respect to ground are easier and and safer to work with.

USING A VIDEO CAMCORDER AS AN IR SCOPE

As mentioned earlier in this chapter, at most all TV imaging tubes are sensitive to infrared light. In fact, some tubes are especially made to pick up

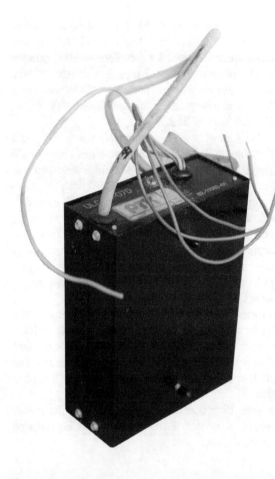

Fig. 19-13. A high-voltage modular power supply. This one generates about 15 kV when supplied with 22 to 24 volts dc. Output is roughly 7.5 kV when supplied with 12 Vdc.

infrared light and are used in night-time security and surveillance. An old black-and-white security camera, available surplus for $100 or less, makes a perfectly good see-in-the-dark viewer. These cameras often use a blue or blue-green filter in front of the image tube to block infrared light. Remove this filter and your scope is ready to go. Of course, unlike the see-in-the-dark viewer described above, you need a separate TV or monitor to view the nighttime scene.

You can use a modern color camcorder to make a night scope. The solid-state imager used in today's camcorders are even more sensitive to infrared light than the older-fashioned TV pickup tubes. Again, some camcorders use an infrared-blocking filter that you must remove if you want to be able to use the unit as a night scope. The best camcorders to use are those with a built-in electronic viewfinder, which is a miniature black-and-white television tube (that usually measures an inch or so across) so you don't need a separate monitor or TV set. The biggest advantage of a camcorder is that you can record the infrared scene. This might be important if you are a licensed private detective and you need to collect official evidence.

Cannibalizing a $1,000 camcorder just to make a night-vision scope might not be to your liking. The next best thing is to cannibalize a $100 camcorder. No such thing, right? Wrong. Fisher-Price makes a kiddie camcorder that retails for about $100 that uses a miniature charge-coupled-device (CCD) image sensor. Though the resolution is rather poor, the image sensor is very sensitive to infrared radiation, and you can even record up to 5 minutes of video on an ordinary audio cassette.

The Fisher-Price PXL 2000 shown in FIG. 19-14 uses a small CCD element measuring approximately 1/8-inch square. The video image picked up by this element is then digitally processed and recorded as a black-and-white image on an audio cassette tape. In order to sufficiently record the wide bandwidth of

Fig. 19-14. The Fisher-Price PXL 2000 "kiddy camcorder." It makes an ideal see-in-the-dark scope, although image quality is poor.

video, the tape is shuttled through the transport at high speeds. A 90-minute audio cassette (45 minutes each side) provides 4 to 5 minutes of video time with the PXL 2000. The camcorder has only an optical viewfinder, so you must use a television set to view the recorded images. I use a 5-inch portable black-and-white set for this purpose. TABLE 19-4 provides the parts list for a night scope setup based on the PXL 2000.

Table 19-4. Parts List for PXL
2000 Camcorder as Infrared Scope

1	Fisher-Price PXL 2000 camcorder, modified as discussed in text
1	Small-screen TV (used as monitor for camcorder)
1	Battery power for TV

As with many other monochrome cameras, the PXL 2000 incorporates a blue-green IR blocking filter over its lens and CCD imager. You must remove this filter to use the camcorder as a night-vision scope. Be warned, however, that removing the filter voids the camcorder's warranty, and with the filter gone, daylight scenes recorded with the camcorder will appear severely washed out.

To remove the IR blocking filter, disassemble the camcorder by removing the two screws toward the front of the unit. Leave the other screws in place. Gently pry apart the two sides of the case until the lens barrel falls out. Actually, this barrel doesn't hold the lens but merely acts as a manual iris control. It is not needed for night-vision work.

The lens of the camcorder is extremely small (you might want to replace it, but that's beyond the scope of this chapter) and is mounted on a threaded plastic bushing. The filter is fitted on the outside of this bushing. To remove it, use a small jeweler's screwdriver to pop the filter out of its holder, as depicted in FIG. 19-15. Save the filter in case you want to use it later on. Replace the screws and the camcorder is now ready.

PROVIDING INFRARED LIGHT

Both the do-it-yourself IR scope and the converted video camera/camcorder require an infrared light source to see at night. This light source emits only infrared radiation, so human eyes can't detect it.

You can construct an IR light source with a flashlight or floodlight outfitted with the proper infrared filter (see TABLE 19-5 for a parts list). These filters are routinely available in the surplus market; as of this writing, Meredith Instruments carries an assortment of plastic and glass filters in many sizes. Other outlets such as Jerryco, Edmund Scientific, and John Meshna & Associates have carried infrared filters. For best results, the diameter of the filter should be the same as the diameter of the flashlight or floodlight. If the filter is too small, you'll have to cover up the edges to prevent white light from spilling out.

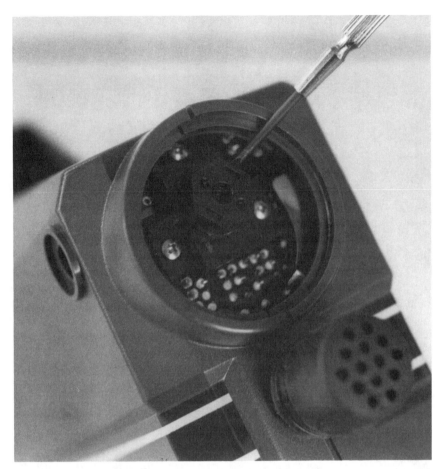

Fig. 19-15. Remove the filter holder from the front of the PXL 2000 camera with a small jeweler's screwdriver. Drop out the filter (save it for later use) and place the holder back on the lens.

Table 19-5. Parts List for Infrared Floodlight

1	Q-beam floodlamp, 350,000 candlepower (or equivalent)
1	5¹/₄-inch diameter infrared glass filter
1	12-volt battery pack (8 amp-hour or better)

Figure 19-16 shows a Q-Beam 350,000 candle power automotive floodlight and a matching 5¹/₄-inch-diameter glass infrared filter (purchased from Meredith Instruments). As it happens, the filter fits the bezel of the floodlight *exactly*. Apparently, the filter was made for this brand of floodlight, which I purchased for about $20 at K-Mart.

Fig. 19-16. The Q-Beam spotlight makes an excellent infrared light source when outfitted with a glass IR filter.

Though the floodlight produces a lot of heat, the infrared filter is made to withstand it. In fact, though you can feel the heat of the infrared radiation produced by the floodlamp, the filter itself remains relatively cool. The filter doesn't block out all visible light. In complete darkness, it's still possible to see a very faint red glow when looking directly at the floodlight. Normally this does not pose a problem.

The Q-Beam floodlight is designed for 12-volt use and comes with a cigarette adapter for plugging into a car lighter. I use the floodlight with a heavy-duty 12-amp-hour rechargeable lead-acid battery so that the floodlight can be operated anywhere.

While floodlights produce copious amounts of light, much of it is contained in a narrow beam. The spot of that beam can cause a "blooming" effect in the infrared converter tube or modified camcorder. If blooming is a problem, add a neutral density filter over the lens of the night scope, or choose a light source that doesn't project such a tight beam.

20

Radiation alert!

Mention the words ''nuclear radiation'' in casual conversation, and most people recoil in fright. Over the years, nuclear radiation has become the bugaboo of modern technology—appreciated for its energetic potential, but feared for its deadly prowess. Driving home these fears are the recent accidents at the Three Mile Island and Chernobyl nuclear power plants, as well as revelations by the government and news agencies of excessive radiation release during the above-ground nuclear tests in the fifties and sixties.

Radiation is certainly nothing to play with. But despite its unpopularity, there are a number of safe, sane, and useful experiments you can conduct with nuclear radiation. You can build a portable Geiger counter to sniff out hidden radiation, watch the tracks of radiation left in a misty atmosphere of alcohol, or test levels of radiation in your home to see if you and your family are exposed to abnormally high levels.

In this chapter, you'll learn what radiation is and how it affects living organisms, how to construct and use a cloud chamber, how to build several types of radiation monitoring equipment, and how to use these monitors to check for radiation.

WHAT IS RADIOACTIVITY?

Though radiation was first discovered by the French chemist Becquerel in 1897, it wasn't fully understood until the 20th Century scientists Bohr, Rutherford, and others developed a theory of the construction of the atom—the smallest unit of matter we know about. As you probably already know, the atom consists of a nucleus or core containing protons and neutrons. Orbiting around this nucleus are one or more electrons, as shown in FIG. 20-1. Nuclear radiation can occur when the atom changes its form, usually by the loss or addition of electrons.

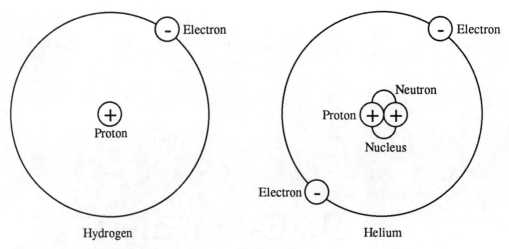

Fig. 20-1. Atomic rendition of the hydrogen and helium atoms.

Of significant importance in the study of radioactivity is the atomic weight and mass number of an atom. The *atomic weight* (designated as Z) represents the number of electrons orbiting the atom's nucleus. Each element has a unique atomic weight. Hydrogen has an atomic weight of 1 because it has just one electron orbiting its nucleus. You can readily see the atomic weights of other elements in a periodic chart.

The *mass number* (designated as M) is the sum of protons and neutrons in the nucleus and is generally written after the name of the element. The mass number indicates an isotope, or a slightly modified form, of a particular atom. For example, although hydrogen has just one electron, hydrogen atoms might have one, two, or no neutrons in the nucleus, as shown in FIG. 20-2. The mass number is typically written after the name of the element, such as uranium-235 or uranium-238. While both of these refer to the element uranium, they are two different isotopes, and as we'll see later in this chapter, they behave quite differently.

Atomic decay

Radiation is the result of atomic decay. The decay occurs because the construction of some isotopes, such as hydrogen-3 (otherwise known as tritium), is fundamentally unstable—the relationship between Z and M (mass number and mass weight) is not in balance. In an attempt to balance itself, tritium atoms will change one of the neutrons into a proton and an electron. The new nucleus consisting of two protons and a neutron (and shown in FIG. 20-3) is the same as helium-3. A by-product of this nuclear change is that one of the electrons is ejected from the atom.

As with chemistry, atomic radiation is often described in a reaction equation. The reaction equation for tritium to its stable helium-3 state is:

$$\text{tritium} \quad \rightarrow \quad \text{helium-3} \quad + \quad 1 \text{ electron}$$

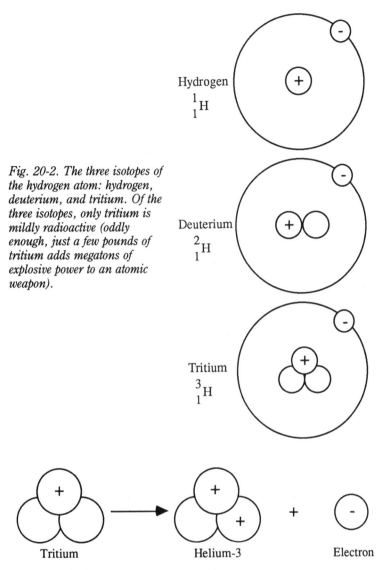

Fig. 20-2. The three isotopes of the hydrogen atom: hydrogen, deuterium, and tritium. Of the three isotopes, only tritium is mildly radioactive (oddly enough, just a few pounds of tritium adds megatons of explosive power to an atomic weapon).

Hydrogen
$_1^1 H$

Deuterium
$_1^2 H$

Tritium
$_1^3 H$

Tritium   Helium-3   Electron

Fig. 20-3. Fusion of tritium atoms to helium, with the release of one electron.

The equation is easy to read: tritium atomically transforms into helium-3 plus the release of one electron.

In the case of tritium, atomic decay leads to an element higher up in the periodic table. Not all atomic reactions work that way. For instance, uranium-238 decays into thorium-234—lower in the table—along with the release of what's called an alpha particle (more on alpha particles shortly). The reaction equation for uranium-238 is:

$$\text{uranium-238} \quad \rightarrow \quad \text{thorium-234} \quad + \quad \text{alpha particle}$$

Note that thorium-234 is also an unstable isotope, so it, too, will undergo an atomic decay reaction. TABLE 20-1 shows the decay chain of uranium-238 as it works its way down the ladder of atomic stability, finally to stop at ordinary lead. You can readily see that the original uranium-238 takes on many forms over billions of years.

Table 20-1. Uranium-238 Decay Chain

Element	Weight/Atomic Number	Main Radiation Emitted
Uranium	238/92	alpha
Thorium	234/90	beta, gamma
Protactinium	234/91	beta, gamma
Uranium	234/92	alpha, gamma
Thorium	230/90	alpha, gamma
Radium	226/88	alpha, gamma
Radon	222/86	alpha, gamma
Polonium	218/84	alpha, beta
Lead	214/82	beta, gamma
Astatine	218/85	alpha
Bismuth	214/83	alpha, beta, gamma
Polonium	214/84	alpha, gamma
Thallium	210/81	beta, gamma
Lead	210/82	alpha, beta, gamma
Bismuth	210/83	alpha, beta
Polonium	210/84	alpha, beta
Thallium	206/81	beta
Lead	206/82	(stable)

You've probably heard the terms atomic *fission* and *fusion*. These terms roughly indicate the type of atomic reaction exhibited in an isotope. Fission takes place when the atoms decay to an element or isotope lower in the periodic table; fusion takes place when the atoms decay to an element or isotope higher in the periodic table.

Half-life

As shown in FIG. 20-4, radioactive isotopes exhibit a *half-life*, or the time it takes for half of the atoms in a given sample of material to decay to another form. For a sample of uranium-238, it takes 4.47 billion years for half of its atoms to transform to thorium-234. However, thorium-234 has a relatively short half-life of 24.1 days, when half of its atoms decay to protactinium-234 (same mass number as thorium-234, but different atomic weight). The half-life of protactinium, in turn, is just 1.17 minutes.

As each isotope is transformed into a new material, it gives off certain radioactive emissions. The three principal types of such emissions are *alpha*, *beta*, and *gamma*.

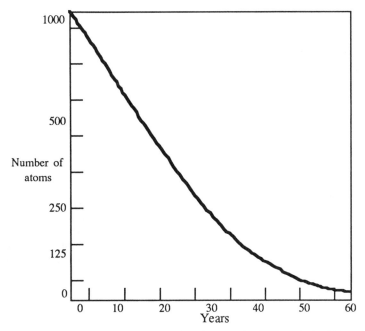

Fig. 20-4. Typical half-life graph of a radioactive sample. With each doubling of years, half the atoms in the sample decay.

Alpha decay

When a nucleus is too heavy—it has too many protons and neutrons compared to the electrons orbiting around it—a compact group called an *alpha particle* is ejected. The alpha particle consists of two neutrons and two protons and is elementally equivalent to helium-4. When an alpha particle is ejected, it leaves the original nucleus four units lighter in mass number and two units lower in atomic weight; that is, two steps down in the periodic table.

Alpha decay is most common among the heavier of elements, such as uranium and thorium, but it doesn't lead directly to a stable nuclei. So that forces the atoms that result from the atomic decay to decay even further, as you saw in TABLE 20-1.

Alpha particles are the most active of all nuclear emissions, with energies of up to 5 million electronvolts (or 5 MeV). An electronvolt is a unit of energy equivalent to the passage of one electron between two separate plates, as shown in FIG. 20-5, when one volt of electricity is applied to those plates. One alpha particle is therefore strong enough to cause 5 million electrons to jump across the plates.

However, even though alpha particles pack a lot of wallop, they don't travel far. In air, alphas travel a distance of no more than an inch or so, and they even

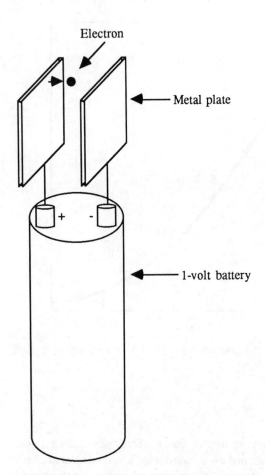

Electron

Metal plate

1-volt battery

Fig. 20-5. A physical representation of the electronvolt. One electronvolt is equal to the energy when one electron jumps from a metal plate to another when charged with a potential of 1 volt.

can't penetrate a sheet of paper. Detecting alpha particles can be a tough proposition because the detecting element can't be covered. As you'll see later in this chapter, however, you can purchase specially made Geiger tubes that use a very thin radiation window to let in alpha particles.

Beta decay

When a nucleus has too many neutrons, one neutron changes to a proton plus an electron (hydrogen-3, or tritium, for example). The electron is ejected from the nucleus and is no longer a part of the isotope. That ejected electron is a beta particle. After the atomic decay, the original nucleus is left with an additional positive charge, so it's one unit higher in Z and one step up the periodic table.

Beta particles have maximum energies of between 0.02 MeV to no more than 5.3 MeV. Unlike alpha particles, betas penetrate several yards in air and can easily go through paper. However, a beta particle can be stopped by even a thin 1/25-inch sheet of aluminum.

Beta decay is the most common form of radioactivity and is often referred to as either *low-level* (or *low-energy*) particles, or *high-level* (*high-energy*) particles. The energy of the particle is determined by its charge in MeV.

Gamma decay

Gamma decay occurs whenever beta decay has not drained enough energy from the atom to provide complete stability. Unlike alpha particles (helium-4 atoms) and beta particles (electrons), gamma decay produces electromagnetic rays, which are just like light or x-rays, except higher up in the electromagnetic spectrum.

Gamma rays are shorter in wavelength than x-rays, and as you probably guessed, they can penetrate through more matter. Gamma rays develop energies of between 0.15 and 2.50 MeV (usually the lower end of this spectrum), but they can penetrate up to 10 inches of solid lead or up to nine feet of concrete (talk about "x-ray vision"!). Gamma radiation has three effects on atoms, as shown in FIG. 20-6. In all cases, it causes an electron to eject from the atom.

TABLE 20-2 shows a comparison of the energy levels and penetration of alpha, beta, and gamma decay. Note that each isotope creates a slightly different level of one or more of these atomic reactions. Much of the gamma radiation produced by radioactive cobalt can be stopped by a lead shield an inch or so thick.

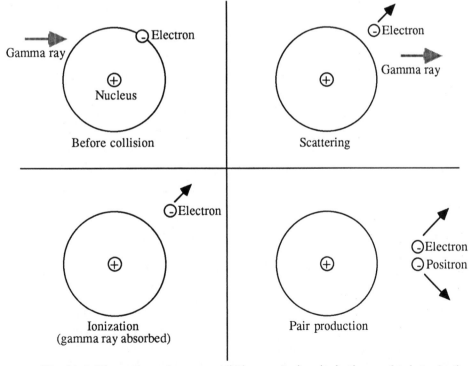

Fig. 20-6. Three effects of gamma radiation: scattering, ionization, and pair production.

Table 20-2. Properties of Atomic Particles

Particle	Rest Mass (in Atomic Mass Units)	MeV Units	Specific Ionization in Air (ions/cm.)	Approx. range Air	Alum.
alpha	4	5	3,000	3.5 cm	0.002 cm
beta	0.0005	1	50	300. cm	0.010 cm
gamma	0	1	Negligible	Not applicable	

In addition to alpha particles, beta particles, and gamma rays, certain radioactive substances emit neutrons. Neutron emission of the pairing of beryllium and polonium, for example, is used to initiate a barrage of neutrons necessary for an atomic explosion.

HOW RADIATION IS MEASURED

Part of the confusion surrounding nuclear radiation arises because of the myriad of ways radioactivity is measured. There are four principal procedures for measuring radiation:

Counts Each radioactive particle intercepted by any of a number of detecting apparatus produces a single click that is used to advance an electronic or mechanical counter. The number of counts in a minute, hour, or even day is used to evaluate the total amount of radiation reaching a particular area. Counting does not generally consider the type of radiation (alpha, beta, gamma, or neutron), nor the source of the radioactivity.

Rad Actual energy absorption is expressed in rads. One rad is 0.01 watt-seconds of energy absorbed per kilogram of body weight. Because one rad represents quite a bit of radiation, measurement is taken as a portion of a rad, typically a millirad (mrad) or thousandths of a rad. Unlike counts, rad measurements typically take into account the type of radiation as well as the source of the radiation.

Rem Different types of radiation and their sources can have varying effects on living tissue. When the radiation is measured against its effect on animal tissue (including human cells), it is generally gauged in rems. While rems and rads are similar, they aren't the same in all situations, so it's important to differentiate the two. As with the rad, one rem is a rather large dose of radiation, so typical measurements are in millirems, or mrem.

Curie One curie (named after Marie Curie, the discoverer of radium) is the amount of any nucleide that undergoes 3.7×10^{10} radioactive disintegrations per second. The curie is a very accurate means of measuring levels of radioactivity. As a comparison, the radioactive source commonly used in ion smoke detectors (discussed more fully later in this chapter) is usually rated at 0.9 to 1.0 microcuries, (one-millionth of a curie). However, the same source might produce tens of thousands of clicks per second and register as 1 to 2 mrads.

RADIATION AND LIVING TISSUE

When alpha particles, beta particles, and gamma rays strike living tissue, they stop or slow down just as if they had struck a piece of paper, a metal shield, or a concrete slab. The energy of the atomic particles or rays must go somewhere, so they are suffused into the cells of the living tissue. The molecules of the cells that have been bombarded by the particles are changed in such a way that the cell can die or can be altered (mutated into another form).

Cells altered by radiation bombardment have been the favorite of Hollywood movie studios. Many late-night movies depict giant men, women, ants, spiders, and other denizens that have been subjected to atomic particles and rays. Invariably, these poor creatures were accidentally exposed to some atomic blast and are out to avenge the scientists and world that caused their suffering.

In reality, cell mutation doesn't cause bugs to grow to the size of sanitation trucks, but it can lead to cancer. Madame Curie, who discovered and worked with the radioactive element radium, eventually died of leukemia, a type of cancer of the blood. And in a number of cases, people and animals exposed to the fallout from the nuclear tests conducted in Nevada during the late fifties and early sixties came down with varying forms of cancer.

As an example, as of this writing, over half of the cast and crew of the 1956 motion picture "The Conqueror," starring John Wayne, Agnes Moorehead, and Susan Heyward have died of cancer (in fact, all three stars and its director died of cancer). That's an unusually high ratio until you think of the likely (but still hotly debated) cause: a good portion of the film was shot in the Nevada and Southwestern Utah deserts where fallout from previous atomic tests had settled. The cast and crew of the movie were continually exposed to low-level radiation that had been impacted in the soil.

Radiation is all around no matter where we live. The most common source of radiation is the sun, which generates prodigious amounts of gamma rays, x-rays, cosmic rays, and numerous other forms of radiation. The earth's magnetic field and its atmosphere prevent most of this radiation from reaching us. On a typical day at sea level, the average person receives approximately 59 millirems (59 mrems) of radiation per year from the sun. Radiation exposure increases at higher elevations—about 1 mrem per 100 feet—so a person living at mile-high Denver might receive about 112 mrems.

Small amounts of natural radiation also surround us. These include radiation from minerals in the soil around the house, radiation from package and home material processing, chest x-rays, and so forth. With all these other contributors included, the average person in the U.S. receives about 200 mrems per year.

There is a great deal of controversy over how much radiation, over a given period of time, is considered safe. While there is a strong argument that says any additional radiation beyond that naturally occurring is potentially dangerous, studies of the physiological effects of radiation show that a single radiation dose of 400 rems (not mrems) will prove fatal to half of the people who receive it. The other half may suffer some form of impairment of function or develop longer-standing cancers.

On a more practical level, several studies have shown there are no *directly observable* effects on humans from a radiation dosage smaller than 10 rems. However, it all depends on the part of the body that receives the radiation. Vital organs such as the eyes and glands seem the most susceptible to the detrimental effects of radiation, so these tissues cannot withstand high exposures.

RADIATION SAFETY

In experimenting with radiation, you might need to collect various radioactive sources, including some potentially hazardous ones. However, unless you somehow get your hands on weapons-grade plutonium or some similar material, you won't be exposed to very dangerous levels as long as you treat the isotope with caution and respect.

Always wear clean, white gloves when handling any radioactive substance. This prevents alphas and most low-level beta particles from reaching your skin and also helps prevent contamination of your skin. Be sure to wear the gloves even when handling materials encased in vials or embedded in plastic. Although the risk of contamination is lower, handling the material with bare hands could still expose your skin to unnecessary radiation.

Carefully wrap all radioactive materials in several layers of aluminum foil. Store your collection of materials in a lead-lined film pouch, the type used by air travelers to prevent fogging of film at airport security gates. Clearly label the pouch "RADIOACTIVE" and store in a safe place.

While the opposite might seem true, isotopes with longer half-lives are generally the safer ones to handle. A long half-life typically means the isotopes are not as energetic as short-lived ones, so they are less radioactive. The difference is that materials with a long half-life continue to decay and give off radiation long after short half-life isotopes have stabilized or turned into another material.

SOURCES OF RADIOACTIVE MATERIALS

If you work at a laboratory or university, lab-grade radiation sources are common and rather easy to obtain. As a home-based gadgeteer, though, you'll have a little more trouble convincing the local Rent-a-Radium to sell you some radioactive samples.

But all is not lost. In almost any household, you can find sources of radiation that you can use in your experiments.

Radium-dial watches   Radium was often used to paint the self-luminous dials and numbers on watches (it is no longer used). If you have an older watch, remove its crystal and you have a source of low-level radium. Radium has a long half-life—of 16,000 years.

Tritium-dial watches   Tritium is now commonly used in self-luminous watches. However, tritium emits very low levels of radiation, none of which passes the plastic or glass crystal of the watch. You might have some luck detecting the tritium radiation by removing the crystal; however, this requires a highly sensitive detector. (The radiation detectors described in this chapter lack this sensitivity.)

Smoke alarms   Almost all smoke alarms use an ionizing chamber equipped with a small dab of americium 241, a human-made element two steps higher on the periodic table from plutonium. While americium 241 is fairly potent stuff, smoke detectors use just one microcurie. Americium-241 has a half-life of about 12 years.

Camping lamp mantles   Though the packaging doesn't say it, mantles for gas camping lamps are impregnated with radioactive thorium. Only use new mantles.

Glazed crockery   Brightly colored crockery used to be (and in some parts of the world still is) glazed with materials containing uranium oxide. Look around your house for bright yellow, orange, or red crockery and test it with one of the radiation detectors described later in this chapter.

Radon gas   Your house itself could be a source of radiation. Elemental radium decays to radon gas, which is slightly radioactive. A good radiation detector (including the Geiger counters described later in this chapter) can pick up the radioactive residue of radon.

Gold jewelry   At one time, some enterprising soul recycled into jewelry some gold that was originally used to encase low-level radioactive samples. The radiation remained even when the gold was melted down and made into something else. Although the radiation count will be low, you could own a piece of gold or gold-plated jewelry that gives off minute levels of radiation.

Small samples of uranium-235 and uranium-238 can often be purchased through lab supply outlets, including some of the chemistry mail-order companies provided in Appendix A. Write for a catalog. Some labs refuse to sell to individuals, so if you're working on your own, you might need to attach yourself to a recognized educational institution or look for another outlet.

Local and mail-order surplus companies occasionally offer new and used radiation monitors (mostly military surplus from the fifties and sixties). Some of these monitors come with a small sample of radioactive material, usually radium. In many cases, the sample is used to calibrate the instrument, so you know how much radioactivity is emitted from the material.

EXPERIMENTS WITH A CLOUD CHAMBER

Radiation particles and rays are far too small for you to see with your eyes or even with the most sophisticated electron microscopes of today. But you *can* see the trails left by these particles with a simple cloud chamber. You have probably experimented with a cloud chamber in physics or chemistry class. Here's how to do it at home.

Ingredients for the cloud chamber include a small plastic bowl with a tight-fitting lid, a block of dry ice, some denatured alcohol, and a radiation source, detailed in TABLE 20-3. A number of local and mail-order lab supply companies provide ready-to-go cloud chamber kits that include everything but the dry ice. Get dry ice at any nearby ice packing or beverage company (particularly those outfits that sell beer in kegs for parties).

Before actually using dry ice, you should be aware of some safety tips regarding its handling. Dry ice is really frozen carbon dioxide (at minus 110

Table 20-3. Parts List for Cloud Chamber

1	5-inch-diameter plastic or glass dish with sealing lid
1	16-inch by 1-inch-wide felt strip
1	Block of dry ice
1	Radiation sample
1	Flashlight or other point source

degrees F). Handle it only with heavy leather gloves and never let it touch your bare skin. Transport it from the ice packing company in a small cooler.

As dry ice melts, it gives off carbon dioxide gas. Although this gas is not toxic, it can displace the oxygen in a closed-up room, so always use dry ice in a well-ventilated room. If you should ever feel light-headed or nauseous, step outside to get some fresh air.

The plastic bowl should be clear except for the bottom, which you should coat (on the underside) with flat black paint. Glue a $1/2$-inch-wide strip of black velvet around the inside rim of the bowl, as shown in FIG. 20-7. When the glue sets, you're ready to go.

Fig. 20-7. Place a $1/2$-inch-wide strip of black velvet in your cloud chamber bowl. Just before using the chamber, soak the felt with alcohol.

Start by placing the dry ice in a pie baking tin. The dry ice should be just slightly larger than the base of the bowl and about an inch thick (you don't need any more, and any less could cause poor results). Thoroughly soak the velvet with undiluted denatured alcohol—the stronger the better. Place the radiation source, such as a sample of uranium oxide, in the middle of the bowl and put the lid securely in place. Position the bowl on the block of dry ice and leave the experiment alone for 10 to 15 minutes.

With the room lights as low as possible, shine a flashlight through the top or edge of the bowl. The inside of the bowl should look misty: the evaporating alcohol has created a saturated atmosphere inside the bowl, or your own self-contained cloud. Look carefully for tiny trails of mist, evidence that a radioactive particle has just ionized the saturated atmosphere, and left a visible trail. You will have better luck if you aim the flashlight towards you so that the trails are back-lit, as shown in FIG. 20-8.

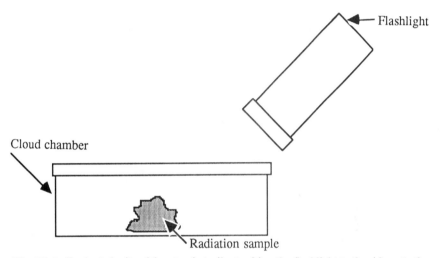

Fig. 20-8. For best clarity of the atomic trails, position the flashlight to the side or to the back of the cloud chamber so that the trails are backlit or sidelit when you look at them.

I have had good results with numerous radioactive samples, including a small chunk of uranium oxide (it came with the cloud chamber kit I purchased) and a radium-tipped pinhead. I obtained poor results with thorium-impregnated camping lamp mantles and an americium-241 disk cannibalized from an old smoke detector. Whenever you change samples, wipe off the excess alcohol from inside the bowl and add more to the velvet strip. Let the chamber settle 10 to 15 minutes before you attempt to view the trails.

BUILD THE GEIGER COUNTER

The most familiar type of radiation detector is the Geiger counter, named after its principal inventor H. Geiger (with E. W. Muller in 1928). At the heart

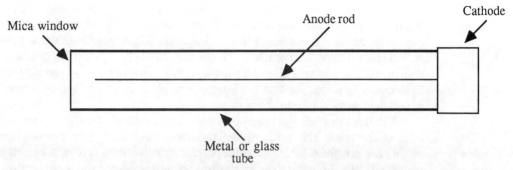

Fig. 20-9. The Geiger-Muller tube is simple in construction, consisting of a metal or glass envelope, an anode rod, a mica window for letting in alpha particles, and a common cathode.

of the Geiger counter is the Geiger-Muller detection tube, which consists of a glass or metal cylinder fitted with electrodes and sealed at low pressure with special gases. A schematic diagram of a typical Geiger-Muller tube is shown in FIG. 20-9.

The G-M tube operates with a high voltage (usually between 500 and 1,000 volts dc) between the anode and cathode electrodes. The exact amount of electricity is just enough to prevent an electrical discharge between the two electrodes. When a radioactive particle enters the tube, it knocks the electrons out of the atoms of the gas. Since the electrons are negatively charged, they are attracted to the anode terminal, and a slight drop in voltage across the anode and cathode terminals occurs. The change in potential in the tube is sufficient so that a sensing circuit connected to the anode can detect it.

When the sensing circuit is connected to an audio amplifier, the drop in voltage is heard as a slight click or pop. You hear a series of clicks when the amount of radiation is high enough to cause a continuous stream of radiation particles to enter the tube.

Geiger counters aren't used as much as they were in the fifties and sixties and have since been replaced in the laboratory with more sensitive solid-state radiation monitors and crystal scintillators (both of these technologies are discussed in more detail later in this chapter). Nevertheless, Geiger counters are sufficiently sensitive for our purposes, and though the Geiger-Muller (G-M) tubes are a little hard to find, the entire circuit is easy to build and use.

G-M tubes can be purchased new or used. Sources for new tubes include Raytheon, Amperex Electronics, and Victoreen. Plan to spend about $35 to $75 for the tube, depending on the size and sensitivity. You don't need anything elaborate, so opt for the less expensive miniature variety. In any case, be sure it is sensitive enough to register the faint clicks from a low-level radiation source such as thorium-impregnated camping lamp mantles. The tube you get should have a mica window on one end. This window is thin enough to let you measure alpha particle radiation.

If you prefer, you might have some luck finding a G-M tube in the used or surplus market. Try at the local electronics surplus outfits, especially those that deal in old military junk. They might have an antiquated radiometer, like the one shown in FIG. 20-10, for sale at a reasonable price. You'll get a better price if the unit doesn't work (after more than 30 years, it probably won't), but make sure the G-M tube is still intact. Though the tubes don't have an unlimited life, odds are they'll still work long after the electron tubes used in the detector bite the dust.

Another approach is to purchase a replacement G-M tube for the Heathkit GM-4 radiation monitor, shown in FIG. 20-11. The monitor includes a high-quality miniature Geiger-Muller tube. Contact Heathkit for a current price list on

Fig. 20-10. A museum piece now but still in working order, this 1950s Radiac set measures alpha, beta, and gamma radiation with amazing accuracy. It was originally manufactured for the military by Admiral.

Fig. 20-11. The pocket-size Heathkit Monitor 4 radiation detector comes in kit form but is also available ready-made from a number of sources, including Edmund Scientific.

replacement parts. Of course, you might also want to build the Heathkit Geiger counter kit; it's relatively inexpensive and its portable size lends itself to many useful applications. I carry it with me on all my trips.

Mail-order surplus is another possible source for Geiger-Muller tubes. I purchased two high-quality tubes from John Meshna & Associates (see Appendix A) for only $10 each. This included the aluminum wands (shown in FIG. 20-12) used to enclose the fragile tubes. However, there's no saying that Meshna will still have the tubes when you call or write, because surplus comes and goes. Check and ask around.

Figure 20-13 shows a schematic of a bare-bones Geiger counter. TABLE 20-4 provides a parts list. Actually, the term ''counter'' is a misnomer because this device doesn't count anything; it merely transforms the radiation. The device works on 6 volts dc and uses a transistor oscillator, step-up transformer, and high-voltage doubler to produce about 750 volts dc. A 4.7-megohm current-limiting transistor drops the voltage to about 500 volts, which is perfect for most Geiger-Muller tubes.

Fig. 20-12. The handheld "wand" of a surplus Radiac set, complete with sensitive Geiger-Muller tube. The wand includes a metal shutter at the end to block alpha particles.

Table 20-4. Parts List for Geiger Counter

R1	8.2 kΩ
R2	1 kΩ potentiometer (5–15 turn recommended)
R3	470 kΩ
R4	4.7 MΩ
C1	470 μF polarized electrolytic
C2 – C5	0.1 μF disc capacitors, 1 kV or higher
C6	50 pF disc capacitor, 1 kV or higher
Q1	2N2055 power transistor
T1	6.3 Vac to 120 Vac step-up/step-down transformer
S1	SPST switch
Misc.	Heatsink for Q1, G-M tube, shielded wire, 6 Vdc power source

All resistors are 5 to 10 percent tolerance, 1/4 watt. All capacitors are 10 to 20 percent tolerance, rated at 35 volts or more, unless otherwise noted.

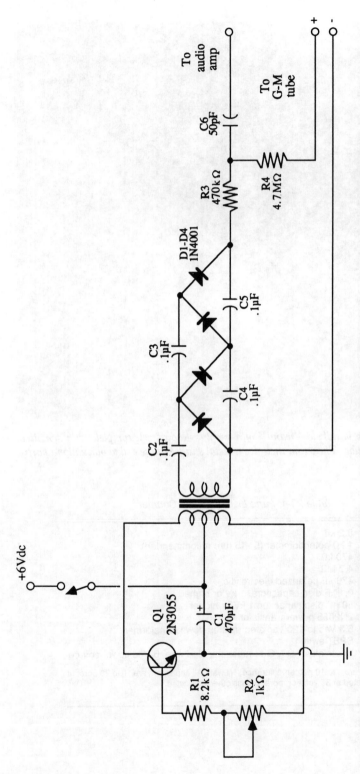

Fig. 20-13. Geiger counters don't require much circuitry. This is about the bare minimum you need and is mostly designed to generate the high voltages required to run the tube. Connect an audio amp to the amp terminals (don't omit C6 or the amplifier will be damaged).

It is extremely important that Geiger-Muller tubes not be subjected to more voltage than they are rated for. If you are unsure of the voltage rating, start low (at about 500 volts), and increase the potential in 50-volt increments while listening for clicks of radiation. When you hear the clicks, the tube has enough juice to monitor the incoming particles.

The basic Geiger counter circuit uses an external battery-powered amplifier. I used a Radio Shack 1-watt pocket amp, but you can use just about anything you have around. You might want to complete the Geiger counter into a self-contained system by adding your own low-power amp. An add-on circuit based around the versatile LM386 IC amplifier is shown in FIG. 20-14 (see the parts list in TABLE 20-5).

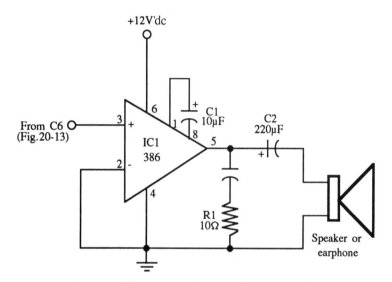

Fig. 20-14. A suitable amplifier for use with the Geiger counter circuit in FIG. 20-13.

Table 20-5. Parts List for LM386 Audio Amplifier

R1	10 Ω
C1	10 μF polarized electrolytic
C2	220 μF polarized electrolytic
IC1	LM386 amplifier IC
SPKR	8 Ω miniature speaker

All resistors are 5 to 10 percent tolerance, 1/4 watt. All capacitors are 10 to 20 percent tolerance, rated at 35 volts or more.

A word of caution: The high-voltage doubler, as well as the output of the Geiger-Muller tube, is at high potential. Though 500 to 600 Vdc won't kill you, getting a shock from it can be unpleasant. Be sure to add the bypass capacitor as shown in the schematics or you run the chance of blowing out the audio amplifier. You are advised to test the circuit with a makeshift audio amp first before connecting it to a more expensive audio amplification system.

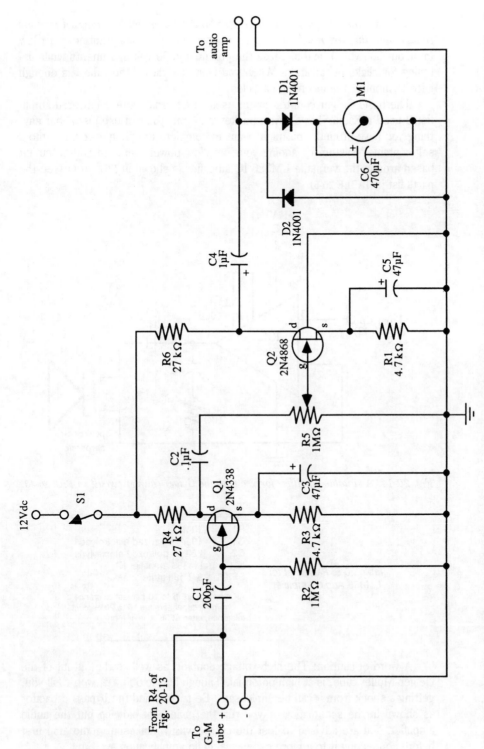

Fig. 20-15. An improved Geiger counter circuit designed to preamplify the clicks caused by radiation and show them on a meter. Use the circuit from FIG. 20-13 as a high-voltage power supply.

Another approach is to use the sensor, amplifier, and meter circuit shown in the schematic in FIG. 20-15. Consult TABLE 20-6 for a parts list. Experiment with the exact voltage at the high-voltage input for best audible results. Note that a certain amount of the high audio-frequency component developed by the oscillator circuit will bleed through to the speaker output. This will sound like a high-pitched shrill over the clicks from the radiation source.

Table 20-6. Parts List for
Advanced Geiger Counter

R1,R3	4.7 kΩ
R2	1 MΩ
R4,R6	27 kΩ
R5	1 MΩ potentiometer
C1	200 pF disc
C2	0.1 μF disc
C3,C5	47 μF polarized electrolytic
C4	1 μF polarized electrolytic
C6	470 μF polarized electrolytic
Q1	2N4338 FET transistor
Q2	2N4868 FET transistor
D1,D2	1N4001 diode
S1	SPST switch
M1	50 mA meter

All resistors are 5 to 10 percent tolerance, 1/4 watt. All capacitors are 10 to 20 percent tolerance, rated at 35 volts or more.

Try the Geiger counter with your assortment of household radiation sources. You can hear the relative levels of radioactivity of the various sources. Note, however, that most Geiger-Muller tubes are oversensitive to the americium 241 used in smoke detectors, so you'll get a very high rating when testing these, but the actual amount of radioactivity won't be as high.

TRANSISTOR RADIATION DETECTOR

You can construct a simple but not-too-sensitive radiation sensor out of an ordinary silicon transistor. Use any npn transistor in metal TO-5 case, such as a 2N3429.

As carefully as possible, remove the metal top using a small pair of pliers. You might need to cut through the metal with wire cutters. Be careful not to nick the transistor substrate inside or you'll ruin the device; you might want to use inexpensive transistors because you are apt to ruin two or three before successfully removing the top.

With the top off, examine the transistor substrate to look for nicks or torn electrodes. Use a $6\times$ or $10\times$ magnifier to look up close. If everything seems okay, connect the transistor to a 1- to 2-watt audio amplifier, as shown in FIG. 20-16. Turn out any room lamps and shield the transistor substrate from the light of the sun. Turn up the amplifier and you should hear a hissing sound.

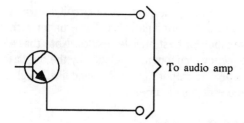

To audio amp

Notes:
1. Leave base unconnected.
2. Reverse connections to amp
   if excessive noise is heard.

Fig. 20-16. After removing the cover of the transistor, connect it to an amplifier as shown. If you hear excessive noise, reverse the connections.

Notice that the hiss grows stronger when sunlight strikes the transistor. Light from an incandescent lamp will cause a loud hum through the amp. The transistor is sensitive to infrared light and gives off a voltage proportional to the amount of light it receives. In the case of the incandescent lamp, the light is pulsed at 120 times per second, which results in the low hum you hear.

Place a strong radiation source close to the transistor. You should hear a faint but audible clicking. A good source to use is an americium-241 sample salvaged from a smoke detector. If you can't hear the clicks, try turning up the amplifier. You might need to try a different transistor.

Bear in mind that this is just a rudimentary form of a solid-state radiation detector. For better results, try the circuit shown in FIG. 20-17 (parts list in TABLE 20-7). It raises the gain from the amplifier several thousand times. As before, keep the transistor away from light sources.

SCINTILLATION RADIATION DETECTOR

One of the most sensitive forms of radiation detectors is the scintillator. The scintillator uses a radiation-sensitive crystal and a photodiode or photomultiplier tube to register alpha, beta, and gamma. The "crystal" can take many forms including Naphthalene, high-silica glass, and other natural and human-made substances. When radiation strikes the crystal, it scintillates—a small amount of light is produced in the material in reaction to the radiation exposure. That light is sensed and amplified by the photomultiplier tube or solid-state sensor.

Crystals and other materials normally used in commercially manufactured scintillators are expensive and difficult to get, so for our project, we'll use an old standby: zinc sulfide. This yellowish material has long been used in glow-in-the-dark paints and plastics. Before the advent of the Geiger counter, zinc sulfide was the compound originally used by Marie Curie as a means to determine the radioactivity of her radium samples. Small amounts of zinc sulfide can be purchased through most lab supply companies at a cost of about 25 cents a gram. You need less than 2 to 3 grams for this project.

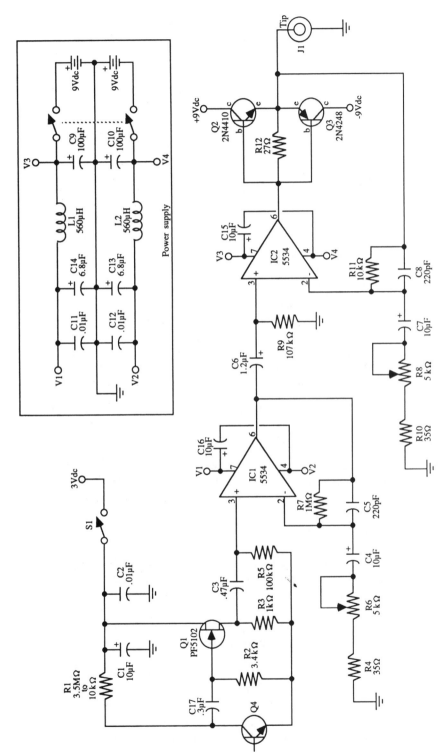

Fig. 20-17. A supersensitive amplifier circuit for use with the transistor radiation detector. It is the same circuit as in Chapter 15 for the long-distance light wave communicator.

Table 20-7. Parts List for Long-Range Receiver

R1	3.5 MΩ to 10 kΩ resistor (see text)
R2	3.4 kΩ
R3	1 kΩ
R4,R10	35 Ω
R5	100 kΩ
R6,R8	5 kΩ potentiometer
R7	1 MΩ
R9	107 kΩ
R11	10 kΩ
R12	27 Ω
C1,C7,C15,C16	10 μF polarized electrolytic
C2,C11,C12	0.01 μF disc
C3	0.47 μF disc
C4	10 μF polarized electrolytic
C5,C8	220 pF disc
C6	1.2 μF polarized electrolytic
C9,C10	100 μF polarized electrolytic
C13,C14	6.8 μF polarized electrolytic
C17	0.3 μF disc
IC1,IC2	5534 low-noise amplifier
Q1	PF5102 FET transistor
Q2	2N4410 npn transistor
Q3	2N4248 pnp transistor
Q4	Sensing transistor (see text)
S1	SPST switch
S2	DPDT switch
J1	Miniature jack (for output)
L1,L2	560 μH choke
Misc.	Two 9-volt transistor battery clips, two 9-volt transistor batteries, case, focusing lenses

All resistors are 5 to 10 percent tolerance, 1/4 watt. All capacitors are 10 to 20 percent tolerance, rated at 35 volts or more.

Start by mixing a gram or two of zinc sulfide with a drop of tap water. Stir until you get a slushy mixture. If the mixture is too runny, add a little more zinc sulfide powder or wait a few minutes for some of the water to evaporate. If it's too dry, add just a small drop of water.

Thoroughly clean both sides of a new microscope slide. With a paintbrush, apply a thin coat of the zinc sulfide mixture over the slide. Be sure the coating isn't too thick. You'll have enough zinc sulfide soup to coat several slides. You might need to smooth out the coating with the edge of an index card.

Let the liquid zinc sulfide dry. It doesn't take long—about 5 to 10 minutes in full sunlight. When completely dry, carefully clean and place another micro-scope slide on top of the zinc sulfide coating or apply a piece of cellophane tape to the coated side. You'll need to clean the ends of the glass so the tape will stick there.

Position a small solar cell directly behind the microscope slide. The solar cell shouldn't be much larger than the slide or the portion of the slide coated with the zinc sulfide. Connect the solar cell to the amplifier circuit shown in FIG. 20-18 (parts list in TABLE 20-8). Alternatively, you can use a photodiode as the sensing element. Position it behind the slide so it touches the glass.

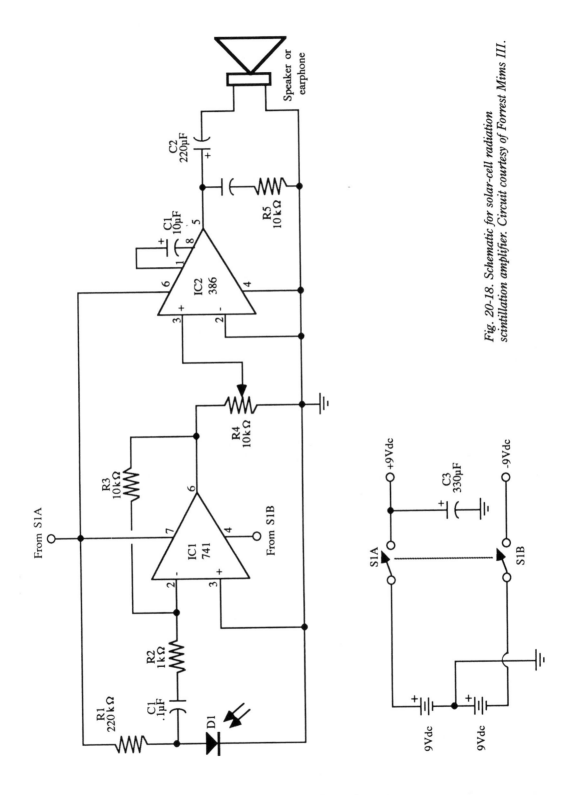

Fig. 20-18. Schematic for solar-cell radiation scintillation amplifier. Circuit courtesy of Forrest Mims III.

Table 20-8. Parts List for
Solar Cell Amplifier Circuit

R1	220 kΩ
R2	1 kΩ
R3	10 kΩ
R4	10 kΩ potentiometer
R5	10 Ω
C1	0.1 μF disc
C2	220 μF polarized electrolytic
C3	10 μF polarized electrolytic
C4	100 μF polarized electrolytic
IC1	LM741 operational amplifier IC
IC2	LM386 audio amplifier IC
Q1	Solar cell
S1	DPDT switch

All resistors are 5 to 10 percent tolerance, $1/4$ watt. All capacitors are 10 to 20 percent tolerance, rated at 35 volts or more.

Fit the microscope slide, solar cell, amplifier, and batteries in a black plastic project box. Cover the slide and solar cell with black electrical tape or thin heat-shrink tubing. Make sure the assembly is lighttight. The object is to prevent any light from striking either the microscope slide or the solar cell. Place the cover on the project box.

Wait at least 20 minutes for the light charge on the zinc sulfide to diminish. Place a radium source (the zinc sulfide doesn't work well with other, weaker radiation sources) close to the slide inside the enclosure. Because the slide is blocked by plastic and electrical tape, alpha particles and most beta particles won't penetrate. The reading you get will be strictly gamma rays.

You should hear clicks as the zinc sulfide intercepts the radiation, turning it into small speckles of light and passing it on to the solar cell. If you hear only a rush of hiss, light could be striking the tube or slide, or the zinc sulfide could still be glowing from its previous exposure to light.

21

Experiments in superconductivity

In the early part of this century, the natural element uranium was largely regarded as a scientific curiosity. Few physicists could pin a worthwhile application on uranium. The most common demonstration of uranium was grinding up a small portion of it and watching it spontaneously combust. The ''explosion'' was about as violent as a soda fizzing in a glass.

Despite these humble demonstrations, uranium was destined to change the world. In refined form, the uranium that crackled and popped on the science teacher's desk would erupt with a force equal to thousands of tons of dynamite. In a further refined form, it would provide electricity to a growing nation and power massive ships at sea. And its rays would penetrate tissue and bones looking for cancerous tumors that were hidden by ordinary x-ray techniques.

The science of superconductivity is now at the same infant stage uranium occupied earlier in this century. Thanks to recent advances in new materials, superconductivity can be easily demonstrated even in the elementary school classroom, and with materials costing less than $20 to $30. But the demonstration—floating a magnet a few tenths of an inch above a black ceramic pellet—doesn't even begin to do justice to the promise of superconductivity. We now stand at a crossroads where the entire future of superconductivity depends solely on applications scientists can dream up for it.

In this chapter, you'll learn what superconductivity is, how it works, and how to perform the almost-magical floating magnet trick. You'll also learn about additional experiments you can perform with superconductors and how this new science holds great promise for those with the vision to look ahead.

A SHORT HISTORY

Superconductivity was first discovered by Heike Kamerlingh Onnes, a Dutch physicist working alone in 1911. Onnes was interested in the effect of extremely cold temperatures on certain metals. Quite by accident, he discovered that a glob of mercury lost all resistance to the flow of electricity when the metal was cooled to around 4 degrees Kelvin (or K), very close to absolute zero.

The loss of resistance to electricity was an important discovery, as it meant that "superconducting" wires could carry power for long distances without loss. Onnes discovered other metals and alloys—such as bismuth and lead—were equally superconductive and behaved the same as mercury.

In all cases, the superconducting action didn't take place until the metal had cooled with the aid of liquid helium to very low temperatures. Because liquid helium is expensive and difficult to manufacture, Onnes' discovery—although important on a scientific level—had no practical applications. The power companies were not interested in making their transmission wires with a superconducting metal, only to cool hundreds of miles of it with a costly liquefied gas.

Since Onnes' time, little progress was made in developing superconductors that worked at higher, more manageable temperatures. Even as recent as 1986, the best superconducting materials still required a frosty 39 degrees K.

All that changed in February 1987, when Ching-Wu Chu at the University of Houston, working on advancements recently reported by Georg Bednorz and Alex Müller at IBM's Zurich Research Laboratory, discovered a method for making a new material that underwent superconductivity at 98 degrees K. Even though 98 degrees K is still bitingly cold (−273 degrees F), the breakthrough is that expensive liquid helium is no longer needed. Rather, the superconductor could be cooled with cheap and plentiful liquid nitrogen. Nitrogen gas liquefies at 77 degrees C, well below the critical temperature required by Chu's superconductor.

Unlike earlier superconductors, which were simply metals like mercury, lead, and bismuth, the superconductors pioneered by Bednorz, Müller, and Chu are *perskovite ceramic discs*. The most common superconducting ceramic is composed of the metals yttrium, barium, and copper in an oxidized form.

During manufacture, the materials of the superconductor are ground up, pressed into a disc shape, and heated in a ceramic-firing furnace. In most instances, the disc is recrumbled, crushed some more, and reheated. This process can be repeated several times to improve the superconducting properties of the material.

Because the superconducting disc is a ceramic rather than a solid metal, it tends to be brittle and absorb water, which makes the disc even more brittle. Water absorption can be a problem when demonstrating superconductivity with a ceramic disc (as you'll see later in this chapter) and should be avoided.

THE PROPERTIES OF A SUPERCONDUCTOR

Today's superconductor materials have two states: nonsuperconducting and superconducting. In the nonsuperconducting state, the material acts like any other matter and possesses no extraordinary properties. In the superconducting state, reached only when exposed to cold temperatures, the material exhibits unusual behavior. Most importantly, the usual resistance to the flow of electricity is gone. Instead of putting in 1 volt and getting out 0.99 volt, the superconductor passes all of the electricity.

Let's put it another way. Suppose the superconducting material is shaped like a donut, as shown in FIG. 21-1. An electrical charge is applied to the donut, and the electrons begin to spin around its circumference. With ordinary resistance, that flow will stop almost immediately after the electrical charge is removed. But with a superconductor, the flow will continue unabated for months or even years.

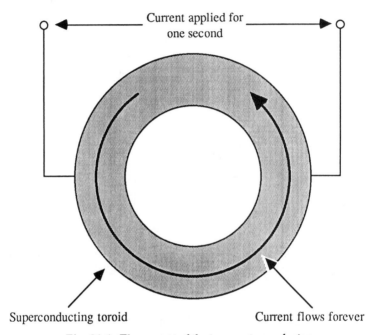

Current applied for one second

Superconducting toroid   Current flows forever

Fig. 21-1. The concept of the torus superconductor.

In fact, this very experiment was performed by Onnes shortly after he began his pioneering work in superconductivity. He first cooled a loop of pure metal to 4 degrees K with liquid helium, applied an electrical charge to it, removed the charge, and then stashed the whole contraption away for one year. When he returned to his experiment, he found the current was still flowing, undiminished.

Coupled with the loss of electrical resistance, materials in their superconducting state can also be subjected to extremely high currents. There is a limit to the amount of current that can be passed through a superconductor, and most of today's superconducting materials are anisotropic (they conduct best in only one direction). But the current-carrying capacity (or current density) is several hundred times higher than that of ordinary materials.

A thin strand of copper wire might melt when passing 5 or 10 amps of current, but a thin superconducting wire can pass several hundred amps of current before breaking down. When subjected to high currents, superconductors first lose their superconductivity. At that point, they are subject to the same effects of current overload as other, nonsuperconducting materials.

A third unique aspect of some (but not all) superconductors is the "magnetic mirror" effect, most often referred to as the *Meissner effect*, named after the scientist who first discovered the phenomenon in 1933. When in the superconducting state, the superconductor acts as a type of mirror to magnetic fields, as depicted in FIG. 21-2. If you placed a small, strong magnet on top of a superconducting disc, the mirrored magnetic fields would repel the magnet, and the magnet floats in midair. This is the principle behind the floating magnet trick so often demonstrated with superconducting discs.

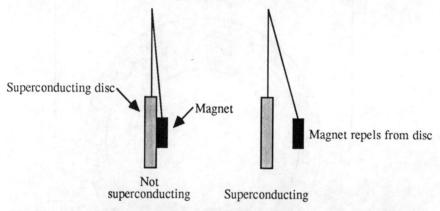

Fig. 21-2. The Meissner, or magnetic mirror, effect causes the magnet to deflect from the superconductor. The deflection is caused by the magnetic lines of force from the magnet "reflecting" off the superconductor.

Not all superconductors are subject to the Meissner effect. Physicists have discovered that the properties of the magnetic mirror occur only with so-called Type I superconductors, which include metals and the commonly available perskovite ceramic discs. The Meissner effect does not occur in Type II superconductors such as niobium and vanadium, some alloys, and semiconductors.

As with current density, there is a limit to the amount of magnetism a superconductor can repel. After a certain point, the magnetism overcomes the material, and the material loses its superconductivity.

MATERIALS YOU NEED

You need only a few materials to begin experimenting with superconductivity and the Meissner magnetic mirror effect.

The superconducting disc These are available from many sources including Edmund Scientific, Jerryco, and Colorado Superconductor. Cost is reasonable, with a basic no-frills disc for about $25.

Small rare-earth magnet The magnet you use for the Meissner experiment must be small but extremely powerful for its size. Ordinary alnico and ceramic magnets are too weak; you need a rare-earth magnet like samarium-cobalt. Good sources for good rare-earth magnets are miniature stereo headphones (be sure the packaging states that rare-earth magnets are used) and those companies offering the superconducting discs.

Plastic tongs Use tongs to handle the magnet and disc. Plastic tongs are required because they don't draw too much cold from the disc and magnet.

Liquid nitrogen You need liquid nitrogen to cool the superconductor disc. You can purchase liquid nitrogen from many welding supply shops, industrial gas companies, and some hospitals. Be sure to read the instructions on handling liquid nitrogen later in this chapter.

Insulating cup An ordinary Styrofoam cup holds the superconducting disc for your experiments and helps contain the liquefied nitrogen. Don't use plastic, ceramic, or glass containers because they could break or crack. Metal containers draw too much of the coldness away from the superconducting disc, requiring you to use additional liquid nitrogen.

Colorado Superconductor offers a kit consisting of a superconductor disc, square rare-earth magnet, plastic tongs, and instruction book. As of this writing, you can obtain the kit directly from Colorado Superconductor or from other mail-order outlets including Jerryco and Edmund Scientific. Consult Appendix A for the addresses to these firms.

WORKING WITH LIQUID NITROGEN

Liquid nitrogen is a colorless, nontoxic liquid composed entirely of nitrogen gas. The nitrogen gas is cooled to liquid form with the use of compression and expansion. Liquid nitrogen is often used in the welding trade for metal fitting, in the medical trade to remove corns and warts, and as a coolant for certain hospital equipment such as nuclear magnetic resonators.

Despite its nontoxicity, liquid nitrogen can be a dangerous material to handle. Its extreme coldness can cause instant frostbite. When the liquid touches objects that are at room temperature, it ''boils'' just like water hitting a hot griddle. Tiny droplets could then strike skin and eyes and cause considerable injury.

Keep the following in mind when handling and using liquid nitrogen.

- Use a stainless steel or glass Thermos bottle to hold the liquid nitrogen (some outlets won't fill these canisters due to breakage and waste; check first). Drill a hole in the Thermos cap to allow the nitrogen vapor to escape. *Never* stopper the container completely because the nitrogen gas can cause the container to violently explode.
- The liquid nitrogen will last about a day in a Thermos. If you want to keep it longer, use a Dewar's flask or other approved container designed for handling refrigerated liquid gas.

- Wear safety goggles and waterproof welder's gloves when handling liquid nitrogen.
- Never allow the liquid to touch your skin or you could receive serious frostbite burns.
- If the liquid touches clothing, immediately grasp the material at a dry spot and pull it away from your body.
- While experimenting, fill liquid nitrogen only in Pyrex glass or metal containers.
- Avoid plastic and regular glass containers because they can shatter on contact with the extremely cold temperatures.
- The evaporating nitrogen gas released as the liquid slowly warms is not harmful (78 percent of the Earth's atmosphere is composed of nitrogen). However, the nitrogen gas can displace oxygen, so always conduct your experiments in a well-ventilated room.
- If you have extra liquid nitrogen left over after your superconducting experiments, simply let it evaporate in its container. If you can't keep the container, simply return it with its contents, or dump the extra liquid nitrogen on cool, soft soil (avoid plants).

CONDUCTING THE MEISSNER EXPERIMENT

Set up the experiment as depicted in FIG. 21-3. You need the ingredients specified in TABLE 21-1 to conduct the experiment. Cut all but the bottom 1/4 inch off of a Styrofoam cup. Place the superconductor disc inside the cup (in addition, be sure to place the cup on a (durable or disposable) surface such as a piece of scrap wood). Gently place the magnet in the center of the disc.

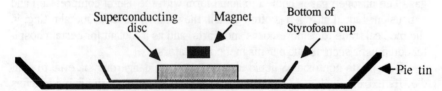

Superconducting disc   Magnet   Bottom of Styrofoam cup

Pie tin

Fig. 21-3. Basic setup for demonstrating the Meissner effect.

1	Styrofoam cup
1	Rare-earth magnet
1	Superconducting disc
1	Plastic tweezers
	Liquid nitrogen, gloves

Table 21-1. Parts List for Basic Meissner Effect Setup

Being careful to pour slowly, douse the disc and magnet with the liquid nitrogen. At first, the liquid will boil furiously and splatter all over the disc, cup, and working surface. As the cup and disc cool, however, the boiling should subside, and the liquid nitrogen will begin pooling at the bottom of the cup and around the superconductor. Fill the cup with liquid nitrogen so that the entire disc is submerged.

As the disc reaches its critical temperature (the point where it changes from a nonsuperconductor to a superconductor), the magnet will levitate a small distance in midair. With the eraser end of a pencil, gently tap the magnet and watch it spin in the air (a square magnet shows rotational movement more readily than a disc magnet). The magnet will slowly come to a stop because of air friction. If the magnet was spinning in a vacuum, it would continue forever or until the superconductor's temperature heated past its critical temperature.

Before long (about 1 to 2 minutes), the liquid nitrogen will evaporate and the superconductor disc will begin to heat up. When the critical temperature is reached—the border between superconductivity and nonsuperconductivity— the magnet will slowly settle down to the disc and the magnetic mirror effect will be lost.

MEASURING THE MAGNETIC MIRROR EFFECT

The superconductor disc does not reflect the entire magnetic field put out by the tiny rare-earth magnet. You can prove this by building the simple gauss meter shown in FIG. 21-4 (see the parts list in TABLE 21-2). The meter uses a linear Hall-effect sensor, manufactured by Sprague (but also available from

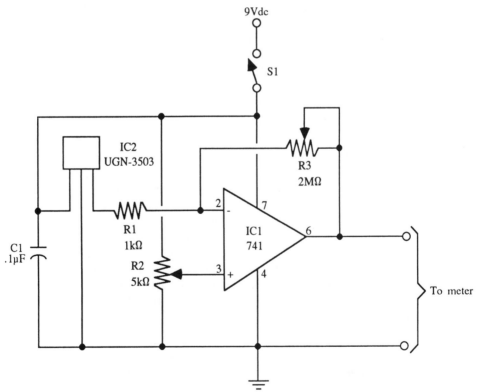

Fig. 21-4. A solid-state magnetometer, or gauss detector, designed around an affordable linear-output Hall effect device. Connect the output of the detector to a digital or analog voltmeter.

Table 21-2. Parts List for Guess Meter

R1	1 kΩ
R2	5 kΩ, 15-turn potentiometer
R3	2 MΩ
C1	0.1 μF disc
IC1	LM741 op amp IC
IC2	Sprague UGN-3503 Hall-effect sensor (or equivalent)
S1	SPST switch

All resistors are 5 to 10 percent tolerance, 1/4 watt. All capacitors are 10 to 20 percent tolerance, rated at 35 volts or more.

other manufacturers), to detect the strength of nearby magnetic fields. The output of the operational amplifier connects to a digital or analog volt-ohmmeter (set to low-range volts). The higher the reading, the greater the magnetic force.

Attach the gauss meter so that the printing on the Hall-effect sensor (the south pole) points upward through a Styrofoam cup. Tape the cut Styrofoam cup you used for the previous Meissner effect experiment end-to-end over the other cup.

Place the magnet on the superconducting disc and note the output of the Hall-effect sensor on the volt-ohmmeter. Now pour on the liquid nitrogen and watch the reading on the meter drop as the magnet is levitated. If the magnetic mirror were complete, the reading on the meter would be zero, or very close to it. In fact, the reading changes only a little, showing that much of the magnetic field from the rare-earth magnet passes through the disc. Enough is reflected back, however, to bob the magnet a few fractions of an inch in the air.

TESTING CONDUCTIVITY

The Meissner effect has many real-world applications such as in bullet trains, magnetic shock absorbers, magnetic bearings, and even unique forms of high-tech art. But these applications are further off than the more rudimentary use of superconductors as a means to transmit electricity without current loss.

The no-frills disc is not really suitable for experimenting with the super conductivity of superconductors. For effective results, you need a four-point probe disc that already has four wires bonded to its surface. These wires allow you to apply a current and test it with a voltmeter. Simply clipping leads to a plain disc yields poor results because of losses in contact resistance. Four-point probe superconducting discs are available from Colorado Superconductor, among others. TABLE 21-3 lists the parts you need to experiment with a four-point superconducting disc.

Table 21-3. Parts List for Four-Point Disc Experiments

1	Four-point superconducting disc (with cup, magnet, liquid nitrogen, etc.)
1	Ammeter
1	Voltmeter
1	Variable dc power supply

The basic hookup diagram for the four-point probe disc is shown in FIG. 21-5. Attach a current-regulated power supply equipped with an ammeter (with a power output of about 500 milliamps to 1 amp) to two of the probes, and attach a digital voltmeter to the remaining two probes. Set the volt-ohmmeter to the 10-microvolt range.

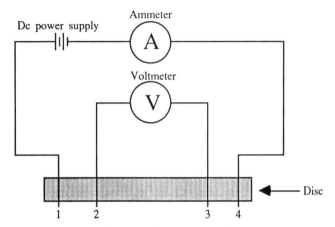

Fig. 21-5. Connection diagram for a four-point superconducting disc.

With the disc at room temperature, turn on the power supply and note the reading on the ammeter and the volt-ohmmeter. Now, immerse the disc in a bath of liquid nitrogen and watch the meter readings jump. You will note that the voltage drop normally associated with conductors will be next to zero when the disc is in its superconducting state. There will be some voltage drop, however, due to the inherent resistance in the hookup wires and test leads.

CARING FOR YOUR SUPERCONDUCTOR DISC

Ceramic superconducting discs don't last forever. As the liquid nitrogen evaporates and the disc warms, condensation collects on the surface of the disc. This moisture can cause the disc to crack and grow increasingly brittle. You can help prolong the life of your disk after you're finished with your experiments by placing it under the heat of a desk lamp for 10 to 15 minutes. Place the disc on a clean Styrofoam cup; avoid paper products like paper towels or facial tissues because the paper can dry onto the surface of the disc, causing even more trouble.

Though the ceramic superconducting discs are relatively safe to handle, they are poisonous if ingested. The superconducting discs are not Necco wafers, so keep them away from kids and adults with a sweet tooth. Never allow children to play with the disc because not only will it break, but they might try to eat parts of the superconductor. For reference, the toxicity of ceramic superconductors is listed in the Rocky Mountain Poison Index. Let your doctor know if you suspect someone has eaten a disc.

Superconducting discs are not biodegradable, so when yours finishes its useful life, don't throw it in the trash or down the garbage disposal. The disc should be disposed of at an official toxic disposal site.

FOR MORE INFORMATION

The topic of superconductivity is an involved and complicated one. If you'd like to learn more, you are encouraged to actually try some of the experiments outlined in this chapter. You might also want to try the more advanced superconducting experimenter's kits offered by Colorado Superconductors and others. These allow you to test other properties of the superconductor including:

- Critical temperature—when the material changes state from superconductor to nonsuperconductor.
- Critical magnetic field—when the material no longer mirrors a nearby magnetic field.
- Critical current—when the material no longer supports the current flowing through it.

For more background on the science and application of superconductivity, read *Superconductivity: The Threshold of a New Technology* by Jonathan L. Mayo (TAB BOOKS, No. 3022), as well as *Superconductivity: Experimenting in a New Technology*, by Dave Prochnow (TAB BOOKS, No. 3132). These books provide additional material on the history, theory, and application of superconductors.

22

BioMetal—a
shape-memory alloy

A metal that has a memory? You bet. As early as 1938, scientists observed that certain metal alloys, once bent into odd shapes, returned to the original form when heated. This property was considered little more than a laboratory curiosity because the metal alloys were weak, were difficult and expensive to manufacture, and they broke apart after just a couple of heating/cooling cycles.

Research into metals with memory took off in 1961 when William Beuhler and his team of researchers at the U.S. Naval Ordnance Laboratory developed a titanium-nickel alloy that repeatedly displayed the memory effect. Beuhler and his cohorts developed the first commercially viable *shape-memory alloy*, or SMA. They called the stuff Nitinol, a fancy-sounding name derived after *ni*ckel *ti*tanium *N*aval *O*rdnance *L*aboratory.

Since its introduction, Nitinol has been used in a number of commercial products—but not many. For example, several Nitinol engines have been developed that operate with only hot and cold water. In operation, the metal contracts when exposed to hot water, and relaxes when exposed to cold water. Combined with various assemblies of springs and cams, the contraction and relaxation (similar to a human muscle) causes the engine to move.

Other commercial applications of Nitinol include pipe fittings that automatically seal when cooled, large antenna arrays that can be bent (using hot water) into most any shape desired, sunglass frames that spring back to their original shape after being bent, and a novel antiscald device that shuts off water flow in a shower should the water temperature exceed a certain limit.

Since regular Nitinol contracts and relaxes with temperature (in air, water, or other liquid), the effectiveness of the metal is limited in many applications where local heat can't be applied. Researchers have attempted to heat the Nitinol metal using electrical current in an effort to exactly control the contraction and relaxation. But because of the molecular construction of Nitinol, hot spots develop along the length of the metal, causing early fatigue and breakage.

In 1985, a Japanese company called Toki Corp. unveiled a new type of shape-memory alloy specially designed to be activated by electrical current. Toki's unique SMA material, called BioMetal, offers all of the versatility of the original Nitinol, with the added benefit of near instant electrical actuation.

While the main application of BioMetal is currently in industry and the R&D lab, you can readily purchase small quantities of this intriguing metal for testing purposes. This chapter details some experiments you can readily perform with BioMetal and gives you some ideas of how to use it in real-world applications.

BASICS OF BIOMETAL

At its most basic level, BioMetal is a strand of nickel titanium alloy wire. Though the BioMetal wire is only 0.15 mm thick—slightly wider than a strand of human hair—it is exceptionally strong. In fact, the tensile strength of Bio-Metal rivals that of stainless steel: the breaking point of such a slender wire is a whopping 6 pounds. Even under this much weight, BioMetal stretches little. In addition to its strength, BioMetal also shares the corrosion resistance of stainless steel.

BioMetal, like all shape-memory alloys, changes its internal crystal structure when exposed to certain higher-than-normal temperatures (this includes the induced temperatures caused by passing an electrical current through the wire). The structure changes again when the alloy is allowed to cool.

More specifically, during manufacture, the BioMetal wire is heated to a very high temperature, which embosses or "memorizes" a certain crystal structure. The wire is then cooled and stretched to its practical limits. When the wire is reheated, it contracts because it is returning to the memorized state.

Although most BioMetal strands are straight, it can also be manufactured in spring form, usually as an expansion spring. In its normal state, the spring exerts minimum tension, but when current is applied, the spring stiffens, exerting greater tension. Used in this fashion, BioMetal becomes an "active spring" that can adjust itself to a particular load, pressure, or weight.

BioMetal has an electrical resistance of about 1 ohm per inch. That's more than ordinary hookup wire, so BioMetal heats up more rapidly when an electrical current passes through it. The more current, the hotter the wire becomes, and the more contracted the strand becomes.

Under normal conditions, a 2- to 3-inch length of BioMetal is actuated with a current of about 450 milliamps. That creates an internally generated temperature of about 100 to 130 degrees C; 90 degrees C is required to achieve the shape-memory change (BioMetal can be manufactured to change shape at most any temperature, but 90 degrees C is the standard value for off-the-shelf material).

Excessive current should be avoided. Why? Extra current causes the wire to overheat, which can greatly degrade its shape-memory characteristics. For best results, current should be as low as possible to achieve the contraction desired. BioMetal contracts by two to four percent of its length, depending on the amount of current applied. Maximum contraction of BioMetal is 8 percent, but that requires heavy current that can, over a period of just a few seconds, damage the wire.

OBTAINING BIOMETAL

BioMetal is not yet a regular staple of the neighborhood electronics store. As of this writing, there are two main sources for BioMetal: directly from Toki, the manufacturer, or mail order from Mondo-Tronics (see Appendix A for addresses).

Both Toki and Mondo-Tronics offer numerous kits you can build with BioMetal such as a robotic hand, a novel "Space Wings" executive toy (the BioMetal slowly flaps a set of mylar wings), and experimenter's packages consisting of BioMetal wire, terminators, electronic actuators, and more. You might want to start out with one of these experimenter's packages until you get used to working with BioMetal. If you develop a commercial application for the material, you can purchase the wire in bulk.

USING BIOMETAL

BioMetal needs little support paraphernalia. Besides the wire itself, you need some type of terminating system, a bias force, and an actuating circuit.

Terminating

The terminators attach the ends of the BioMetal wires to the support structure or mechanism you are moving. Because BioMetal expands as it is heated, using glue or other adhesive will not secure the wire to the mechanism. Soldering is not recommended as the extreme heat of the soldering can permanently damage the wire. The best approach is to use a crimp-on terminator. These and other crimp terminators are available from Toki (either in the experimenter's kit or separately).

You can make your own crimp-on connectors using 18-gauge or smaller solderless crimp connectors (the smaller the better). Although these connectors are rather large for the thin 0.15 mm BioMetal, you can achieve a fairly secure termination by folding the wire in the connector and pressing firmly with a suitable crimp tool. Be sure to completely flatten the connector. If necessary, place the connector in a vise to flatten it all the way.

Bias force

If you apply current to the ends of a BioMetal wire, it just contracts in air. To be useful, the wire must be attached to one end of the moving mechanism and biased (as shown in FIG. 22-1) at the other end. Besides offering physical support, the bias offers the counteracting force that returns the BioMetal wire to its limber condition once current is removed from the strand.

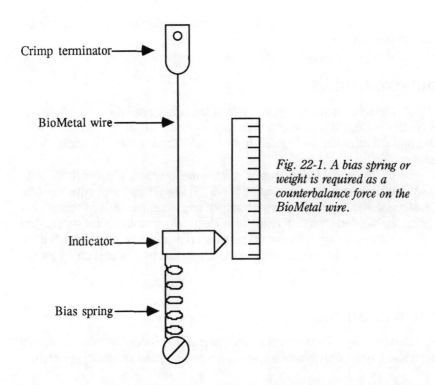

Crimp terminator

BioMetal wire

Fig. 22-1. A bias spring or weight is required as a counterbalance force on the BioMetal wire.

Indicator

Bias spring

Actuation

BioMetal can be actuated with a 1.5-volt penlight battery. Because the circuit through the BioMetal wire is almost a dead short, the battery delivers almost its maximum current capacity. But the average 1.5-volt alkaline penlight battery has a maximum current output of only a few hundred milliamps, so the current is limited through the wire. You can connect a simple on/off switch in line with the battery, as detailed in FIG. 22-2, to contract or relax the BioMetal wire.

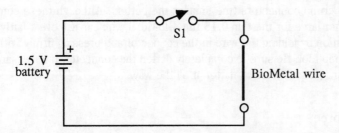

S1

1.5 V
battery

BioMetal wire

Fig. 22-2. A simple switch in series with a 1.5-volt penlight battery forms a simple BioMetal driving circuit. The low current delivered by the penlight battery prevents damage to the strand of BioMetal.

The problem with this setup is that it is wasteful of battery power, and if the power switch is left on for too long, it can lead to some damage of the BioMetal strand. A more sophisticated approach uses a 555 timer IC that automatically shuts off the current after a short period of time. The schematic in FIG. 22-3 shows one way of connecting a 555 timer IC to control a length of BioMetal. TABLE 22-1 provides a parts list for the 555 BioMetal circuit.

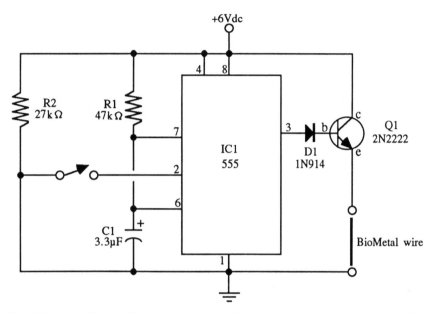

Fig. 22-3. A 555 timer IC is the heart of an ideal driving circuit for BioMetal. The 555 removes the current a short moment after you release activating switch S1.

Table 22-1. Parts List for 555 BioMetal Driver

R1	47 kΩ
R2	27 kΩ
C1	3.3 μF polarized electrolytic
Q1	2N2222 transistor
D1	1N914 diode
Misc.	BioMetal wire, with terminators

All resistors are 5 to 10 percent tolerance, 1/4 watt. All capacitors are 10 to 20 percent tolerance, rated at 35 volts or more.

In operation, pressing momentary switch S1 activates the wire, and it contracts. Release S1 immediately, and the BioMetal stays contracted for an extra fraction of a second and then releases as the 555 timer shuts off. Since the total on time of the 555 is dependent on the length of time S1 is held down plus the 1/10th of a second delay, you should depress the switch only momentarily.

You can vary the on time by changing the value of R1. For flexibility in experimenting, change R1 to a 1-megohm potentiometer, and add an extra 10 kΩ resistor in-line with it, as shown in FIG. 22-4 (parts list in TABLE 22-2). Rotate the pot to dial in other delay periods.

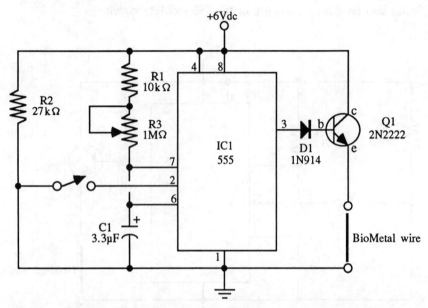

Fig. 22-4. An improved BioMetal driver, incorporating a potentiometer to alter the on time of the 555 timer chip.

R1	10 kΩ
R2	27 kΩ
R3	1 MΩ potentiometer
C1	3.3 µF polarized electrolytic
Q1	2N2222 transistor
D1	1N914 diode
Misc.	BioMetal wire with terminators

All resistors are 5 to 10 percent tolerance, 1/4 watt. All capacitors are 10 to 20 percent tolerance, rated at 35 volts or more.

Table 22-2. Parts List for
Enhanced 555 Timer Biometal Driver

The circuit in FIG. 22-5 automatically pulses the BioMetal wire once every 3 to 30 seconds (BioMetal can be cycled several thousands of times per second, but in actual practice, the oscillations shouldn't be more rapid than about 10 Hz). This circuit, with parts list in TABLE 22-3, also uses the versatile 555 timer IC to let you control the rate of oscillation just by turning potentiometer R1.

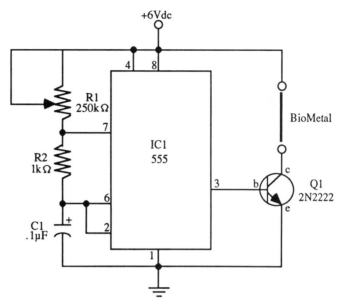

Fig. 22-5. Schematic showing a 555 timer IC connected in astable mode, delivering a series of short pulses to the BioMetal wire.

Table 22-3. Parts List for Astable BioMetal Circuit

R1	250 kΩ potentiometer
R2	1 kΩ
C1	0.1 μF disc
Q1	2N2222 transistor
Misc.	BioMetal wire with terminators

All resistors are 5 to 10 percent tolerance, 1/4 watt. All capacitors are 10 to 20 percent tolerance, rated at 35 volts or more.

BioMetal mechanisms

With the BioMetal properly terminated and actuated, it's up to you and your own imagination to think of ways to use it. Figure 22-6 shows a typical application of using BioMetal in a pulley configuration. Apply current to the wire, and the pulley turns, giving you rotational motion. Large-diameter pulleys turn very little when the BioMetal tenses up, but small-diameter ones turn an appreciable distance. Use the driven pulley as the hub of a larger gear system if you need to spin bigger wheels.

Another application uses the block-and-tackle approach (see FIG. 22-7). Here, a length of BioMetal is attached to a rigid arm. On this arm are two idler pulleys. Opposite the arm is a moving jib equipped with three pulleys. You can readily see what will happen when the BioMetal wire is actuated: the strand will

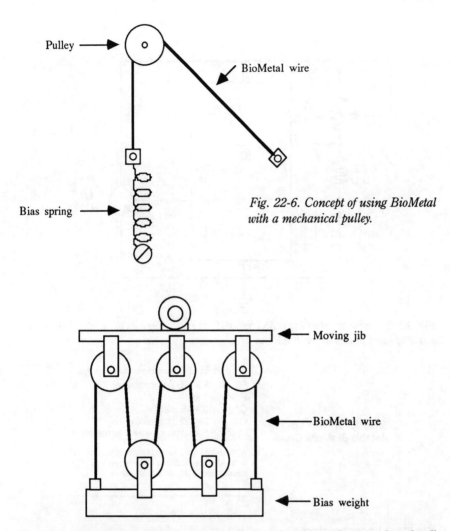

Pulley

BioMetal wire

Fig. 22-6. Concept of using BioMetal with a mechanical pulley.

Bias spring

Moving jib

BioMetal wire

Bias weight

Fig. 22-7. A block-and-tackle arrangement can greatly enhance the strength and pull distance of a strand of BioMetal.

contract and draw the jib closer to the rigid arm. The more pulleys you have, the greater the distance of travel. Note that the two ends of the BioMetal strand must be electrically insulated from one another.

Figure 22-8 shows a length of BioMetal wire used in a lever arrangement. Here, the BioMetal strand is attached to one end of a bell crank. On the opposite end is a bias spring. Applying juice to the wire causes the bell crank to move. The spot where you attach the drive arm dictates the amount of movement obtained when the BioMetal contracts. As shown in FIG. 22-9, place the drive arm close to the BioMetal, and you get little movement. Place the arm on the opposite end, and it moves much further. In both instances, the BioMetal contracts the same amount.

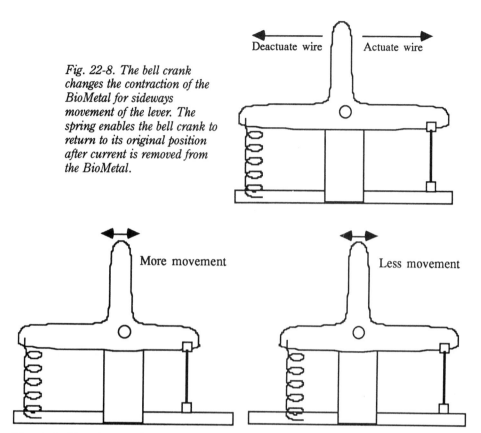

Fig. 22-8. The bell crank changes the contraction of the BioMetal for sideways movement of the lever. The spring enables the bell crank to return to its original position after current is removed from the BioMetal.

Fig. 22-9. Changing the distance between pivot and BioMetal wire alters the amount of movement of the bell crank lever.

The miniature hardware designed for model RC airplanes can be most useful in constructing these mechanisms. Any well-stocked hobby store should have a full variety of bell cranks, levers, pulleys, wheels, gears, springs, and other odds and ends to make your work with BioMetal more enjoyable.

23

High-tech
espionage devices

In the James Bond movies, 007 has at his disposal a specialized department of Her Majesty's Secret Service called Q-branch, headed by an eccentric known only by the name "Q." It was the job of Q and his minions to come up with useful gadgets for 007 to use in his various international exploits. In the James Bond movies especially, these gadgets took on almost magical qualities: homing devices to track a moving vehicle from miles away, explosive-firing pens, even powerful electromagnetic wristwatches that could deflect the path of a bullet.

The real world of espionage has its toys and gadgets, but they aren't anything like what James Bond carries. High-tech espionage technology involves superlisteners to eavesdrop on conversations blocks away or mini FM transmitters that relay the sounds of one room to another.

This chapter details some inexpensive and practical "espionage" gadgets you can build. These trinkets are presented more as an exercise in miniaturization and high-gain amplification than they are as real spy tools, so keep the following in mind: using electronic means to eavesdrop into other people's conversations or otherwise intrude into someone else's privacy is against the law in most states. Use these projects carefully and only when you have the full permission of every party involved.

The usefulness of these projects goes beyond mere "spying" tools. For example, the electronic ear is great for listening to and recording outdoor wildlife, especially birds and small animals that won't let you get too close. And you can use the mini FM transmitter as a one-way intercom for electronically checking up on baby while you're doing the dishes or are out in the laundry room.

ELECTRONIC EAR

Imagine being able to hear whispers at 500 feet or conversations at a quarter of a mile away. The electronic ear brings distant sounds closer to you so you can listen in with your headphones or record for posterity on a portable cassette player. Two versions of the electronic ear are presented: one uses a parabolic dish to focus sound waves to a common point, and the other uses a long tube as a highly directional "shotgun" microphone. Both use the same easy-to-build superamplifier circuit.

Audio amplifier circuit

The audio amplifier circuit shown in FIG. 23-1 is based on an LM387 dual preamp IC. The parts list for the circuit is provided in TABLE 23-1. The two pre-amps are used independently to boost the sound amplification in steps. The first stage increases the amplification from the microphone some 47,000 times. The microphone used in the project is an ordinary electret condenser type with a built-in FET preamp transistor. The mike used in the prototype was the two-terminal variety (as opposed to three-terminal), available at a number of electronics surplus outfits.

The second stage of the preamp IC is used in a novel sound-limiting scheme. To understand the importance of audio limiting, imagine you're listening to some faraway sounds through a set of headphones. Suddenly, a jokester nearby fires a cap gun in front of the microphone. Without limiting, your ears would be subjected to a great burst of noise as the amplifier boosts the signal as high as it can. Given the right circumstances, your eardrums could be damaged if exposed to such high-volume sounds.

The limiter permits audio signals of only 1 volt peak-to-peak or less to pass through to the headphone amplifier. That means that signals greater than one volt peak-to-peak are cut off by the feedback mechanism constructed around the second preamp stage. A 220 pF capacitor, C6, is attached to the anode end of D1 to prevent oscillation. If you experience oscillation in your circuit, increase the value of this capacitor to 330 pF or 470 pF.

The second half of the amplifier circuit consists of a headphone amp. This amp consists of a 2N2222 transistor and adds enough current to the signal for use with 8-ohm dynamic headphones. Alternatively, you can connect the output of the LM387 amplifier to the line input of a tape recorder. Use a 10 kΩ potentiometer as a level control and be sure to use shielded cable to reduce unwanted hiss. In fact, you'll want to use shielded cable for both the mike input and all amplified signal outputs.

Because you might want to use this same amplifier circuit in many projects, build the project using 1/8-inch audio jacks for the microphone input and sound output, as shown in FIG. 23-2. An added switch lets you turn the circuit on and off.

The LM387 amplifier circuit uses readily available components and can be built using almost any circuit construction technique. I used a universal solder board to put the prototype together, and fit everything in a 2³/4-by-5¹/8-by-1³/4-inch plastic project box. As of this writing, the LM387 amp is available in kit form from Allegro Electronics Systems (see Appendix A for the address). This mail-order firm headed by fellow gadgeteer Anthony Charlton supplies a miniature printed circuit board and all parts except the project box.

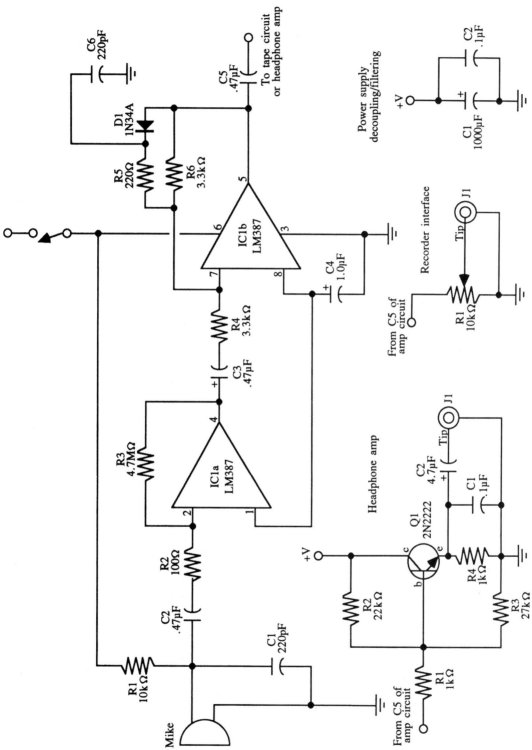

Fig. 23-1. Schematic for the low-noise audio amplifier/limiter circuit designed for the parabolic dish and tube listeners detailed later in this chapter. Circuit courtesy Anthony Charlton.

Table 23-1. Parts List for LM 387 Amplifier

R1	10 kΩ
R2	100 Ω
R3	4.7 MΩ
R4,R6	3.3 kΩ
R5	220 Ω
C1	220 pF disc
C2,C3,C5	0.47 µF disc
C4	1.0 µF polarized electrolytic
IC1	LM387 integrated amplifier IC
D1	1N34A
S1	SPST switch
Mike	Electret condenser microphone

All resistors are 5 to 10 percent tolerance, 1/4 watt.
All capacitors are 10 to 20 percent tolerance, rated
at 35 volts or more.

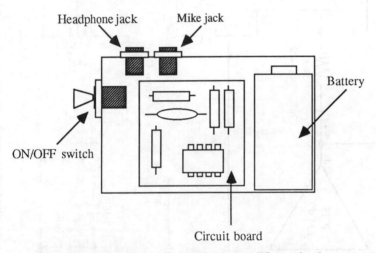

Fig. 23-2. Suggested parts layout for the amplifier project box.

Test the circuit before building the rest of the electronic ear. Connect an electret condenser microphone to the input and a pair of headphones to the output. Don't put the headphones on yet. You can wear the headphones, but place the earpieces on above your ears on the sides of your head instead of directly over your ears for now.

Turn the circuit on and listen for the telltale squealing sound of an oscillating circuit. If you don't hear a squealing, proceed to the next step. If oscillation does pose a problem, you might need to increase the value of C6, as detailed previously, or increase R5. Note that increasing R5 from 220 ohms to a higher value will reduce or eliminate oscillations, but it will also decrease the gain of the circuit. Don't replace R5 with a potentiometer because the construction of a pot can add significant noise to the amplified circuit. Keep trying new values until you get one that works.

If oscillation is still a problem, you can reduce gain (and hence the oscillations) by replacing R3, a 4.7-megohm resistor, with a 1.5-megohm resistor. Decreasing the value of R3 also decreases the overall gain of the circuit.

Assuming the circuit is working and not oscillating, place the headphones over your ears and point the microphone away from you. Speak softly and you should hear a greatly amplified version of your voice. When constructed properly, the LM387 amplifier can detect your breathing from several feet away, fans in other rooms, or even small bugs flying around. The sound level increases when you add a parabolic dish or tube.

You might find that full-cover headphones provide better sensitivity and privacy than the small portable-radio or Walkman types. Because the full-cover headphones don't let much sound escape, there is less risk of audio feedback as the already-amplified sound from the headphones reaches back to the microphone.

Parabolic dish

The most effective way of tuning in to distant sounds is to use a parabolic dish. A parts list for the complete dish (minus amplifier, detailed above) is provided in TABLE 23-2. I used an 18-inch-diameter dish purchased surplus from Edmund Scientific. Other dishes, both metal and plastic, will do. For example, those round aluminum snow sleds make excellent parabolic dishes, as do solar collectors and dishes for receiving microwave pay TV (MDS TV).

Table 23-2. Parts List for Electronic Ear

1	Spun aluminum or plastic parabolic dish
1	10-inch-length 12-gauge wire
2	#12 crimp-on connector
1	Project box, $2^{3}/4$ by $5^{1}/8$ by $1^{3}/4$ inches
Misc.	Hardware, battery clips, batteries, tripod mounting hardware

The microphone must be positioned at the focal point of the dish. While you can calculate the focal point of a parabolic dish using algebra, an easier method is to position the dish in full sunlight and move a small piece of paper in and out of the center until you see a small, bright spot. Measure the distance of the paper to the base of the dish and you have the focal point. If your dish is not bright metal, temporarily cover it with aluminum foil. Smooth the foil to remove excess ripples and ridges.

Figure 23-3 shows one way to mount the electret condenser microphone to a standoff. This standoff, made from a $1/8$-inch diameter welding rod (a metal coat hanger works just as well) can be adjusted to fine tune the focal point. Simply loosen the bolts, turn the nuts one direction or another, and tighten. Note that the microphone is physically insulated from the standoff using double-sided tape. The springiness of the tape helps damp vibrations from the dish to the microphone. A completed microphone and standoff is shown in FIG. 23-4. The black tape was added only for cosmetic purposes.

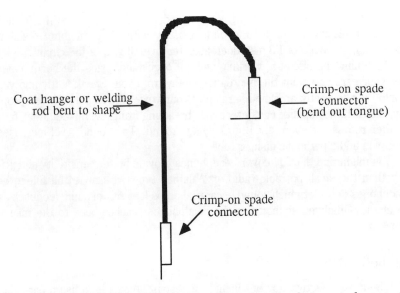

Coat hanger or welding rod bent to shape

Crimp-on spade connector (bend out tongue)

Crimp-on spade connector

Fig. 23-3. The microphone standoff: bend a wire so that the crimp-on spade connector (with the tongue bent out) equals the focal length of the parabolic dish.

Attach the amplifier to the side of the dish using hardware or double-sided tape. The prototype uses a single bolt that attaches the amplifier and microphone standoff to the dish.

Although not necessary, you might want to paint your parabolic electronic ear flat black, especially if you plan to use it in daylight. The dish focuses light as well as it does sound, and you could disturb your target by shining bright, focused light into its eyes. Add a 1/4" 20 bolt and sleeve coupling to the side of the dish to provide a means to secure it to a tripod. A completed parabolic electronic ear, attached to tripod, is shown in the photograph in FIG. 23-5.

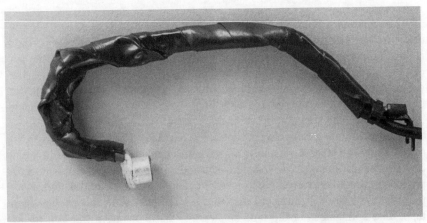

Fig. 23-4. The complete microphone standoff, with electret condenser microphone attached to tongue of the crimp-on connector with double-sided foam tape.

Fig. 23-5. The complete parabolic "big ear" dish mounted on a camera tripod.

Tube listener

The parabolic dish focuses sound waves to a common point. It is the most effective means of listening to faraway sounds because the surface area of the dish captures (or collects) a great deal of sound. The shape of the dish focuses all the captured sound into one spot, where the microphone is placed.

In addition to collecting sound, the parabolic dish discriminates against those sounds that occur off axis. This helps tune out all sounds except those occurring directly in front of the dish. This same directionality, a property of the parabolic dish, is also used to receive signals from distant satellites, or to beam telephone calls from mountaintop to mountaintop.

Yet another method of achieving directionality is to place the receiving microphone deep inside a long tube. Only those sound waves exactly parallel to the length of the tube can travel far enough inside to reach the microphone. Tube listeners don't collect as much sound waves as the parabolic types, but they are generally more directional and less expensive to build. A parts list for the tube listener is given in TABLE 23-3.

Table 23-3. Parts List for Tube Listener

1	40-inch length of 4-inch PVC or ABS plastic pipe
1	3-inch round speaker (replaces Mike in TABLE 23-1)
Misc.	Foam backing for speaker, mounting hardware

Start with a 40-inch length of 4-inch PVC or ABS plastic pipe. You can obtain the pipe from most any hardware or home-improvement store. Pipe supply and plumbers stores have more to choose from, so be sure to check them out as well. Also purchase one 4-inch end cap to fit onto one end of the pipe.

The tube electronic ear uses a 3-inch-round speaker instead of an electret condenser microphone. The speaker needs to be physically isolated from the tube to reduce the effects of vibration. Construct the isolator by cutting a piece of 2-inch-thick foam and cut it to $4^1/2$ inches in diameter. Punch or cut out a small circle in the center of the foam as shown in FIG. 23-6.

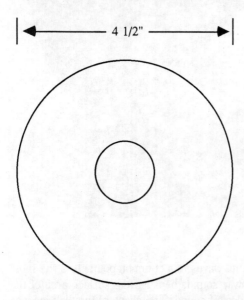

4 1/2"

Fig. 23-6. Cutaway diagram for the speaker foam.

Solder an 8-inch length of shielded cable to the terminals of the speaker. Press the magnet of the speaker through the small hole and slip the foam and speaker into one end of the tube. The front of the speaker should be facing into the tube so that if you look into the other end, you'll see the cone of the speaker deep inside the tube. Solder the other end of the speaker wire to a $1^1/8$-inch plug.

Mount the LM387 amplifier (assuming you've installed it in a suitable box) as shown in FIG. 23-7. Drill a hole in the end cap for the speaker wire, and pass it through to the amplifier. Test the tube electronic ear in the same fashion described earlier for the parabolic dish.

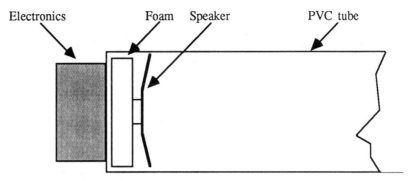

Fig. 23-7. Assembly suggestion for tube listener. Mount the speaker in the foam and place the foam at one end of a length of PVC pipe. Attach the amplifier (in project box) to an end cap and place the end cap over the pipe.

TELEPHONE RECORDING DEVICE

You might have occasion to record your telephone conversations (you're conducting an interview for a magazine, for example), but you don't want to get into the mechanics of turning a recorder on every time you make or receive calls.

The simple schematic in FIG. 23-8 lets you automatically tape messages with an ordinary cassette recorder (see the parts list in TABLE 23-4). The recorder is activated only when you are on the phone. The circuit works only with recorders with a remote control input.

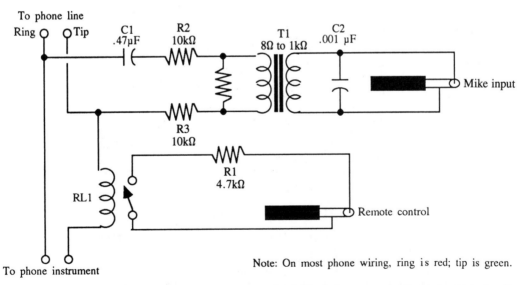

Note: On most phone wiring, ring is red; tip is green.

Fig. 23-8. Schematic of the telephone conversation interceptor circuit.

Table 23-4. Parts List for Phone Listener

R1	4.7 kΩ
R2,R3	10 kΩ
C1	0.47 μF disc
C2	0.001 μF disc
T1	8 kΩ-to-1 kΩ impedance-matching transformer
RL1	SPST reed relay

All resistors are 5 to 10 percent tolerance, 1/4 watt. All capacitors are 10 to 20 percent tolerance, rated at 35 volts or more.

To use the recording device, connect the circuit to the phone line as shown, and attach the output plugs to the respective jacks of a cassette recorder. Keep the recorder turned on and in the RECORD position at all times. It will only activate on when you are on the phone.

The first portion of the circuit senses when the phone is off the hook. When the phone is on the hook (no call), the reed relay is not activated and the relay contacts are broken. The tape recorder doesn't run. When you pick up the phone, the reed activates and pulls in the contacts, thus activating the recorder.

The second portion of the circuit is the interface between phone line and the microphone input of the recorder. The transformer couples the impedance of the phone line to the mike input.

Before you use the telephone recording device, take note: Under most circumstances, you are not allowed to record a conversation unless both parties (you and the person you are talking to) are aware that the chat is being recorded. This can be as simple as ''I'm recording this call for my records'' at the beginning of the conversation to a more complex ''beep'' every 10 or 15 seconds. Note that the circuit can be used only with the telephone that is connected to it; it won't activate the recorder if you use an extension phone.

When building the circuit, you can hard-wire the phone cord into the circuit or use modular jacks. These jacks are available at Radio Shack or any hardware or home-improvement store. Remember that phone lines are color coded (or at designated ''tip'' and ''ring''), so make sure to match the colors or the circuit won't work. For your reference (and as long as your phone was installed according to standards), the ''tip'' connection is the green wire and the ''ring'' connection is the red wire.

EAVESDROPPING WITH RADIO TRANSMITTERS

In just about every spy book ever written, the hero checks his hotel room for a hidden ''bug,'' a radio transmitter that sends everything he says to the bad guys. Radio bugs are surprisingly simple to build and, because they use so few parts, can be packed into small, easily concealed packages. The range is limited—usually no more than 100 or so feet—but that should be sufficient for most applications.

You can build your own FM radio transmitters from scratch, using components that for the most part are easily obtainable. You can also salvage a useful FM transmitter from a number of inexpensive sources:

- Purchase a kiddie wireless mike and tear out the microphone and transmitter circuitry. The circuit is built small so it fits into a microphone enclosure. The signal is received over an unused FM frequency. The cost is about $8.
- Purchase a low-power toy walkie-talkie. Tear out the guts of the walkie-talkie and replace the bulky speaker with a microphone. Jam the send button so that the unit transmits only, and you have an instant FM bug. The real benefit is that walkie-talkies have ranges of up to $1/4$ mile, but you do need a matching unit to receive the signal. The cost is $10 to $12.
- Purchase an FM transmitter kit such as the EKI 1044. The kit can be tuned for best reception on an FM station. The cost is $5 to $10.

Salvaging an FM transmitter from a wireless microphone

The easiest method of salvaging an FM transmitter is to tear apart a toy wireless microphone. The microphone typically includes an electret condenser mike, transmit circuitry (tunable across most of the FM band), and a 9-volt battery.

Begin by removing the battery (if installed), wind sock, and any labels. If the enclosure is held together by screws, loosen them to take the microphone apart. If no screws are used, pry apart the two halves of the enclosure with a small screwdriver. You won't be saving the plastic enclosure, so don't worry about destroying it.

Carefully pick the components out of the enclosure and lay them on your work table. The wires connecting the microphone and battery to the main circuit are usually fragile, so be sure you don't pull on them too hard. Analyze the construction of the transmitter board and your intended application and modify the components accordingly. For example, you'll probably want to save space by attaching the microphone directly to the transmitter circuit.

What kind of battery does the circuit use? Many require a 9-volt transistor battery, which you can place under the transmitter board (add a layer of insulating tape to prevent shorts). Other wireless microphones operate on one or two penlight cells. These are bulkier than a 9-volt transistor battery (especially when installed in holders), but you can often substitute them for smaller power cells. Use 1.5- or 3-volt mercury or lithium batteries, for example. These might not last as long as the penlight cells, but they will allow you to build much smaller bugs. A sample FM bug constructed from a toy wireless microphone is shown in FIG. 23-9.

To use the bug, apply power to it (switch it on or clip in a new battery). Place the transmitter about 5 feet from an FM radio. Slowly turn the radio knob until you hear a distinct drop or void in normal signals. They could indicate that you've found the frequency of the wireless microphone. Say a few words into the mike to see if the sound comes through. If not, keep dialing.

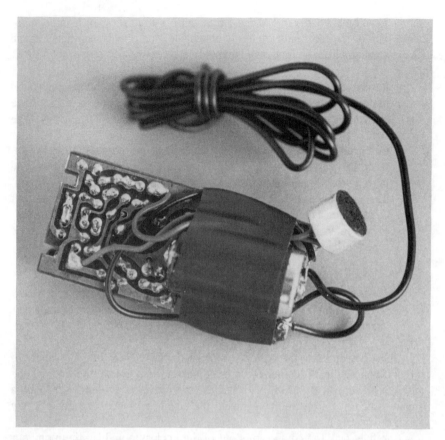

Fig. 23-9. The innards from a toy FM wireless microphone make a good compact "bug."

Sometimes, the transmitter is adjusted to the same frequency as a strong, local radio station. In this case, your voice might not get through or you might hear terrible interference. Try adjusting the output frequency on the transmitter (if so equipped), and redial the radio.

Salvaging an FM transmitter from a walkie-talkie

Kiddie wireless microphones have a very restricted range, typically less than 30 or 40 feet. If you need more range, try a 100-milliwatt walkie-talkie. The best types to use are the kind equipped with a flexible whip antenna because the aerial is more compact and easier to manage.

Begin by removing the back from the walkie-talkie, usually by removing one or two screws. Depending on the model of walkie-talkie you use, inside you'll find a relatively small transmit/receive board, a battery clip, a speaker, and an antenna. Carefully unsolder the speaker connections (at the speaker) and remove the antenna and circuit board. The circuit board, such as the one in FIG. 23-10, will include a press-to-talk button and switch. Since you want to use the

Fig. 23-10. A walkie-talkie, with back and tranceiver board removed.

walkie-talkie only as a transmitter, you can remove the button and jam the switch in the "talk" position.

Alternatively, you can remove the switch and short out the proper connections to force the walkie-talkie into the transmit mode, but the push-to-talk switch has many terminals so you must decode the action of the switch before attacking the board with your soldering iron. By removing the switch, you make the FM bug even smaller.

In place of the speaker that you removed, attach a two-terminal electret condenser microphone to the walkie-talkie circuit. Electret condenser mikes are polarity sensitive, so try to trace which speaker wire goes directly to ground (use a volt-ohmmeter and watch for 0 ohms when connecting the terminals between the ends of the wire and a known ground point, like the black battery lead). If you can't ascertain the proper polarity, temporarily connect the microphone and try transmitting. If you don't hear anything or the sound is very weak, reverse the connections.

Many toy walkie-talkies have a volume control that also incorporates the on/off switch. You don't need the volume control, so you can remove it. Add a miniature on/off toggle switch in place of the control. A salvaged walkie-talkie circuit board, with whip antenna and 9-volt battery intact, is shown in FIG. 23-11.

Fig. 23-11. A walkie-talkie transceiver board trimmed of all its nonessentials, and packed small enough to fit in a vest pocket.

Unlike a wireless microphone, transmitted signals from a walkie-talkie must be received by another walkie-talkie or CB radio. As most toy walkie-talkies are sold in pairs, you probably already have the receiver. Test the bug by turning on both transmitter and receiver. Place the receiver on a table a few feet away and talk into the microphone of the transmitter. If all is working properly, you'll hear your voice on the receiving end.

Note that you can substitute the antenna of the walkie-talkie—whether it is the telescoping or whip kind—with a 16-inch length of insulated hookup wire. When you "plant" the bug, be sure the wire is oriented the same way as the receiving antenna. When possible, the antennas should point vertically to provide the greatest range.

Building from scratch

The schematic in FIG. 23-12 (parts list in TABLE 23-5) shows a low-power FM bug you can build with ordinary parts. You need to build the circuit on a printed circuit board. This requires you to design the layout of the board following the schematic and then etch and drill it prior to soldering the components. Making printed circuit boards is beyond the scope of this book, so if these techniques are new to you, check out the available literature listed in Appendix B.

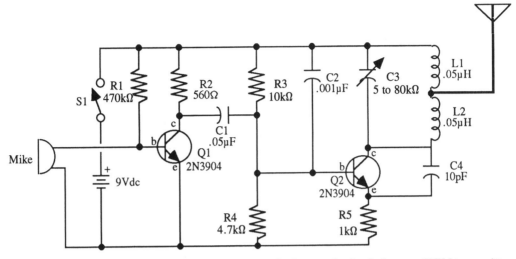

Fig. 23-12. Schematic diagram for the do-it-yourself FM transmitter.

Table 23-5. Parts List for FM Bug

R1	470 kΩ
R2	560 Ω
R3	10 kΩ
R4	4.7 kΩ
R5	1 kΩ
C1	0.05 µF disc
C2	0.001 µF disc
C3	5 pF-to-60 pF variable capacitor
C4	10 pF disc
Q1,Q2	2N3904 npn transistor
L1,L2	0.05 µF coil
Mike	Electret condenser microphone
S1	SPST switch

All resistors are 5 to 10 percent tolerance, 1/4 watt. All capacitors are 10 to 20 percent toler- ance, rated at 35 volts or more.

To use the circuit, connect a 9-volt transistor battery to the battery clip and turn switch S1 to the OFF position. Place the board a few feet from an FM radio and orient the antenna vertically. Turn the radio on and rotate the dial until no station can be heard. Then turn the transmitter on by flicking switch S1. With a plastic (*not* metal) TV alignment tool, rotate variable capacitor C2 until you hear a hiss or rush in the radio. Talk into the microphone to see if your voice comes through. If not, keep rotating C2 until your voice comes through clear.

You will find that the transmitter puts out many harmonics of the main broadcast frequency, so you can often pick up the sounds of the transmitter at several points along the band. However, only one spot will yield the cleanest, most powerful signal. Note that frequency on the radio dial and write it directly on the circuit board.

24

Projects in robot design

The word *robot* is defined as a mechanical device that is capable of performing human tasks or behaving in a humanlike manner. To the robotics experimenter, however, *robot* has a completely different meaning. A robot is a special brew of motors, solenoids, wires, and assorted electronic odds and ends, a marriage of mechanical and electronic gizmos.

Taken together, the parts make a half living but wholly personable creature that can vacuum the floor, serve drinks, protect the family against intruders and fire, entertain, educate, and more. In fact, there's almost no limit to what a well-designed robot can do.

In just about any science, it is the independent experimenter who first establishes the basic ideas and technologies. Robert Goddard experimented with liquid-fuel rockets during World War I; his discoveries paved the way for modern-day space flight. In the mid 1920s, Scotsman John Logie Baird experimented with sending pictures of objects over the airwaves. His original prototypes, which transmitted nothing more than shadows of images, opened the door to today's explosive television and video marketplace.

Robotics, like rocketry, television, and countless other technology-based industries, started small. But growth and progress have been slow. Robotics is still a cottage industry, even considering the special-purpose automatons now in wide use in the car industry. The science of personal robotics—the R2-D2 and C-3PO kind of ''Star Wars'' fame—is even smaller, an infant in a brand new family on the block. Hence, for the robotics experimenter, there is plenty of room for growth. There are a lot of discoveries yet to be made.

In this chapter, you'll learn how to build a rudimentary robot from the ground up—starting with the basic frame, and going on to the propulsion system, power pack, and control network. Actually, what sounds rather complex is really pretty simple: you can build the ''Scooterbot'' that appears in the following pages in less than a few hours at a cost of under $25. And as you'll see in the next two chapters, Scooterbot is an excellent springboard for advanced robotics projects.

BUILDING SCOOTERBOT

Scooterbot is a small robot built from channel aluminum, mending plates, nuts and bolts, and a few other odds and ends. All parts are commonly available at most any hardware or home improvement store. The parts list for Scooterbot appears in TABLE 24-1.

Table 24-1. Parts List for Scooterbot

Frame

2	11-inch-long aluminum or steel shelving standards
2	5³/4-inch-long aluminum or steel shelving standards
4	1¹/2-by-³/8-inch flat corner irons
8	1/2-inch-by-8/32 stove bolts, nuts, lockwashers

Motor and Mount

1	Surplus Big Trak motor (or two dc gear reduction motors)
1	5³/4-inch-long, 1¹/4-inch-wide galvanized nail mending plate
2	1-inch-by-8/32 stove bolts, nuts, flat washers, toothed lockwashers
1	Four-cell D battery holder

Support Caster

1	5³/4-inch-long, 1¹/4-inch wide galvanized nail mending plate
1	1¹/4-inch swivel caster
2	1/2-inch-by-8/32 stove bolts, nuts, toothed lockwashers, flat washers (as spacers)

Motor Control Switch

1	Small electronic project enclosure
2	DPDT momentary switches, with center off
Misc.	Hookup wire

Framework

Build the frame of the Scooterbot from a single 3-foot length of channel aluminum or steel shelving standards. These items are routinely available at most any hardware or home improvement stores. The prototype used aluminum shelving standards; you can use steel standards or extruded aluminum channel.

Cut the pieces using a hacksaw and miter box, as shown in FIG. 24-1. Be sure to cut precise 45-degree angles and that the pieces are as close to the specified length as possible. A deviation of as little as ¹/8 inch will cause the frame to be off square, and the robot might not roll in a straight line.

Using 1¹/2-inch-by-³/8-inch flat corner irons, as shown in FIG. 24-2, attach the pieces like a picture frame. The flat corner has two holes on each leg; drill only one matching hole in the channel or standard stock to match. Assemble using 8/32-by-¹/2-inch pan-head stove bolts, and secure with toothed lock-washers and nuts, but do not tighten the nuts until the entire frame is assembled. Using a carpenter's square, align the frame as close to square as possible.

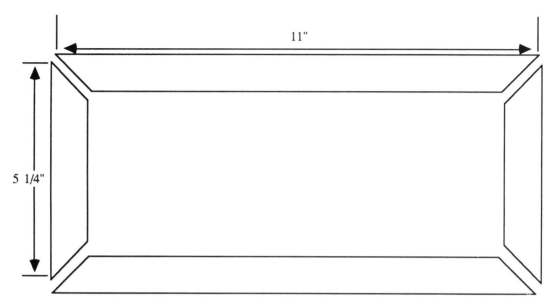

Fig. 24-1. Cutting diagram for the Scooterbot frame. Note the orientation of the mitered ends.

If a corner will not square up, try reversing one of the pieces, or file down the ends for a better match. Tighten the nuts and bolts when you are satisfied with the alignment.

Motor mount

The prototype Scooterbot made use of a dandy surplus geared motor system. This all-in-one drive system, shown in FIG. 24-3, was originally designed for the Milton-Bradley BigTrak toy and contains two motors and two gear reduction systems, not to mention a novel magnetic clutch arrangement to assure that the motors turn at the same speed when the toy is propelled directly forward.

Thousands of these motors appeared in the surplus market a few years ago, but alas, their popularity among hobby robot builders has made them somewhat hard to find. However, you can still find the motors at many local electronics surplus outfits. If you can't locate one locally, try H&R Sales (see Appendix A for the address), who offers a specially manufactured version of the BigTrak motor. If all else fails, you can still build the Scooterbot with almost any other lightweight, geared hobby motor.

The motor attaches to the frame with a $1^1/4$-inch-by-$5^3/4$-inch mending plate, as shown in FIG. 24-4. The plate was cut to size from another that measured 12 inches long. Secure the motor at the center flanges of the unit with two $8/32$-by-1-inch stove bolts and $8/32$ nuts. Use washers as needed. Secure the plate to the frame using $8/32$-by-$1/2$-inch bolts and nuts. Attach 5- or 6-inch rubber wheels to the motor shafts. The shafts of the motors are notched, and you can easily slot the hubs of the wheels to match.

A

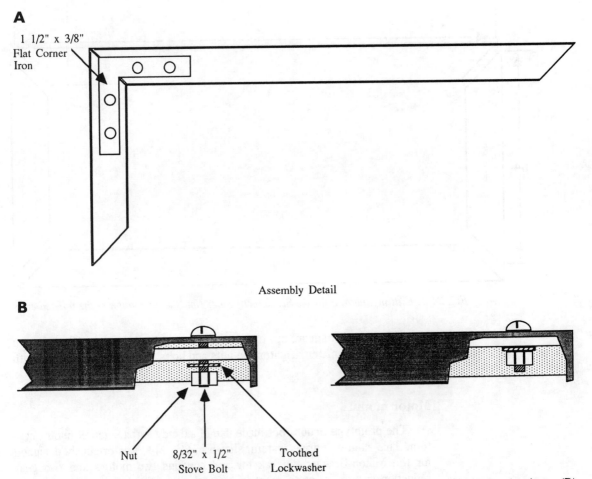

1 1/2" x 3/8"
Flat Corner
Iron

Assembly Detail

B

Nut   8/32" x 1/2"   Toothed
Stove Bolt   Lockwasher

Fig. 24-2. Construction detail of the frame. (A) Using 1¹/₂- by ³/₈-inch flat corner irons to secure the pieces; (B) Assembly detail using ¹/₂-inch-by-⁸/₃₂ hardware.

Support caster

The Scooterbot uses the two-wheel drive tripod arrangement. You need a caster on the other end of the frame to balance the robot and provide a steering swivel. The 1¹/₄-inch swivel caster is not driven and doesn't do the actual steering. Driving and steering are taken care of by the drive motors.

Refer to FIG. 24-5. Attach the caster using another piece of 1¹/₄-inch by 6-inch mending plate. Secure the plate to the frame with ⁸/₃₂-by-¹/₂-inch bolts and ⁸/₃₂ nuts. Using the baseplate of the caster as a drilling guide, drill two holes in the mending plate. Secure the caster with ⁸/₃₂-by-¹/₂-inch bolts and nuts. Be sure to add a few washers or another nut on the top side of the mending plate, as depicted in the figure, because if you don't, the end of the bolt will interfere with the swivel motion of the caster. Alternatively, you can cut the bolt to length.

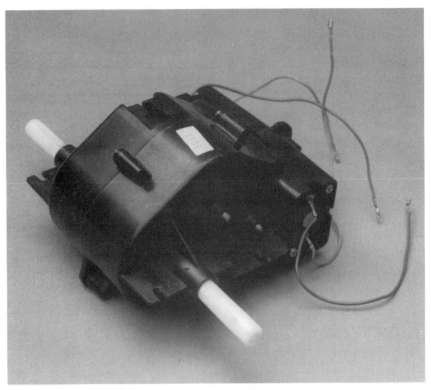

Fig. 24-3. The motor used in the prototype Scooterbot. This dual motor is still sometimes available in surplus and was originally built for use in the Milton-Bradley BigTrak toy.

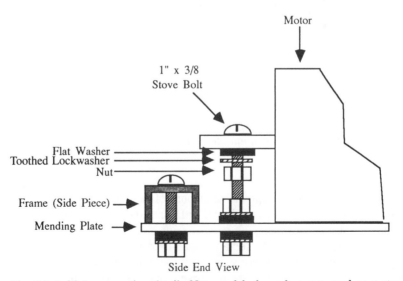

Fig. 24-4. Motor mounting detail. Nuts and lock washers are used as a spacer to secure the bolt against the motor flange and the mending plate.

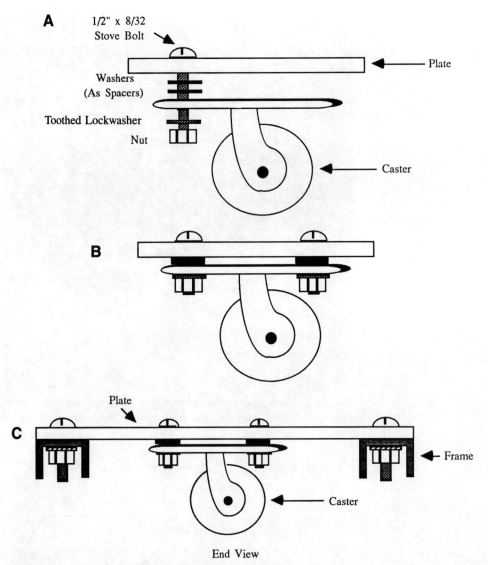

A 1/2" x 8/32
Stove Bolt

Plate

Washers
(As Spacers)

Toothed Lockwasher

Nut

Caster

B

C

Plate

Frame

Caster

End View

Fig. 24-5. Mounting the caster to the Scooterbot. (A) Hardware assembly detail; (B) Assembled caster with spacers; (C) The mending plate, with caster, attached to the frame of the robot.

Battery holder

The motors require an appreciable amount of current, so the Scooterbot should really be powered by heavy-duty C or D size cells. The smaller AA cells just won't cut it.

The prototype Scooterbot used a four-cell D battery holder. The holder is nearly 6 inches wide, so it fits nicely on top of the frame. Drill holes in the corners of the holder and secure it to the base using $6/32$-by-$1/2$-inch pan-head stove bolts and $6/32$ nuts. Be sure the heads of the bolts do not interfere with any of the batteries.

WIRING SCOOTERBOT

The basic test version of the Scooterbot uses a manually wired switch control. Chapters 25 and 26 detail how to control Scooterbot with a computer and your own voice.

The wiring diagram in FIG. 24-6 allows you to control the movement of the Scooterbot in all directions. This simple two-switch system uses commonly available double-pole, double-throw (DPDT) switches. The switches called for in the circuit are spring loaded so they return to a center-off position when you let go of them.

Once you've tested the operation of the Scooterbot with the switch harness, you can build the computer interface and command your robot by electronic control.

To prevent the control wire from interfering with the operation of the robot, attach a piece of heavy wire (the bottom rail of a coat hanger will do) to the

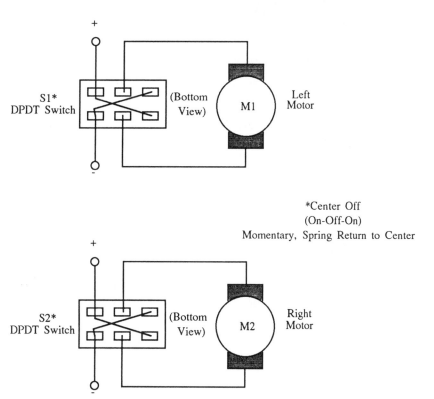

Fig. 24-6. Use this schematic to build a control pendant for the Scooterbot. The switches are momentary DPDT with center off.

Fig. 24-7. The completed Scooterbot, with batteries and wheels. Note the whip used to keep the control wires from tangling with the robot.

caster plate and lead the wire up it. Use nylon wire ties to secure the wire. The finished Scooterbot is shown in FIG. 24-7. Note that a small piece of perforated circuit board was used to solder the motor, battery, and control wires together. Alternatively, you can use wire nuts to bundle the wires together.

TEST RUN

You'll find that the Scooterbot is an amazingly agile robot. It turns in a distance a little longer than its length (about 14 inches in the prototype) and has plenty of power. There is room on the front and back of the robot to mount additional control circuitry. You can also add control circuits and other enhancements over the battery holder. Just be sure that you can remove the circuit(s) when it comes time to change or recharge the batteries.

25

Computer control of robots

In "The Wizard of Oz," the Scarecrow laments "If I only had a brain." He imagines the wondrous things he could do and how important he'd be had he more than straw filling his noggin!

In a way, your robot is just like the Scarecrow. Without a computer to control it, your robot can only be so smart. The Scooterbot introduced in Chapter 24 operates under your direct control and is really nothing more than a motorized car. You can build a computer from scratch for use in your robot by using various IC building blocks to make a reasonably smart brain for your automaton. There are numerous advantages to constructing your own computer for your robot, but a far easier and more immediately rewarding way is to use the personal computer that's sitting on your desk at home or in the office.

With little circuitry, you can use your personal computer to control your robot. If the computer is small enough, like the the old Timex/Sinclair 1000 or Commodore 64, you might even be able to install it temporarily on your robot.

Most personal computers come with, or have as an option, a parallel printer port. This interface port is intended primarily for connecting the computer to printers, plotters, or some other computer peripherals. With a few ICs and some rudimentary programming, it can also be used to directly control your robots.

This chapter reveals how to use your computer—specifically an IBM PC or clone—to automate your Scooterbot creation. Why the IBM PC? It's a common computer, often found in businesses and homes, and it provides a great deal of flexibility. And if you decide to graduate to bigger and better robots, ready-made IBM PC compatible *motherboards* are available that you can use as easily as slipping the board into your automaton.

INTERFACE CIRCUITRY

Too bad . . . you can't just plug the Scooterbot into your computer. A computer has no direct means to control a robot (or most any real-world device for that matter, such as a light dimmer, home security alarm, or kitchen appliance). To connect the robot and computer together, you need suitable interface electronics. This can take many forms, including relays and transistors, as detailed here.

Of primary importance to Scooterbot builders is starting the motors and controlling their rotation. It's fairly easy to change the operation and rotational direction of a dc motor. The Scooterbot discussed in Chapter 24, Projects in robot design, dealt with using a double-pole double-throw (DPDT) switch to control the direction of drive motors. Two such switches were used, one for each of the drive motors. The DPDT switches used here have a center-off position. When in the center position, the motors receive no power, so the robot does not move.

You can use the control switches for experimenting, but you'll soon want to graduate to more automatic control of your robot. Fortunately, that's not hard, either. There are a number of ways to accomplish electronic or electrically assisted operation and direction control of motors. None are really better than the others; they have advantages and disadvantages. Let's see what they are.

Relay control

Perhaps the most straightforward approach to controlling motors is to use relays. It might seem rather daft to install something as old-fashioned and cumbersome as relays in a high-tech robot, but you should try the other techniques before making up your mind. You'll find that while relays may wear out in time (after a few hundred thousand switchings), they are less expensive than the other methods, easier to implement, and actually take up less space.

Basic on/off motor control can be accomplished with a single-pole relay. Rig up the relay so that current is broken when the relay is not activated. Turn on the relay, and the switch closes, thus completing the electrical circuit. The motor should then turn.

How you activate the relay isn't the important consideration here. You could control it by a pushbutton switch, but that's no better than the manual switch method above. Relays can easily be driven by digital signals. Figure 25-1 shows the complete driver circuit for a relay-controller motor. Logical 0 (low) turns the relay off; logical 1 (high) turns it on. The relay can be operated from any digital gate, including a computer or microprocessor port. The transistor is used because most digital ports (parallel printer ports included) lack sufficient current drive to directly operate a relay.

Controlling the direction of the motor is only a little more difficult. This requires a DPDT relay wired in series after the on/off relay described above (see FIG. 25-2). With the contacts in the relay in one position, the motor turns clockwise. Activate the relay and the contacts change positions, turning the motor counterclockwise. Again, you can easily control the direction relay with

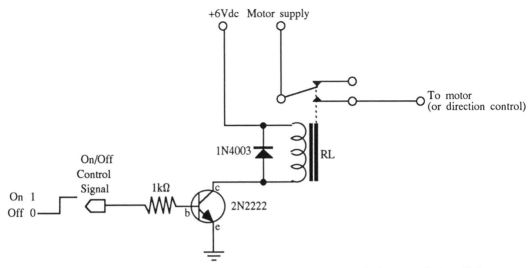

Fig. 25-1. Basic hookup for on/off relay control of small dc motors.

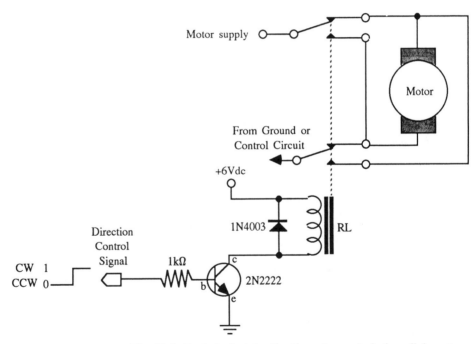

Fig. 25-2. Basic hookup for direction relay control of small dc motors.

digital signals. Logical 0 makes the motor turn in one direction (let's say forward), and logical 1 makes the motor turn in the other direction. Both on/off and direction relay control is shown combined in FIG. 25-3. A parts list is provided in TABLE 25-1.

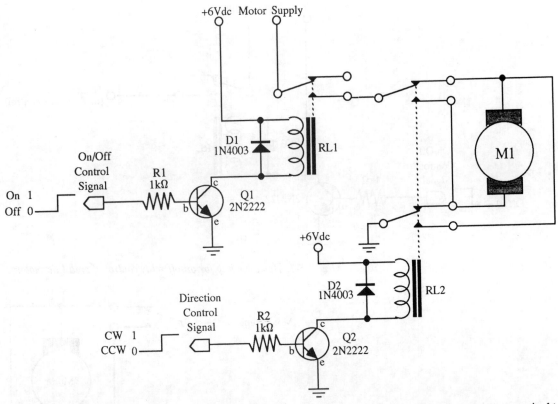

Fig. 25-3. Both on/off and direction control combined in one circuit. Only two digital control signals are required to operate the motor.

R1,R2	1 kΩ
Q1,Q2	2N2222 npn transistor
RL1	SPST relay
RL2	DPDT relay
D1,D2	1N4003 diode
Misc.	Heatsinks for transistors

Table 25-1. Parts List for Relay Motor Control

All resistors 5 to 10 percent tolerance, 1/4 watt.

You can quickly see how to control the operation and direction of a motor using just two data bits from a computer. Since most robot designs incorporate two drive motors, you can control the movement and direction of your robot with just four data bits. That's exactly what we'll be doing later in this chapter.

When selecting relays, make sure the contacts are rated for the motors you are using. All relays carry contact ratings, which vary from about 0.5 amp to over 10 amps, at 125 volts. Higher capacity relays are larger and might require bigger transistors to trigger them (the very small reed relays can often be triggered by digital control without adding the transistor). For most applications, including Scooterbot control, you don't need a relay rated higher than 2 or 3 amps.

Transistor

Transistors provide true solid-state control of motors. For the purpose of motor control, you use the transistor as a simple switch.

An excellent transistor motor control scheme is shown in FIG. 25-4 (parts list in TABLE 25-2). This is a familiar sight to electronic buffs. The "H" network is wired in such a way that only two transistors are on at a time. When transistor 1 and 4 are on, the motor turns in one direction. When transistor 2 and 3 are on, the motor spins the other way. When all transistors are off, the motor remains still.

Note the resistor used to bias the base of each transistor. These are necessary to prevent the transistor from pulling excessive current from the gate controlling it (computer port, logic gate, whatever). Without the resistor, the gate would overheat and be destroyed. The actual value of the resistor depends on the voltage and current draw of the motor as well as the characteristics of the

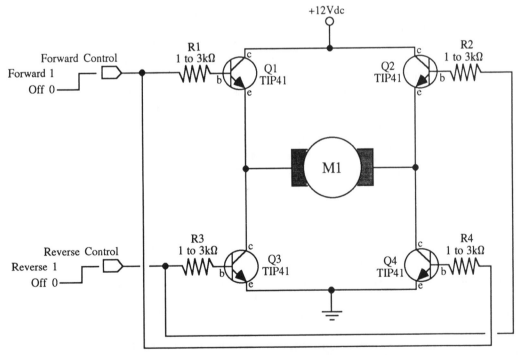

Fig. 25-4. Motor control (power plus direction) using transistors. For best operation, mount the transistors on heatsinks. The transistors specified are fine for small hobby motors, or up to about 6 volts dc and between 800 and 1,000 milliamps.

R1 – R4	1 kΩ to 3 kΩ
Q1 – Q4	TIP41 npn power transistor
Misc.	Heatsinks for transistors

All resistors 5 to 10 percent tolerance, 1/4 watt.

Table 25-2. Parts List for Transistor Motor Control

particular transistors used. For ballpark computations, the resistor is usually in the 1 kΩ to 3 kΩ range.

The choice of transistors really makes little overall difference as long as they comply to some general guidelines. First, they must be capable of handing the current draw demanded by the motors. The drive motors for the Scooterbot draw about 1 ampere continuously, so the transistors you choose should be able to handle this. This immediately rules out the small signal transistors, which are rated for no more than 500 or 600 mA.

A good npn transistor for medium-duty applications is the TIP31, which comes in a TO-220 style case. For high-power jobs, the npn transistor that's almost universally used is the 2N3055 (get the version in the TO-3 case because it handles more power).

The driving transistors should be located off of the main circuit board, preferably directly on a large heatsink or at least on a heavy board with clip-on or bolt-on heatsinks attached to the transistors. Use the proper mounting hardware when attaching transistors to heatsinks.

Remember that with most power transistors, the case is the collector terminal. This is particularly important when there are more than one transistor on a common heatsink and they aren't supposed to have their collectors connected together. It's also important that the heatsink is connected to the grounded metal frame of the robot. You can avoid any extra hassle by using the insulating washer provided in most transistor mounting kits.

The power leads from the battery and to the motor should be 12- to 16-gauge wire. Use solder lugs or crimp-on connectors to attach the wire to the terminals of TO-3 style transistors. Don't tap power from the electronics for the driver transistors; get it directly from the battery.

Power MOSFET

Wouldn't it be nice if you could use a transistor without bothering with biasing resistors? Well you can, as long as you use a special brand of transistor, the power MOSFET. The *MOSFET* part stands for *m*etal *o*xide *s*emiconductor *f*ield *e*ffect *t*ransistor. The *power* part means you can use them for motor control without worrying about burning the MOSFETs, or the controlling circuitry, up in smoke.

MOSFETs look a lot like transistors, but there are a few important differences. First, like CMOS ICs, it is entirely possible to damage a MOSFET device by zapping it with static electricity. When handling it, always keep the protective foam around the terminals. Further, the names of the terminals are different than transistors. Instead of base, emitter, and collector, MOSFETs have a gate, source, and drain. Broadly speaking, the gate is the same as the base of a transistor, and the source and drain are the same as the emitter and collector, respectively. You can easily damage a MOSFET by connecting it in the circuit improperly. Always refer to the pinout diagram before wiring the circuit, and double-check your work.

A commonly available power MOSFET is the IRF-511 (available at Radio Shack and through most mail-order electronics companies). It comes in a TO-220 style transistor case and can control several amps of current (when on a suitable heatsink). A practical circuit that uses MOSFETs is shown in FIG. 25-5. A parts list is given in TABLE 25-3. Note the similarity between this design and the transistor design in FIG. 25-4.

Here, a 4011 NAND CMOS gate has been added to provide positive-action control. When the control signal is low, the motor turns clockwise. When the control signal is high, the motor turns counterclockwise. There's more on this flexible design in the next section.

About the only real problem with MOSFETs is their price. You can purchase 2N3055 power transistors for less than 75 cents if you look long enough, but it's hard to find power MOSFETs for under $2.00. Seeing how you need four of them for each motor, it's obvious that costs can mount quickly.

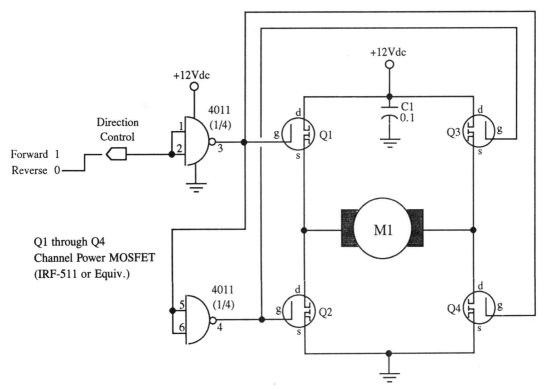

Fig. 25-5. Motor control (power plus direction) using power MOSFET devices.

| Q1 – Q4 | IRF511 (or equiv.) power MOSFET |
| IC1 | 4011 CMOS quad NAND gate IC |

Table 25-3. Parts List for MOSFET Motor Control

CONNECTING TO THE PARALLEL PORT

In the IBM PC, expansion input/output (I/O) boards, like a parallel printer or RS-232 serial board, are plugged into the main motherboard. Slots in the motherboard accept these auxiliary cards, and you are relatively free to put any type of card in any slot.

The PC *addresses*, or accesses, its various I/O ports by using a unique address code. Each device or expansion board in the computer has an address unique to itself, just as you have a home address that no one else in the world shares with you.

The parallel port that comes on the monochrome display adapter has a starting address of 956. This address is in decimal, or base-10 numbering form; some programming languages require that the address be given in hexadecimal (''hex''), or base-16, form. In hex, the starting address is 3BCH (the address is really 3BC; the H means that the number is in hex). The parallel port contained on an I/O expansion board, like a multifunction card, has a decimal address of 888 (or 378H hex) or 632 (278H). Usually, you specify the address of the port when you install the board.

Parallel ports in the IBM PC are given the *logical* names LPT1:, LPT2:, and LPT3:. Every time the system is powered up or reset, the ROM BIOS (*basic input/output system*) chip on the computer motherboard automatically looks for parallel ports at these I/O addresses—3BCH, 378H, and 278H, in that order (it skips 3BCH if you don't have a monochrome card/printer port installed). The logical names are assigned to these ports as they are found.

TABLE 25-4 is a chart that shows the port addresses for all the possible parallel ports in the IBM PC and AT. The logical port names are often used by applications software instead of the actual addresses. In robotics, there is more flexibility using the addresses, so we'll stick with them.

Table 25-4. Parallel Port Addresses

Adapter	Data Output	Status	Control
Mono card (PC, XT, AT)	3BCH, 956D	3BDH, 957D	3BEH, 958D
PC/XT Printer Adapter PC jr. Printer Adapter AT S/P card (as LPT1:)	378H, 888D	379H, 889D	37AH, 890D
AT S/P card (as LPT2:)	278H, 632D	279H, 633D	27AH, 634D

H suffix = Hex
D Suffix = Decimal

The parallel port on the IBM PC is a 25-pin connector, often referred to as a DB-25 connector. Cables and mating connectors are in abundant supply, making it easy for you to wire up your own peripherals. You can buy connectors that crimp onto 25-conductor ribbon cable, or others are designed for direct soldering.

Figure 25-6 shows the pinout designations for the connector (shown with the end of the connector facing you). Note that only a little more than half of the pins are in use. The others are not connected inside the computer or are grounded to the chassis. TABLE 25-5 shows the meaning of the pins.

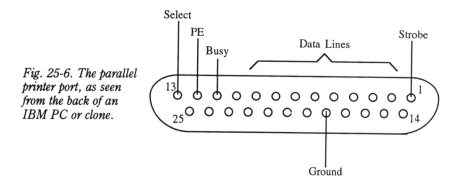

Fig. 25-6. The parallel printer port, as seen from the back of an IBM PC or clone.

Take a closer look back at the chart in TABLE 25-4. Notice that not one address is given, but three. The so-called "starting" address is used for data output. The data output is comprised of eight bits, something on the order of 01101000, as shown in FIG. 25-7. There are 256 possible combinations of the eight bits. In a printer application, this means that the computer can send out the specific code for up to 256 different characters. The data output pins are numbered 2 through 9. The other two addresses are for status and control, some of which are shown in FIG. 25-8.

An important control pin is pin number 1. This is the STROBE line, used to tell the peripheral—like a printer—that the parallel data on lines 2 through 9 is ready to be read. The STROBE line is used because all the data might not arrive at their outputs at the same time. It is also used to signal a change in state.

The output lines are latched, meaning that whatever data you place on them stays there until you change it or turn off the computer. During printing, the STROBE line toggles from high to low and then high again. You don't have to use the STROBE line when commanding your robot (we won't bother with it for the Scooterbot), but it's a good idea for bigger robots. Other control lines are:

- Auto form feed (not always implemented)
- Select/deselect printer
- Initialize printer
- Printer interrupt (not always implemented)

Table 25-5. Parallel Port Pinout Functions

Pin #	Function (Printer Application)
1	Strobe
2	Data bit 0
3	Data bit 1
4	Data bit 2
5	Data bit 3
6	Data bit 4
7	Data bit 5
8	Data bit 6
9	Data bit 7
10	Acknowledge
11	Busy
12	PE (out of paper)
13	Printer on line
14	Auto linefeed after carriage return
15	Printer error
16	Initialize printer
17	Select/deselect printer
18 – 25	Unused or grounded

Control

Bit	Function
0	Low = normal; High = output of byte of data
1	Low = normal; High = auto linefeed after carriage return
2	Low = initialize printer; High = normal
3	Low = deselect printer; High = select printer
4	Low = printer interrupt disables; High = enabled
5 – 7	Unused

Status

Bit	Function
0 – 2	Unused
3	Low = printer error; High = no error
4	Low = printer not on-line; High = printer on-line
5	Low = printer has paper; High = out of paper
6	Low = printer acknowledges data sent; High = normal
7	Low = printer busy; High = printer ready

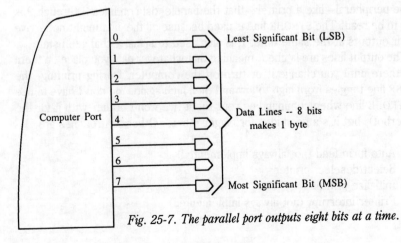

Fig. 25-7. The parallel port outputs eight bits at a time.

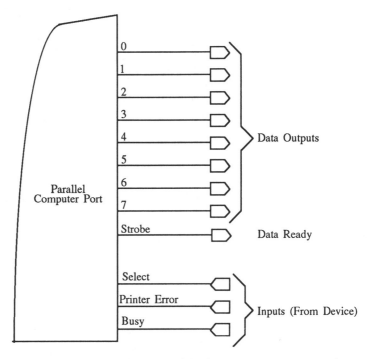

Fig. 25-8. The minimum parallel port: eight data outputs, a STROBE (data ready) line, and inputs from the printer or plotter including SELECT, PRINTER ERROR, and BUSY.

The status lines are the only ones that feed back into the computer. There are five status lines, but not all parallel ports support them every one. They are:

- Printer error
- Printer not selected
- Paper error
- Acknowledge
- Busy

The acknowledge and busy lines are really the same as far as printers are concerned, but you can use the two separately in your own programs (one helpful tidbit: when the BUSY line is low, the ACK line is high). The Scooterbot has no use for the input or status lines on the PC parallel port, but the information is presented here for your edification.

USING THE PORT TO OPERATE A ROBOT

The simplest way to operate your robot via computer is to connect each of the data output lines to one of the interface circuits presented earlier in this chapter. You can control up to eight motors or other devices in this fashion.

Let's say that you only have two motors connected to the interface and are using lines 0 and 1 (pins 2 and 3 respectively). To turn on motor #1, you must activate the bit for line 0; that is, make it high. To do this, output a *bit pattern* number to the port using the BASIC OUT command.

The OUT command is a lot like the BASIC POKE command, except it is used to send data to an I/O port, not to memory. The OUT command consists of just two variables: port address and value. The two are separated by a comma. To send data to the data output, type:

OUT 888, *x*

In place of the *x*, put the decimal value of the binary bit pattern. TABLE 25-6 shows the first 10 binary numbers and the bit pattern that makes them up.

Table 25-6. Decimal/Binary
Equivalents

Decimal Number	Binary
0	0000
1	0001
2	0010
3	0011
4	0100
5	0101
6	0110
7	0111
8	1000
9	1001
10	1010

Normally, you initialize the port at the beginning of the program by outputting decimal 0, or all logic 0s. The line of code for this is:

OUT 888, 0

You'll note that in every other decimal number, bit 0 (on the right-hand side) changes from 0 to 1. To activate just the #1 motor, choose a decimal number where only the first bit changes. There is only one number that meets the criteria: it is decimal 1, or 00000001. So type:

OUT 888,1

The motor turns on. To turn it off, send a decimal 0 to the port. Now, how might you turn on both motor #1 and #2? Look for the binary bit pattern where the first and second bits are 1 (it's decimal 2), and output it to the port.

Not only do you need to turn the motors on and off, you need to indicate a direction. The Scooterbot then requires four data bits:

- Data bit 0 turns the left motor on and off. A logical 0 means off; a logical 1 means on.
- Data bit 1 turns the right motor on and off. A logical 0 means off; a logical 1 means on.
- Data bit 2 controls the direction of the left motor. A logical 0 means forward; a logical 1 means backward.
- Data bit 3 controls the direction of the right motor. A logical 0 means forward; a logical 1 means backward.

Assuming you are using relay control, the overall schematic for the Scooterbot motor control is shown in FIG. 25-9. A parts list is provided in TABLE 25-7.

Software for controlling the Scooterbot needn't be elaborate. You can easily decode the output of the parallel port to operate the Scooterbot because you only need to worry about the first four bits. If you can count from 0 to 15 in binary, you can write a simple program to control the Scooterbot.

Here is how the motor control is decoded (note that the direction bit makes no difference if the motor is off, but for clarity, we'll consider the bit to be 0 for forward):

Left Motor	Direction	Right Motor	Direction	Bit Notation	Decimal
Off	N/A	Off	N/A	0000	0
On	F	Off	N/A	0001	1
Off	N/A	On	F	0010	2
On	F	On	F	0011	3
On	B	Off	N/A	0101	5
Off	N/A	On	B	1010	10
On	B	On	B	1111	16

OTHER SOFTWARE APPROACHES

There are plenty of other ways to get data to the printer port. If LPT1: is set up as the printer device, you can use the LPRINT CHR$(x) statement to output data. As usual, in place of x you put the decimal number that equals the bit pattern you want placed on the data output lines. Note that the STROBE line is automatically activated for each byte and that you don't have direct control of the status and control lines of the port.

Also, the LPRINT command ends each line with a carriage return and line feed, which could be interpreted by your robot as something like "kill master" or "wreck livingroom." You can prevent carriage returns and line feeds from occurring by placing a semicolon after each LPRINT statement line. When the program ends, however, BASIC sends out a carriage return and line feed.

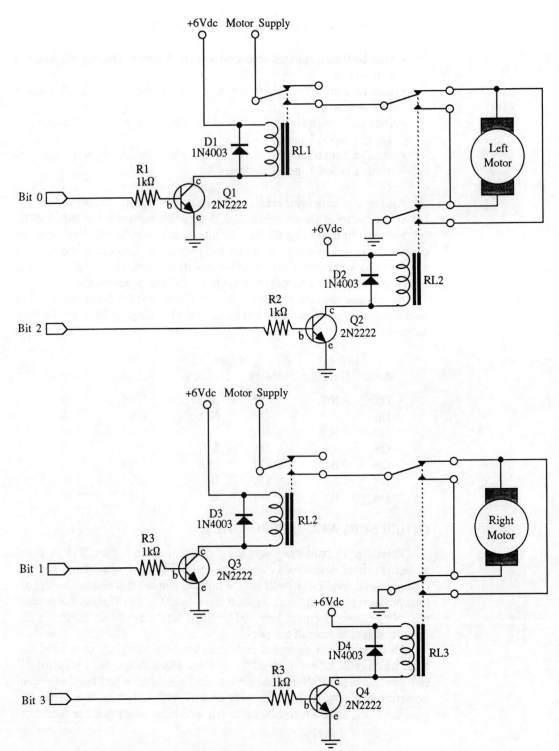

Fig. 25-9. Relay circuit for interfacing the Scooterbot to an IBM PC parallel port.

Table 25-7. Parts List for
Scooterbot Relay Motor Control

R1,R2	1 kΩ
Q1,Q2	2N2222 npn transistor
RL1	SPST relay
RL2	DPDT relay
D1,D2	1N4003 diode
Misc.	Heatsinks for transistors

All resistors 5 to 10 percent tolerance, $1/4$ watt.

Another approach is to use the OPEN command, as in:

OPEN "LPT1" as #1

You can then use the PRINT# command statement. Again, a carriage return/line feed pair is sent when using PRINT#; you can suppress it by ending the PRINT# statement line with a semicolon.

26

Voice control
of robots

The robots of science fiction films are always commanded by human voice. That requires speech recognition, which even in today's push-button world is a tough achievement. Yet with just a handful of affordable electronic parts, you can give your robot ears and control its basic locomotion functions—turn right, turn left, go, stop, and so forth—with your voice.

The voice control project detailed in this chapter centers around the VCP200 speaker-independent word recognizer, currently available for under $10 from Radio Shack. This chip, when connected to a microphone, amplifier, and interface electronics, lets you control your robot just by speaking short commands. Because the VCP200 chip is speaker-independent, anyone can talk to your robot.

ABOUT SPEECH RECOGNITION

Computerized speech recognition is a complex science. It involves breaking down individual components of speech and analyzing them against a framework of known words. When a match is found, the computer (or robot) obeys by carrying out the command.

Speech recognition taxes even the most powerful computers to the limit. It requires fast processing time and prodigious amounts of memory to store known words. One of the most serious shortcomings of speech recognition is that most systems are *speaker-dependent*, meaning the computer recognizes the speech patterns from only one person. Before using the speech recognition system, the user must "train" it by speaking a series of command words—like "enter," "move right," and so forth. The voice pattern of these command words are then analyzed and stored on disk as speech files for future reference. During speech recognition, the computer scans through these stored speech files until it finds a match.

Because speech patterns are different for each person, the computer won't likely find a match if someone other than the "trainer" speaks to it. For example, let's say you train the computer to understand the word "gadgeteer." The computer simply stores a digitized version of this word in its memory banks, and that digitization depends entirely on your inflection, pronunciation, and other factors that describe your speech patterns.

The one advantage to speaker-dependent voice recognition systems is that you merely say a word and match the word to a specific command, but there's a limit to the number of words in the vocabulary: most speaker-dependent voice recognizers can store up to 200 or 300 words.

Speaker-independent systems trade the advantage of a large vocabulary for broad voice pattern recognition. The speaker-independent voice recognition system understands relatively few words—seldom over eight or ten—but it can understand almost anyone who says them. The words are chosen so that they sound as different as possible, for example "Pepsi," "Diet Coke," "Seven Up," and so forth. Most speaker-independent voice recognizers are single-chip affairs built specially for the purpose. Unlike speaker-dependent systems—which are typically personal-computer based—speaker-independent voice recognizers are housed in a single IC, such as the VCP200 speaker-independent word recognizer chip. This chip understands five basic motion commands in its main operating mode: GO, LEFT TURN, TURN RIGHT, REVERSE, and STOP. You select the operating mode by bringing a control pin high or low.

The VCP200 connects directly to a microphone and IC amplifier. No extra analog-to-digital circuits are required; the chip itself contains the circuitry necessary to transform analog voice signals to digital patterns. In operation, the VCP200 performs a spectral analysis of the incoming speech signal over the range of 300 Hz to 5.5 kHz. This analysis identifies the various phonemes (parts of speech) contained in the signal and matches them with stored voice prints. A pinout diagram for the VCP200 word recognizer chip is shown in FIG. 26-1.

CIRCUIT DESCRIPTION

The schematic in FIG. 26-2 shows the complete voice recognition system for commanding the Scooterbot introduced in Chapter 24, Projects in robot design. A parts list for the circuit is shown in TABLE 26-1. An electret condenser microphone plugs into the input of the LM324 op amp IC. This IC amplifies and clips the voice signal and pipes it to the VCP200 word recognizer chip.

Depending on the word spoken, the chip activates one of five control lines. Activation is active low, meaning that normally the lines are at a high logic state. When the chip decodes a word it understands, it brings one of its control lines low.

LEDs attached to each control line indicate when that line is on. The control lines are also attached to a set of relays. These relays control the dc motors of the Scooterbot. Buffer inverters are used to steer the control signals in the proper direction and to invert the active-low control signals to active high.

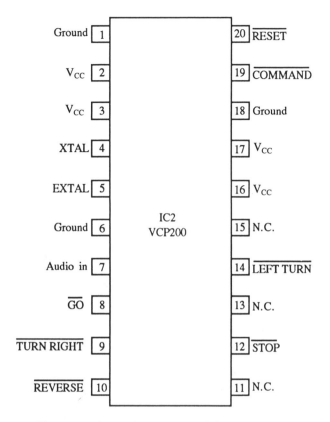

Fig. 26-1. Pinout diagram for the VCP-200 speaker-independent word-recognizer chip.

You can build the circuit using almost any construction technique. The schematic for the LM324 op amp and VCP200 ICs are directly from the Radio Shack application notes for the word-recognizer chip. Included in the application notes is a printed circuit board layout you can use. Add a second board for the inverter IC, driver transistors, and relays. Because the Scooterbot motors draw less than a few amps, the relays can be the small PCB-mounted type. The relays indicated in the parts list are rated for 6 Vdc, but you can also use 5 Vdc relays without worry of overvoltage.

USING THE VOICE COMMAND UNIT

The VCP200 has trouble operating in noisy environments. The quieter the room, the more chance your spoken commands will be carried out correctly. For best results, hold the microphone in your hands instead of mounting it on the robot. Ambient noise from the room and motors will certainly cause trouble. Position the microphone about an inch away from your mouth and speak directly into it. You may find that speaking louder improves the accuracy of the recognized commands.

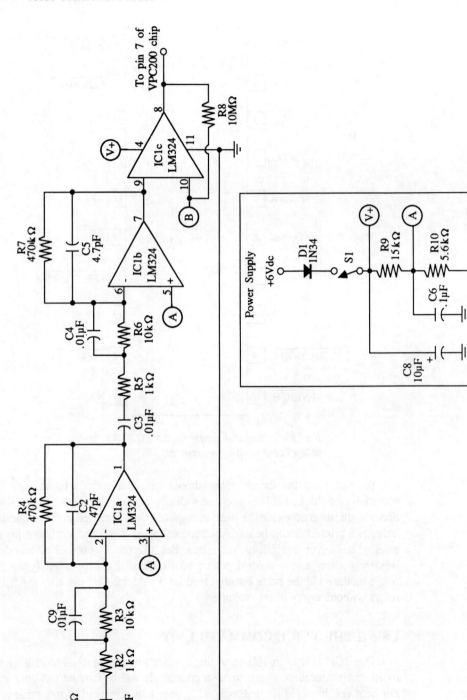

Fig. 26-2. Schematic diagram for interfacing the VCP-200 speech chip to a microphone and Scooterbot relay motor control. (A) Audio amplifier and clipper; (B) VCP-200 hookup diagram; (C) logic interface for controlling the Scooterbot relays.

B

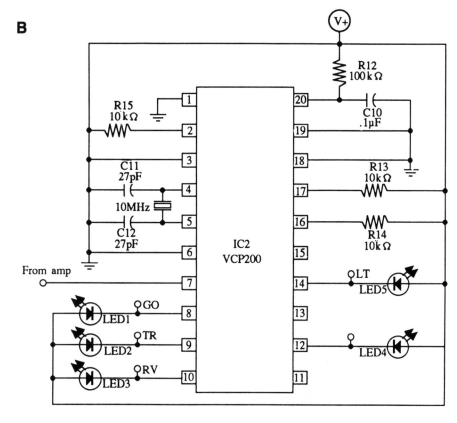

C

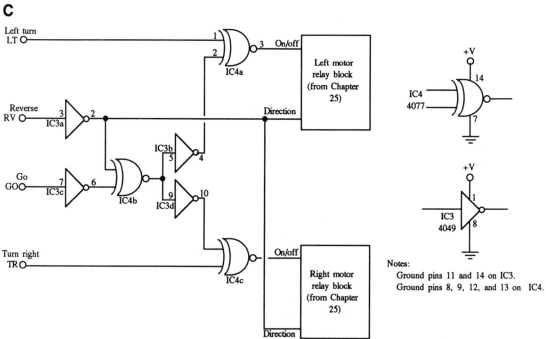

Notes:
Ground pins 11 and 14 on IC3.
Ground pins 8, 9, 12, and 13 on IC4.

Table 26-1. Parts List for Speech Amplifier

R1	2.2 kΩ
R2,R5	1 kΩ
R3,R6,R13,R14,R15	10 kΩ
R4,R7	470 kΩ
R8	10 MΩ
R9	15 kΩ
R10	5.6 kΩ
R11	4.7 kΩ
R12	100 kΩ
C1	0.22 μF disc
C2	47 pF disc
C3,C4,C9	0.01 μF disc
C5	4.7 pF disc
C6,C7,C10	0.1 μF disc
C8	10 μF polarized electrolytic
C11,C12	27 pF disc
IC1	LM324 amplifier IC
IC2	VCP200 speaker independent voice recognition IC
IC3	4049 hex inverter buffer IC
IC4	4077 exclusive OR gate IC
D1	1N34 diode
LED1 – 5	Light-emitting diodes
S1	SPST switch

All resistors are 5 to 10 percent tolerance, $1/4$ watt. All capacitors are 10 to 20 percent tolerance, rated at 35 volts or more.

To use the voice-commanded Scooterbot, turn its power switch on and set it down in the center of a large room (indoors is better than outdoors, especially on windy days). The Scooterbot understands the following words:

- GO—Activates both right and left motors and propels the robot forward.
- STOP—Deactivates all relays and stops the robot.
- REVERSE—Activates both right and left motors and reverses their direction; the robot moves backward.
- LEFT TURN—Activates just the left motor so the robot turns left. Note the sequence of words: ''left turn.''
- TURN RIGHT—Activates just the right motor so the robot turns right. Note the sequence of words: ''turn right.''

Say the commands in any order. Each new command changes the operation or direction of the robot. Say each command clearly and succinctly, and allow enough time for the relays and motors to react. Remember, to turn left, say ''left turn''; to turn right, say ''turn right.'' The two commands switch the sequence of words around so that the VCP200 chip can better differentiate between the two.

If the robot careens out of control, calmly say ''stop'' until the robot halts. You might find you need a period of adjustment before the Scooterbot understands all of the commands you give it. Be patient and you will discover the correct inflections to ensure nearly 100 percent reliability.

ADDING VOICE BROADCAST CAPABILITIES

You can dispense with the tethered microphone cable in a number of ways. An inexpensive solution is to electrically connect the speaker output of a toy walkie-talkie to the microphone input of the LM324 op amp. Use another walkie-talkie to issue commands.

For best results, keep the push-to-talk button to the sending walkie-talkie depressed at all times. This prevents the VCP200 from listening to static when the RF carrier from the transmitting walkie-talkie is broken. Add a switch to the speaker terminals on the transmitter walkie-talkie so that you can disconnect it at will. This prevents the VCP200 from responding to unintentional commands.

Another solution is to use an infrared (IR) audio link. Mono versions of IR links are available for under $40 (you don't need the stereo version). Or, you can construct one of your own. Consult back issues of the various electronics magazines for projects on infrared audio links. One such project is included in the August 1988 issue of *Radio-Electronics*. The project is aimed at making wireless speakers, but you can easily adapt it to infrared control of your Scooterbot. Connect the receiver to the Scooterbot and attach a microphone to the sender (you might need to add a small amplifier to boost the microphone signal).

Appendix A

Sources

This appendix lists dozens of sources for electrical and mechanical components suitable for your gadgeteering habit. Many mail-order companies specialize in a particular type of product or services; when appropriate these are noted in the listings. Ask for a catalog before ordering. When writing or calling, be sure to say you read about them in *Gadgeteer's Goldmine!*

COMPONENTS, PARTS, AND SYSTEMS

A.C. INTERFACE, INC.
17911 Sampson Ln.
Huntington Beach, CA 92647
Representative for Stanley high-output LEDs.

ADVANCED FIBEROPTICS CORP.
7650 East Evans Rd., Ste. B
Scottsdale, AZ 85260
(602) 483-7576
Edu-Kit fiberoptic kit, fiberoptic components, and systems.

ALCO ELECTRONICS
1830 N. 80th Place
Scottsdale, AZ 85257
Electronic parts and boards—RF, computer parts, semiconductors, and unusual parts.

ALL ELECTRONICS CORP.
P.O. Box 567
Van Nuys, CA 91408
(800) 826-5432; (818) 904-0524 (in CA)
Retail and surplus components, switches, relays, keyboards, transformers, computer-grade capacitors, cassette player/recorder mechanisms, etc. Mail-order and retail stores in L.A. area. Regular catalog.

ALLEGRO ELECTRONICS SYSTEMS
3E Mine Mountain
Cornwall Bridge, CT 06754
(203) 672-0123
Mail order. Laser, xenon strobe, and high-voltage equipment and plans. Catalog available.

ALLIED ELECTRONICS
401 E. 8th St.
Ft. Worth, TX 76102
(800) 433-5700
Electronic parts outlet; catalog and regional sales offices. Good source for hard to find items.

ALLKIT ELECTRONICS
434 W. 4th St.
West Islip, NY 11795
Electronic components, grab bags, and switches—all at attractive prices.

ALLTRONICS
15460 Union Ave.
San Jose, CA 95124
(408) 371-3053
Mail order and store in San Jose, CA. New and surplus electronics.

ALPHA PRODUCTS
7904-N Jamaica Ave.
Woodhaven, NY 11421
(800) 221-0916 (orders); (203) 656-1806 (info); (718) 296-5916 (NY orders)
Mail order. Stepper motors and stepper motor controllers (ICs and complete boards), process control devices.

AMERICAN DESIGN COMPONENTS
62 Joseph St.
Moonachie, NJ 07074
(201) 939-2710
New and surplus electronics, motors, and computer gear.

AMERICAN SCIENCE CENTER, INC.
601 Linden Place
Evanston, IL 60602
(312) 475-8440
As the parent company to the surplus outfit Jerryco, American Science Center sells new and surplus lab and optic components.

ANALYTIC METHODS
1800 Bloomsbury Ave.
Ocean, NJ 07712
(201) 922-6663
Mail order. Integrated circuits, LEDs, surplus parts, computer cables.

ANALYTICAL SCIENTIFIC
P.O. Box 675
Helotes, TX 78023
(512) 684-7373
Lab supplies and equipment. Offers wide selection of chemicals, which can be purchased by both educational institutions and individuals.

ANCHOR ELECTRONICS
2040 Walsh Ave.
Santa Clara, CA 95050
(408) 727-3693
Electronic components. Catalog available.

ANTIQUE ELECTRONICS SUPPLY
P.O. Box 1810
Temoe, AZ 85281
(602) 894-9503
Mainly a good source for old-time radio parts, but also handy for the general gadgeteer. Offers high-voltage caps, high-voltage insulation, heavy-duty potentiometers, and more.

BARRETT ELECTRONICS
5312 Buckner Dr.
Lewisville, TX 75028
Mail-order electronic surplus: components, power supplies, computer-grade capacitors.

BCD ELECTRO
P.O. Box 830119
Richardson, TX 75083-0119
Electronic components of all types. Some surplus, including laser systems. Very good prices. Catalog available.

BNF ENTERPRISES
119 Foster St.
P.O. Box 3357
Peabody, MA 01961-3357
Surplus parts such as flyback transformers, high-voltage caps, and lasers.

BIGELOW ELECTRONICS
P.O. Box 125
Bluffton, OH 45817-0125
New and surplus electronics, hardware, and HAM gear. Catalog available.

BISHOP GRAPHICS
5388 Sterling Center Dr.
P.O. Box 5007EZ
Westlake Village, CA 91359
(818) 991-2600
EZ Circuit PCB making supplies.

BURDEN'S SURPLUS CENTER
1015 West "O" Street
P.O. Box 82209
Lincoln, NE 68501-2209
(800) 228-3407
Surplus and new mechanical gear including motors, pneumatics, and electrics.

C & H SALES
2176 E. Colorado Blvd.
Pasadena, CA 91107
(800) 325-9465
Electronic and mechanical surplus, including some optics like prisms, mirrors, and lenses. Reasonable prices for most items; regular catalog. Their walk-in store in Pasadena has more items than those listed in the catalog, so if you're in the area be sure to drop in.

CIRCUIT SPECIALISTS
Box 3047
Scottsdale, AZ 85257
(602) 966-0764
New electronic goodies including Sprague Hall-effect devices and motor control chips (stepper, half-bridge, full-bridge). Full complement of standard components at good prices.

COMPUTER PARTS MART
3200 Park Blvd.
Palo Alto, CA 94306
(415) 493-5930
Mail-order surplus. Good source for stepper motors, lasers, power supplies, and incremental shaft encoders (pulled from equipment). Regular catalog.

COMPUTER SURPLUS STORE
715 Sycamore Dr.
Milpitas, CA 95035
(408) 434-0168
Surplus electronics. Mostly computers, printers, and power supplies, but also carries components and sometimes even lasers. Walk-in store in Northern California and limited mail order. If you don't see what you want in their flyer, contact them and tell them your needs.

CONSUMERTRONICS
2011 Crescent Dr.
P.O. Drawer 537
Alamogordo, NM 88310
Booklets and plans on an assortment of gadgeteering topics including high-voltage and radiation detection.

DATAK CORP.
3117 Paterson Plank Rd.
North Bergen, NJ 07047
(201) 863-7667
Supplies for making your own printed circuit boards including the popular direct-etch dry transfer method.

DIGI-KEY CORP.
701 Brooks Ave. South
P.O. Box 677
Thief River Falls, MN 56701-0677
(800) 344-4539
Mail order. Discount components—everything from crystals to integrated circuits to resistors and capacitors in bulk. Catalog available.

DOKAY COMPUTER PRODUCTS
2100 De la Cruz Blvd.
Santa Clara, CA 95050
(408) 988-0697; (800) 538-8800 and (800) 848-8008 (in CA)
Electronic parts (most popular linear, TTL, and CMOS devices) and computer gear (including computer chips). Catalog available.

DOLAN-JENNER INDUSTRIES, INC.
P.O. Box 1020
Woburn, MA 01801
Fiberoptic components.

EDMUND SCIENTIFIC CO.
101 E. Gloucester Pike
Barrington, NJ 08007-1380
(609) 573-6250
Mail order. New and surplus motors, gadgets, and other goodies for laser build-ing. Includes laser tubes, power supplies, complete systems, holography and optics kits, lenses, and much more. Regular catalog (ask for both the Hobby and Industrial versions).

ELECTRONIC GOLDMINE
P.O. Box 5408
Scottsdale, AZ 85261
(602) 451-7454
New and surplus electronics kits and parts such as simulated laser kit (high out-put LED), piezo discs, EKI kits, and wide assortment of resistors and capaci-tors.

ERAC CO.
8280 Clairemont Mesa Blvd., Ste. 117
San Diego, CA 92111
(619) 569-1864
Mail-order and retail store. PC-compatible boards and sub-systems, surplus computer boards, computer power supplies and components.

FAIR RADIO SALES
1016 E. Eureka St.
P.O. Box 1105
Lima, OH 45802
(419) 223-2196; (419) 227-6573
An old and established retailer of surplus goods of all types, particularly military communications. A boon to the amateur radio enthusiast and gadgeteer. Look for old radiation monitors, high-voltage capacitors, motors, and more.

FOBTRON COMPONENTS
2102 W. Artesia Blvd.
Torrance, CA 90504
(213) 769-5677
Laser components including tubes and power supplies, optics, and mirrors. Flyer available.

FORDHAM RADIO
260 Motor Parkway
Hauppauge, NY 11788
(800) 645-9518
(516) 435-8080 (in NY)
Electronic test equipment, tools, power supplies. Catalog available.

G.B. MICRO
P.O. Box 280298
Dallas, TX 75228
(214) 271-5546
Mail-order components: 3.12 MHz crystals, 300 kHz crystals, UARTs, popular TTL, CMOS, and linear ICs, construction parts, and more.

GENERAL SCIENCE & ENGINEERING
P.O. Box 447
Rochester, NY 14603
(716) 338-7001
Laser components, power supplies, laser power supply kits, optics, high-output Stanley LED, variable-rate flash kits, long-range communications link kits, surplus goodies of all types.

GIANT ELECTRONICS INC.
19 Freeman St.
Newark, NJ 07105
(800) 645-9060; (201) 344-5700
New and surplus electronics, motors, computer peripherals, and more.

GIL ELECTRONICS
P.O. Box 1628
911 Hidden Valley
Soquel, CA 95073
Electronic parts, books, new and surplus computer components.

GUIDEBOOKS
2642 Hope St.
Oceanside, CA 92056
Plans, kits, and newsletters on laser, robotics, and general gadgeteering subjects.

H&R CORP.
401 E. Erie Ave.
Philadelphia, PA 19134
(215) 426-1708
Surplus mechanical components. Excellent source for dc motors including Big Trak dual motor for Scooterbot. Regular catalog.

HALTED SPECIALITIES CO.
3500 Ryder St.
Santa Clara, CA 95051
(800) 4-HALTED; (408) 732-1854
Mail-order and retail stores in Northern California. New and surplus components, PC-compatible boards and subsystems, power supplies, lasers, optics, all popular ICs, transistors, etc.

HALTEK ELECTRONICS
1062 Linda Vista Ave.
Mountain View, CA 94043
(408) 744-1333; (415) 969-0510
Electronics surplus of all types. Also occasionally handles laser goodies.

HAL-TRONIX INC.
12671 Dix-Toledo Highway
P.O. Box 1101
Southgate, MI 48195
(313) 281-7773
Mail order. Surplus computers, computer components, PC-compatible boards
and subsystems.

HEATHKIT
P.O. Box 1288
Benton Harbor, MI 49022
(800) 253-0570
Perhaps the best source of professionally produced electronic kits. Heathkit
offers laser projects, Geiger counters, and more. Also available are testing gear
(assembled and kit form) and computers.

HIGH VOLTAGE PRESS
P.O. Box 532
Claremont, CA 91711
Books on Tesla coil design and other high-voltage apparatus.

HOSFELT ELECRONICS, INC.
2700 Sunset Blvd.
Steubenville, OH 43952
(800) 524-6464
New and surplus electronics at attractive prices. Catalog available.

HUBBARD SCIENTIFIC
P.O. Box 104
Northbruck, IL 60062
(312) 272-7810
Radiation cloud chamber kits.

INFORMATION UNLIMITED
Box 716
Amherst, NH 03031
(603) 673-4730
Plans and kits for lasers, high-voltage devices, plasma globes, and lots of other
high-tech gadgets. Good source for reasonably priced new laser tubes. All kits
professionally produced with excellent step-by-step plans.

INTERNATIONAL MEDCOM
7497 Kennedy Rd.
Sebastapol, CA 95472
(800) 225-3825 (in CA); (800) 257-3825
Digital radiation monitor (kit or ready-built).

JAMECO ELECTRONICS
1355 Shoreway Rd.
Belmont, CA 94002
(415) 592-8097
Mail order. Components, PC-compatible boards and subsystems. Catalog.

JDR MICRODEVICES
1224 S. Bascom Ave.
San Jose, CA 95128
(408) 995-5430
Large selection of new components, wire-wrap supplies, PC-compatible boards
and subsystems. Mail-order and retail stores in San Jose area.

JERRYCO INC.
607 Linden Place
Evanston, IL 60202
(312) 475-8440
Regular catalog lists hundreds of surplus mechanical and electronic gadgets for
robots. Good source for motors, rechargeable batteries, switches, solenoids,
lots more. Gadgeteers need Jerryco.

J.I. MORRIS CO.
394 Elm St.
Southbridge, MA 01550
(617) 764-4394
Small components and hardware; miniature screws, taps, and nuts/bolts.

JOHN J. MESHNA, JR. CO.
P.O. Box 62
East Lynn, MA 01904
(617) 595-2272
New and surplus merchandise, including radiation monitors and Geiger-Muller
tubes. Catalog available.

LINDSAY PUBLICATIONS, INC.
P.O. Box 12
Bradley, IL 60915
(815) 468-3668
Numerous hard-to-find new and reprinted books of interest to every gadgeteer.
You'll especially appreciate the books on old-time Tesla coil and high-voltage
device design. Although some of the books are old (many dating back to the
beginning of the century) and the plans seriously outdated, the information is
still relevant. Catalog available. Ask for the Technical and Electrical catalogs.

MARTIN P. JONES & ASSOC,
P.O. Box 12685
Lake Park, FL 33403-0685
(407) 848-8236
All sorts of goodies including electronic components (transistors, capacitors, resistors, ICs, etc.), high-voltage capacitors, optical stuff, and more. Catalog available.

METROLOGIC INSTRUMENTS INC.
Laser Products Division
143 Harding Ave.
Bellmawr, NJ 08031
(609) 933-0100
Lasers and supplies, ideal for school. Catalog available.

MCM ELECTRONICS
858 E. Congress Park Dr.
Centerville, OH 45459-4072
(513) 434-0031
Test equipment, tools, and supplies. Catalog available.

MEADOWLAKE
25 Blanchard Dr.
Northport, NY 11768
TEC-200 direct-transfer film for making quick integrated circuits from artwork in magazines. Try this stuff out! All you need is a photocopier (or access to one).

MEREDITH INSTRUMENTS
P.O. Box 1724
Glendale, AZ 85311
(602) 934-9387
Laser surplus including tubes, power supply, optics, components, more. Stock comes and goes, so call first to make sure they have what you want. Flyer available; very good prices and friendly service.

MERRELL SCIENTIFIC
1665 Buffalo Rd.
Rochester, NY 14624
(716) 426-1540
Lab equipment and supplies. Offers wide assortment of chemicals, but they won't sell mail order to individuals (schools only).

MICRO MART, INC.
508 Central Ave.
Westfield, NJ 07090
(201) 654-6008
Electronic components (ICs, transistors, etc.). Their grab bags are better than average and prices are low.

MONDO-TRONICS INC.
1014 Morse Ave., Ste. 11
Sunnyvale, CA 94089
(408) 734-9877
BioMetal wire, terminators, kits, and plans.

MOUSER ELECTRONICS/TEXAS DISTRIBUTION CENTER
2401 Hwy. 287 North
North Mansfield, TX 76063
(817) 483-4422
OR
MOUSER ELECTRONICS/CALIFORNIA DISTRIBUTION CENTER
11433 Woodside Ave.
Santee, CA 92071
(619) 449-2222
Mail order. Discount electronic components. Catalog available.

MWK IINDUSTRIES
1440 State College Blvd., Building 3B
Anaheim, CA 92806
(800) 357-7714; (714) 956-8497
Lasers tubes, power supplies, optics, and more.

OCTE ELECTRONICS
Box 276
Alburg, VT 05440
(514) 739-9328
New and surplus electronic stuff at reasonable prices. OCTE carries a lot of complete components, such as cable converters and power supplies, but individual parts, like batteries and connectors, are also usually available. Aimed primarily at the surplus dealer but samples of merchandise are available on a one-piece basis.

PRECISON ELECTRONICS CORP.
605 Chestnut Street
Union, NJ 07083
(800) 255-8868; (201) 686-4646 (in NJ)
Electronic components—linear ICs, transistors, hard-to-find ICs (mostly for TV), and more.

RADIO SHACK
One Tandy Center
Fort Worth, TX 76102
Nation's largest electronics retailer. Many popular components, though short on popular ICs. Good source for general electronic needs, including some fiberoptic components. As of this writing, stocks VCP200 speaker-independent word recognizer chip. Catalog available through store.

R&D ELECTRONICS
1202H Pine Island Rd.
Cape Coral, FL 33909
(813) 772-1441
Mail order. New and surplus electronic, components, switches, ICs.

R&D ELECTRONIC SUPPLY
100 E. Orangethorpe Ave.
Anaheim, CA 92801
(714) 773-0240
Mail order. Power supplies, computer equipment, test equipment.

SHARON INDUSTRIES
1919 Hartog Road
San Jose, CA 95131
(408) 436-0455
Mail-order and retail store. New and surplus electronic components, ICs, computers, and PC-compatible boards and subsystems.

SILICON VALLEY SURPLUS
4222 E. 12th
Oakland, CA 94601
(415) 261-4506
Surplus electronic, computer, and mechanical goodies. Mail-order and retail store (in Oakland).

SMALL PARTS
6891 NE Third Ave.
P.O. Box 381966
Miami, FL 33238-1966
(305) 751-0856
A potpourri of small parts ideally suited for miniature mechanisms. Not cheap, but good quality.

SPECTRA LASER SYSTEMS
P.O. Box 6928
Huntington Beach, CA 92615
Laser light show consultation.

STAR-TRONICS
P.O. Box 683
McMinnville, OR 97128
Surplus electronics at very affordable prices. Stocks wide variety of capacitors (including high voltage), TV flyback transformers, and resistors.

STOCK DRIVE PRODUCTS
55 S. Denton Ave.
New Hyde Park
New York, NY 11040
Gears, sprockets, chains, and more. Available through local Stock Drive distributor. Catalog and engineering guide available.

SYNERGETICS
Box 809
Thatcher, AZ 85552
(602) 428-4073
Columnist's Don Lancaster's company, with "hacker's help line." Lancaster offers free technical advice (within limits), a collection of his old and new books (his TTL and CMOS cookbooks are classics), and several "info packs" of useful programming, graphics, and technical stuff.

TESLA BOOK CO.
P.O. Box 1649
Greenville, TX 75401
(214) 454-6819
Books on and about Nikola Tesla. Heavy emphasis on Tesla coils.

UNICORN ELECTRONICS
10010 Canoga Ave., Unit B-8
Chatsworth, CA 91311
(800) 824-3432; (818) 341-8833
All popular electronic parts including resistors, capacitors, ICs, and components. Some surplus. All at good prices.

UNITED PRODUCTS, INC.
1123 Valley
Seattle, WA 98109
(206) 682-5025
Mail order. Motors, computer components, test equipment.

WINDSOR DISTRIBUTORS
19 Freeman St.
Newark, NJ 07105
(800) 645-9060; (201) 344-5700
Mail order. Surplus electronics.

SEMICONDUCTOR MANUFACTURERS

Most semiconductor manufacturers maintain a network of regional sales and distribution offices. If you would like more information on a particular product or would like to place an order, contact the manufacturer at the address below and ask for a list of dealers, distributors, and representatives in your area. Most firms will work with individuals; sales might be subject to minimum orders and/or service charges.

ADVANCED MICRO DEVICES
901 Thompson Place
Sunnyvale, CA 94088
(408) 732-2400

BURR-BROWN
P.O. Box 11700
Tucson, AZ 85734
(602) 746-1111

CHERRY SEMICONDUCTOR CORP.
2000 South Country Trail
East Greenwich, RI 02818-0031
(401) 885-3600

DATA GENERAL CORP.
4400 Computer Dr.
Westborough, MA 01581
(617) 366-1970

EG&G RETICON CORP.
345 Potrero Ave.
Sunnyvale, CA 94086
(408) 738-4266

EXAR INTEGRATED SYSTEMS, INC.
750 Palomar Ave.
Sunnyvale, CA 94088-3575
(408) 732-7970

FAIRCHILD
10400 Ridgeview Ct.
Box 1500
Cupertino, CA 95014
(408) 864-6250

FUJITSU MICROELECTRONICS, INC.
3320 Scott Blvd.
Santa Clara, CA 95054-3197
(408) 727-1700

HARRIS SEMICONDUCTOR
P.O. Box 883
Melbourne, FL 32901
(305) 724-7000

HITACHI AMERICA LTD.
2210 O'Toole Ave.
San Jose, CA 95131
(408) 435-8300

INTEL
3065 Bowers Ave.
Santa Clara, CA 95051
(408) 987-8080

INTERSIL
10600 Ridgeview Ct.
Cupertino, CA 95014
(408) 996-5000

MAXIM INTEGRATED PRODUCTS
510 N. Pastoria Ave.
Sunnyvale, CA 94086
(408) 737-7600

MICRO SWITCH
11 West Spring St.
Freeport, IL 61032
(815) 235-6600

MOTOROLA
5005 E. McDowell Rd.
Phoenix, AZ 85008
(602) 244-7100

NATIONAL SEMICONDUCTOR
2900 Semiconductor Dr.
Santa Clara, CA 95051
(408) 721-5000

NCR
8181 Byers Rd.
Miamisburg, OH 45342
(513) 866-7217

NEC ELECTRONICS, INC.
401 Ellis St.
Mountain View, CA 94039-7241
(415) 960-6000

PLESSEY SOLID STATE
9 Parker St.
Irvine, CA 92718
(714) 472-0303

PRECISON MONOLITHICS, INC.
1500 Space Park Dr.
Santa Clara, CA 95052-8020
(408) 727-9222

RAYTHEON SEMICONDUCTOR
350 Ellis St.
Mountain View, CA 94039-7016
(415) 968-9211

RETICON
245 Potrero Ave.
Sunnyvale, CA 94086
(408) 738-4266

ROCKWELL INTL.
4311 Jamboree Rd.
P.O. Box C
Newport Beach, CA 92658-8902
(714) 833-4700

SGS SEMICONDUCTOR
1000 E. Bell Rd.
Phoenix, AZ 85022
(602) 867-6100

SHARP ELECTRONICS CORP.
10 Sharp Plaza
Paramus, NJ 07652

SIGNETICS
811 E. Arques Ave.
Sunnyvale, CA 94088-3409
(408) 991-2000

SILCONIX
2201 Laurelwood Rd.
Santa Clara, CA 95054
(408) 988-8000

SPRAGUE ELECTRIC
70 Pembroke Road
Concord, NH 03301

SPRAGUE SOLID STATE
3900 Welsh Rd.
Willow Grove, PA 19090
(215) 657-8400

TEXAS INSTRUMENTS
Literature Response Center
P.O. Box 809066
Dallas, TX 753800
(214) 232-3200

TOSHIBA AMERICA, INC.
2692 Dow Ave.
Tustin, CA 92680
(714) 832-0102

ZILOG, INC.
210 Hacienda Ave.
Campbell, CA 95008-6609
(408) 370-8000

ADDITIONAL MANUFACTURERS

AMPEREX ELECTRONICS CORP.
Box 418
230 Duffy Ave.
Hicksville, NY 11802
Geiger-Muller tubes.

COLORADO SUPERCONDUCTOR, INC.
P.O. Box 8223
Fort Collins, CO 80526
(303) 491-9106
Superconducting discs and kits.

EASTMAN KODAK COMPANY
Photographic Products Group
Rochester, NY 14650
Film for holography.

EKI ELECTRONIC KITS INTERNATIONAL, INC.
22732 Granite Way-B
Laguna Hills, CA 92653
Electronic kits including FM transmitter, variable-rate strobe light, strobe flare,
and power supplies.

GENERAL SCANNING INC.
500 Arsenal Street
P.O. Box 307
Watertown, MA 02272
(617) 924-1010
Precision galvanometers suitable for laser light shows.

GLENDALE OPTICAL CO.
130 Crossways Park Dr.
Woodbury, NY 11797
Laser safety goggles.

INTERLINK ELECTRONICS
535 N. Montecito St.
Santa Barbara, CA 93101
(805) 965-5155
Conductive inks and paints.

IR SCIENTIFIC
P.O. Box 110
Carlisle, MA 01741
(617) 369-7118
See-in-the-dark scopes, kits, and image converter tubes.

J.C. WHITNEY CO.
P.O. Box 8410
Chicago, IL 60680
Ignition coils, reproductions of Model-T ignition coil.

LASER DRIVE INC.
5465 William Flynn Hwy.
Gibsonia, PA 15044
(412) 443-7688
Laser power supplies

MELLES GRIOT/OPTICS
1770 Kettering St.
Irvine, CA 92714
(714) 261-5600
Precision optics.

MELLES GRIOT/GAS LASERS
2551 Rutherford Rd.
Carlsbad, CA 92008
(619) 438-2131
He-Ne lasers.

PENNWALT
P.O. Box 799
Valley Forge, PA 19482
(215) 666-3500
Kynar piezoelectric film, kits, and technical information.

RAYTHEON INC.
465 Center Street
Quincy, MA 02169
Geiger-Muller tubes.

REX RESEARCH
P.O. Box 1258
Berkeley, CA 94701
Clippings from old magazines and newspapers on unusual scientific discoveries and phenomena. Clippings are organized by topic such as "Tesla Coils," "Electrostatics," and "Sound." Reasonable prices, but photocopies sometimes illegible.

SPECTRA-PHYSICS INC.
Laser Analytics Division
25 Wiggins Ave.
Bedford, MA 01730
(617) 275-2650
He-Ne lasers.

TOKI US INC.
18662 MacArthur Blvd. Ste. 200
Irvine, CA 92715
(714) 476-1206
Manufacturer of BioMetal shape-memory alloy metal. Sells bulk and sample wire, terminators, and design kits.

VICTOREEN INC.
10101 Woodland Ave.
Cleveland, OH 44104
Geiger-Muller tubes.

OTHER IMPORTANT ADDRESSES

AMERICAN NATIONAL STANDARDS INSTITUTE (ANSI)
1430 Broadway
New York, NY 10018
Sets and maintains standards; useful data on laser practices and specifications.

LASER INSTITUTE OF AMERICA
5151 Monroe St.
Toledo, OH 43623
(419) 882-8706
Provides publications of interest to users, manufacturers, and sellers of lasers. Several good publications including *Laser Safety Guide, Laser Safety Reference Book,* and *Fundamentals of Lasers.* Ask for a current price list and availability of titles.

CENTER FOR DEVICES AND RADIOLOGICAL HEALTH (CDRH)
HF2-312
8757 Georgia Ave.
Silver Spring, MD 20910
(301) 427-8228
A department of the Food and Drug Administration (FDA) that regulates the commercial manufacturer and use of lasers. If you manufacture, sell, or demonstrate laser systems, you must comply with minimum standards.

INTERNATIONAL TESLA SOCIETY
330-A West Uintah Street, Ste. 215
Colorado Springs, CO 80905-1095
Clearinghouse on subjects relating to Nikola Tesla and the Tesla coil.

TESLA COIL BUILDERS ASSOCIATION
RD3 Box 181
Glens Falls, NY 12801
Association and newsletter on Tesla coil design and construction, pioneered by Harry Goldman.

U.S. DEPARTMENT OF COMMERCE
Patent and Trademark Office
Washington, DC 20231
Provides general information on patents. For information on specific patents, it's better to visit a library that offers copies of old patents and patent applications or consult a patent attorney.

Appendix B

Further reading

Here is a selected list of magazines and books that can enrich your understanding and enjoyment of all facets of gadgeteering.

MAGAZINES

Computer Shopper
5211 S. Washington Ave.
P.O. Box F
Titusville, FL 32781
Monthly magazine (some call it a bible) for computer enthusiasts. Probably more ads than articles, *Computer Shopper* carries updated listings of swap meets and bulletin boards as well as classified advertising from small surplus dealers.

Electronics Today
1300 Don Mills Rd.
Toronto, Ontario M3B 3M8
Canada
General interest electronics magazine with emphasis on hobby how-to articles.

Hands-on Electronics
500 Bi-County Blvd.
Farmingdale, NY 11735
Monthly magazine put out by the editors of *Radio-Electronics*. The articles and construction projects are aimed at beginning electronics enthusiasts.

Laser Focus
1001 Watertown St.
Newton, MA 02165
A monthly trade magazine for the laser industry. Available by paid or audited (free to qualified readers) subscription. Check a good public library for back issues.

Modern Electronics
76 North Broadway
Hicksville, NY 11801
Monthly magazine for electronics hobbyists. Don't miss the regular columns by hobby electronic guru Forrest Mims III. Many of the editors used to be involved with *Popular Electronics* before that magazine changed over to computer-only coverage (and then ceased publication). Check back issues of *Modern Electronics* for Don Lancaster's fascinating columns (he moved his Hardware Hacker Column to *Radio-Electronics* in late 1987).

Nuts & Volts
P.O. Box 1111
Placentia, CA 92670
Monthly "magazine" that contains only advertising—both display ads and classifieds. Often carries ads for new and used electronics equipment and lists upcoming swap meets (both computer and electronic). Subscription price is fairly low.

QST
225 Main St.
Newington, CT 06111
Monthly magazine aimed at the amateur radio enthusiast but often carries articles of interest to the gadgeteer.

Spectra
P.O. Box 1146
Pittsfield, MA 01202
A monthly trade journal that focuses on lasers, fiberoptics, and related topics. Available by paid or audited (free to qualified readers) subscription. Check a good public library for back issues.

Radio-Electronics
500 Bi-County Blvd.
Farmingdale, NY 11735
Monthly magazine for electronics hobbyists. Be sure to read the column by Don Lancaster of Synergetics. Although he seldom addresses lasers in specific, his tips and hands-on help are invaluable for general electronics experimentation. He also provides sources for hard-to-find parts and information.

BOOKS

Basic Digital Electronics—2nd Edition
Ray Ryan and Lisa A. Doyle
TAB BOOKS, No. 3370
Introduction to the principles of digital theory and practice.

Beginners Guide to Reading Schematics—2nd Edition
Robert J. Traister and Anna L. Lisk
TAB BOOKS, No. 3632
How to read and interpret schematic diagrams.

*Build Your Own Laser, Phaser, Ion Ray Gun, and Other Working
     Space-Age Projects*
Robert E. Iannini
TAB BOOKS, No. 1604
A guide to making high-tech gadgets, with some insight and designs for laser-based projects. Iannini, who runs a mail-order firm, presents six laser projects plus a few others such as infrared light detection) that can be used with laser systems. A companion book by the same author, *Build Your Own Working Fiberoptic, Infrared, and Laser Space-Age Projects* (TAB, No. 2724) offers more laser-based designs.

Circuit Scrapbook
Forrest Mims III
McGraw-Hill
More of Mims' Popular Electronics columns—these from 1979 to 1981. Several good circuits and designs that can be adapted for laser work. Also check out the sequel, *Circuit Scrapbook II* (Howard W. Sams Co.). This volume includes several chapters on experimenting with solid-state laser diodes. Be sure to read the sections on laser diode handling precautions.

CMOS Cookbook
Don Lancaster
Howard W. Sams Co.
A classic in its own time, the *CMOS Cookbook* presents useful design theory and practical circuits for many popular CMOS chips. The companion book, *TTL Cookbook,* is equally as helpful.

Engineer's Mini-Notebook
Forrest Mims III.
Radio Shack book series
The *Engineer's Mini-Notebook* publications are a series of small books written by Forrest Mims that cover a wide variety of hobby electronics such as using the NE555 timer to optoelectronics circuits to op amp circuits, and more. The entire set is a ''must have,'' and besides, they're cheap.

Fiberoptics and Laser Handbook
Edward L. Safford and John A. McCann
TAB BOOKS, No. 2981
Theory and general applications of fiberoptics and lasers.

44 Power Supplies for Your Electronic Projects
Robert J. Traister and Jonathan L. Mayo
TAB BOOKS, No. 2922
Forty-four complete power supplies designed for general electronics projects.

Fundamentals of Optics
Jenkins and White
McGraw-Hill
An in-depth and scholarly look at optics. Broken down into logical subjects of
geometrical, wave, and quantum optics.

Handbook of Microcomputer Interfacing
Steve Leibson
TAB BOOKS, No. 3101
How to connect outside devices and circuits to a computer or microprocessor.

Holography Handbook
Unterseher, Hansen, and Schlesinger
Ross Books
Perhaps the best book on amateur holography. It even comes with a white light
reflective rainbow hologram. Although the artwork is too "folksy" for my taste,
the designs are technically sound. If you are interested in holography, you need
this book.

How to Build a 40,000 Volt Induction Coil
Walt Noon
Lindsay Publications
Smashing booklet on the construction of a 40 kV induction coil using an automo-
tive ignition coil.

Troubleshooting and Repairing Electronic Circuits—2nd Edition
Robert L. Goodman
TAB BOOKS, No. 3258
General troubleshooters guide—both analog and digital.

IBM PC Connection
James W. Coffron
Sybex Books
A good beginner's guide on connecting the IBM PC (or compatible) to the out-
side world. Information on circuit building, programming, and troubleshooting.

Kirlian Aura, The
Stanley Krippner and Daniel Rubin, Ed.
Doubleday & Co., Anchor Books
History and background on the Kirlian electrophotography technique.

Laser Cookbook, The
Gordon McComb
TAB BOOKS, No. 3090
My earlier book on an assortment of projects useful to the laser enthusiast.

Lasers—The Light Fantastic—2nd Edition
Clayton L. Hallmark and Delton T. Horn
TAB BOOKS, No. 2905
An introduction to the mechanics and applications of lasers. Interesting chapters on laser gyroscopes, quantum mechanics, and lasers in space.

Optics
N.V. Klein
John Wiley & Sons
A highly technical text about optics with lots of formulas and math on optics design.

Principles & Practice of Laser Technology
Hrand M. Muncheryan
TAB BOOKS, No. 1529
An introduction to lasers and laser applications. Heavy on the industrial side of laser use.

Programmer's Problem Solver, for the IBM PC, XT, & AT
Robert Jourdain
Brady
A technical book on the inner-workings of the IBM PC, with special emphasis on programming in BASIC, Assembly, and machine code. Extensive section on parallel ports.

Robot Builder's Bonanza
Gordon McComb
TAB BOOKS, No. 2800
My earlier book on a compendium of projects useful to the robot experimenter.

Silliconnections: Coming of Age in the Electronic Era
Forrest M. Mims III
McGraw-Hill
A history and personal autobiography of the electronics revolution. Mims, a writer and inventor, contributed to the development of the first personal computer and was employed at the Air Force Weapons Laboratory in New Mexico where he worked with early solid-state diodes. This book includes fascinating material of how the *National Enquirer* paid Mims to build a lightbeam listening device using a laser and phototransistor receiver to tap into the private conversations of the late Howard Hughes.

Index